Reasons of Conscience

**Reasons of Conscience:
The Bioethics Debate in Germany**

Stefan Sperling

The University of Chicago Press :: Chicago and London

Stefan Sperling earned his PhD in medical anthropology from Princeton University and has held postdoctoral fellowships at the Harvard Kennedy School and at the Department of History of Science at Harvard University. He has taught at Harvard University, Humboldt University of Berlin, and Deep Springs College in California.

The University of Chicago Press, Chicago 60637
The University of Chicago Press, Ltd., London
© 2013 by The University of Chicago
All rights reserved. Published 2013.
Printed in the United States of America
22 21 20 19 18 17 16 15 14 13   2 3 4 5

ISBN-13: 978-0-226-92431-1 (cloth)
ISBN-10: 0-226-92431-9 (cloth)
ISBN-13: 978-0-226-92432-8 (paper)
ISBN-10: 0-226-92432-7 (paper)
ISBN-13: 978-0-226-92433-5 (e-book)
ISBN-10: 0-226-92433-5 (e-book)

Library of Congress Cataloging-in-Publication Data
Sperling, Stefan.
  Reasons of conscience: the bioethics debate in Germany / Stefan Sperling.
     pages; cm.
  Includes bibliographical references and index.
   ISBN 978-0-226-92431-1 (cloth: alk. paper)—ISBN 0-226-92431-9 (cloth: alk. paper)—
ISBN 978-0-226-92432-8 (pbk.: alk. paper)—ISBN 0-226-92432-7 (pbk.: alk. paper)—
ISBN 978-0-226-92433-5 (e-book)—ISBN 0-226-92433-5 (e-book) 1. Bioethics—Germany.
I. Title.
   QH332.S64 2013
   174.2—dc23   2012023056

♾ This paper meets the requirements of ANSI/NISO Z39.48-1992 (Permanence of Paper).

Contents

	Pretext	1
	Building, *Bildung*	4
	The Visible Public Sphere	6
	Creating Readers	10
	Normativity—Look It Up!	12
	Grappling with Bioethics	14
1	A Tale of Two Commissions	19
	Two Ethical Visions	20
	Building an Ethical Imperative—The Ethics Lag	21
	Veilings and Unveilings	23
	New Kanzler, New Kanzleramt	24
	Parliamentary Ethics—The Enquete Kommission	26
	New Ethics—The Nationaler Ethikrat	28
	Looking Back—The Enquete Kommission in History	31
	Looking Around—The EK and the NER	33
	Can the Nationaler Ethikrat Be Ethical?	37
	Ethics Commissions as *Saalordner*	44
	"This Is Not Bioethics"—"Bioethics Is a Dirty Word"	49
	The Bundestag Comes to Life—*Sternstunde des Parlaments*	52
	Conclusions	58
2	Disciplining Disorder	61
	Learning to See the Right Things	61
	Becoming an Ethical Insider	62

The First Day	64
Ethics Made Transparent	65
First Impressions	68
A Place for Disability	70
Dienstweg	73
Du und Sie—More Ways of Creating Insider-ness	74
Writing Bioethics	75
Grammar of Democracy	81
"What *Are* the Ethical Aspects of Organ Transplantation?"	85
Translation—The Semi-Legitimate Outsider Attempts to Produce a Legitimate Text	87
Glossary—Marking Science, Unmarking Law	89
The Beginning of Life	91
Conflict of Objectivities	92
Paper Wars	94
A Visit to the Media	95
The Nationaler Ethikrat Goes Public	99
Karlsruhe—Merging Law and Art	101
The Last Day of the Commission	105
Rules, and Rules on Following Rules	106
Leaving the Field—An Outsider Again	108

3   Transparent Fictions   111

Toward an Ethnography of Transparency	112
Transparency Today	114
Crafting Citizens through *Bildung*	119
Democracy Made Transparent at the German Hygiene Museum	120
Place—A Pedagogical Training Ground	123
Participants—Who Are the Citizens?	127
Process—Education in Citizenship	131
The Citizens Speak, but Have Not Heard Clearly	136
Expert Reactions	139
Conclusions	141

4   Conscientious Objections   145

Constitutions of Glass—Transparent, or Merely Fragile?	147
Constituting Conscience	148
Kant's Conscience	152
Native Theories of Conscience—Kant as Germany's Moral Gold Standard	155

	Public and Private Reason	157
	Beamte—Delegated Conscience Then and Now	158
	Tortured Conscience	164
	Conscience and Resistance	169
	Conscientious Objectors	171
	Conscientious Abortions	178
	Constraints on Conscience	181
	Conclusions	190
5	A Failed Experiment	193
	Abwicklung und Aufarbeitung	196
	One *Volk*, One History?—Writing History Together	201
	Making East Germany Transparent—And Seeing an *Unrechtsstaat*	206
	Obsessive Transparency	209
	Transparency on Display—The Stasi in Museums	211
	Learning to See Themselves as Victims	214
	How German Was It?	216
	Mauerschützen—Suspending the *Rechtsstaat*/Erasing East German Conscience	221
	Abortion	229
	East Germany in the Enquete Kommission *Recht und Ethik*	232
	Bioethics and the East German Public Sphere	234
	Coda—A Very Private Place	242
6	Stem Cells, Interrupted	247
	Ethical Imports at Last	247
	"No Embryo Shall Die for German Research"	249
	Ethics Becomes Law	250
	Converting Ethics into Reason	251
	Reading the Law	253
	The Cutoff Date—An Unenforceable Line	254
	Prohibited yet Permitted	255
	Ethical *German* Research	256
	The ZES and the RKI Reconfigure Science and Ethics	257
	Inside the ZES	261
	Jürgen Hescheler	266
	Wolfgang Franz	271
	Conclusion	275
	Reading Borges, Reading Germany	276

Transparency—Text and Context	277
Potentialities—Setting Limits as an Ethical Act	280
Taboo—*Dammbruch*	284
Law and Memory—*Recht und Unrecht*	287
Acknowledgments	291
Notes	295
Bibliography	311
Index	317

Pretext

A pretext, as the Oxford English Dictionary defines it, is that which is put forward to cover the real purpose or object; the ostensible reason or motive of action; an excuse, pretense, specious plea. I use the word here to draw attention to the malleability of perception and the cognitive shifts that occur between the observation of phenomena and their subsequent assumption of factuality. This pretext precedes my text. Read as an introduction to an ethnography of bioethics, it calls attention to the contingency of impressions and moral perceptions before they are codified into text. It is an ironic reminder that the context for any text is as selectively chosen, and hence as contingent, as the text itself.

One morning in early February of 2002, I walked down the central seven-story-high hall of a new glass building in the center of Berlin, the Paul-Löbe-Haus, which contains many of the offices and meeting rooms of the German parliament.[1] Behind me an enormous wall of glass opened onto an expansive lawn, beyond which sits the cube-shaped Kanzleramt, the official seat and residence of the chancellor. In front of me a similarly impressive wall of glass opened onto a large construction site where more government offices were in the making. The path I took through the building was slightly lowered, so that the meandering river Spree, which divided this building from those future offices, appeared to come right up to the glass. The tour boats that on warmer days add life to the quiet waters seemed almost ready to float into the hall itself. The building had just been completed, and its keys had been handed to the Bundestag barely one month earlier.

On this winter morning the cold, dry air made me shiver, intensifying the sense of awe evoked by this vast, sacred space of democratic decision making. Light streamed through the outside wall at a low angle, bounced off the smooth floor and the glass interior, and filled the entire cathedral-sized hall with a radiant glow. Slender bridges connected the upper-story offices on the two sides of the hall. The internal glass elevators moving up and down in silence signified frictionless movement between different administrative levels, and the glass telephone booths hinted at the transparency of all official communication.

Lining the lowered central walkway were two rows of metal letters whose large size and wide spacing kept me from taking them in all at once. Indeed, their physical arrangement suggested that meaning was to be found not in the words themselves, but in their very materiality and their inaccessibility to the stationary observer. Viewed from any point in the hall, these letters trailed off in both directions. Only by walking down the whole length of the hall would the reader come to see the words in their entirety, thus incorporating motion, time, and space into the act of understanding.

Walking along the letters that morning, I saw that the first set of words was a quote taken from Ricarda Huch (1864–1947), a German poet and historian who obtained a doctorate in Switzerland in 1891, when German universities were still closed to women. Her biographers have nearly forgotten her literary creations and remember mainly her courageous resistance to National Socialism. Huch had been elected to the prestigious Prussian Academy of Arts and Sciences, but had refused to swear loyalty to the new regime, and in spring 1933 she left the institution in protest. After World War II she helped rebuild German democracy. The quote, in my translation,[2] reads:

> For what is human life? Like drops of rain that fall from the heavens down upon the earth, we go through time, driven here and there by the winds of fate. The winds and fate as well have their immutable laws according to which they move; but what does the drop which they sweep before them know of this? It rushes through the air along with the others, until it can vanish in the sand. But heaven collects all the drops up again and pours them out again, and collects and pours again and again, always the same ones, but different nevertheless.

The row of words on the opposite side of the walkway was a quote taken from Thomas Mann (1875–1955), the Nobel Prize–winning author who in the Nazi period emigrated to the United States, where he taught at Princeton University and acquired American citizenship before taking up residence in politically neutral Switzerland. In his numerous literary works Mann explored the ideological conflicts that preoccupied Europe during his lifetime and their resonances with the tensions that inhere in our modern selves. Mann's political views were conservative under the rule of Kaiser Wilhelm II, but became liberal during the Weimar Republic and openly critical of the Nazis as they gained power. This quote, also in my translation, reads:

> What then was life? It was warmth, the warm by-product of form-maintaining insubstantiality, a fever of matter, that accompanied the process of unstoppable disintegration and reconstitution of unsustainably entangled and unsustainably artistically constructed protein molecules. It was the Being of that-which-cannot-actually-be, of that which, only in this intercalated and feverish process of degeneration and renewal, balances with sweetly-painfully-precise effort on the tip of Being. It was not matter, and it was not spirit. It was between them both.

In the United States, I might have expected to find one of the foundational texts of American liberty inscribed in the floor of an official space like this. In Germany, where almost unspeakable discontinuities have defined political life, the apparently neutral texts that constitute *Bildung* and *Kultur* provide a continuity that politics cannot. Rather than giving purpose or spurring action, the two quotes are designed to make one pause and reflect, not on the organization of politics, but on the very meaning of life.[3]

On this particular morning I saw a young man kneeling over one of the metal letters. Using both hands, he was rubbing a piece of cloth slowly back and forth along the letter while pushing down with the entire weight of his body. His devotion to the labor of polishing the inscription in that cathedral of democracy seemed almost religious.

The young man's dedication would be invisible a few days later, when the walkway and its quotations were covered by two red carpets spanning the hall's entire length. They had been laid down for an official reception for the annual Berlinale film festival. For an evening, a celebration of cinema's transient celluloid projections would obliterate the polished prose of these set-in-stone reflections on the relation between individual and society, between biological and political life, and between the transient and the eternal. But as we will see, this juxtaposition of firmness and fluidity, and the tension between materiality and imagination, were explicitly designed into this workplace of German parliamentary democracy.

Building, Bildung

To each side of the magnificent central hall of the Paul-Löbe-Haus are four tall cylindrical structures containing the round meeting rooms of various parliamentary committees. These cylinders are clearly visible from both inside and outside the building, whose full-front glass facades allow the public to observe the sessions of its elected representatives. A monthly magazine published by parliament proudly dubbed the building the Motor of the Republic (*Motor der Republik*), thereby giving verbal expression to the link between a quietly humming V8 engine, which the building vaguely resembles, and the power and efficiency that many Germans hope will drive the country's political system, like its auto industry, forward to new levels of achievement.

Inside these circular meeting rooms, round tables surrounded by identical chairs eradicate visible hierarchies among committee members. Like King Arthur's knights, they can come together and engage in Habermasian dialogue with other free and equally situated participants. An outer ring of chairs against the wall for those supernumeraries who may not speak indicates, however, that not all are invited into the idealized public sphere.

From the Paul-Löbe-Haus a tunnel leads to the Reichstag, and I emerged from what felt like an underground parking garage into the basement of the historic seat of parliament. In the basement of the Reichstag, not accessible to the general public but open to me as a temporary member of the parliamentary administration, is an art installation by the Paris artist Christian Boltanski entitled *Archive of the German Representatives* (*Archiv der Deutschen Abgeordneten*). Small metal boxes are stacked one on top of another from floor to ceiling in two rows, leaving a narrow and dimly lit passage between them. The boxes, reminiscent of bank safe deposit boxes in which one keeps one's most prized possessions, are neatly labeled with the names of all the representatives elected between 1919, the year of the Weimar constitution, and 1999, the year of parliament's move from Bonn to Berlin. While there are boxes for the elected representatives of the National Socialist Party, the names of representatives elected in East Germany are not included. The stacked boxes form two solid walls that stand as the symbolic foundation of parliament, but signs of rust and decay on the thin metal also provide troubling reminders of the fragility of the democratic tradition in Germany. Black ribbons mark each of the boxes representing a parliamentarian who was murdered by the National Socialists. In the center of the installation, a single black box represents the years from 1933 through 1945, the dark center of German democracy when all other parties were dissolved and no elected parliament represented the people. Those twelve years are at once the legacy that the modern German state must come to terms with and the black box that must not be opened if that democracy is to be preserved. Symbolizing the taint that will forever attach to the German state, the black box also legitimates reborn Germany by contrasting with its present transparency.

Going upstairs again I entered the enormous Reichstag lobby, where gigantic glass walls opened into the main hall of parliament. The plenary hall looked almost circular, with the parliamentarians' seats arranged in the half-circle of an amphitheater. When a speaker takes the stage, he or she walks down to the lowered podium and then faces the sovereign body, while the chancellor and his or her ministers sit behind the speaker on the right, and the prime ministers of the states (*Länder*) who constitute the Bundesrat are seated behind on the left. Visitors' galleries and an *Ehrentribüne* for guests of honor extend into the hall above the representatives' heads.

From the Reichstag another tunnel leads to the Jakob-Kaiser-Haus, another new building containing offices for the parliamentarians and a popular canteen. In this aseptic underground passage, which is unmarked and leads through several doors that look as if they conceal storage areas, there are also wider spaces into which short moving walkways are set. Passersby can see a piece of the original tunnel, which looks barely big enough for one bent-over person to squeeze through, as well as a Lufthansa electronic check-in terminal that makes it unnecessary for parliamentarians traveling light to wait in line at Tegel Airport.

The tunnels to and from the Reichstag once served darker purposes. Historians of Nazi Germany claim that on February 27, 1933, Nazis entered the building through a secret tunnel connected to the house of Reichstag president Hermann Göring. They set the fire that was blamed on the Communists and that enabled Hitler to announce far-reaching restrictions on democracy the very next day.

In the Jacob-Kaiser-Haus, I was again surrounded by light and glass. As I went upstairs and exited onto Wilhelmstrasse, the themes of law and legibility followed me out the door. Textuality, codification, law, and *Rechtsstaat* are built, signaled, and celebrated together. Turning toward the Reichstag again, and following the river downstream, I passed a tall row of nineteen contiguous glass plates that rise up straight from the ground. They are perhaps two inches thick and function as a fence around another transparent government building. Etched on these plates are the first nineteen articles of the Basic Law in their unamended 1949 version. These first articles specify the basic rights the state guarantees its citizens, such as the inviolability of human dignity (Article 1), the right to life (Article 2), the equality of all before the law (Article 3), the freedoms of faith, conscience, and confession (Article 4), and the freedoms of opinion, information, press, art, and scientific inquiry (Article 5). Every now and then I could see passersby stopping to read these rights of theirs, reflecting perhaps on their meaning and relevance in their day-to-day lives. This time I overheard a couple talking to each other about how, literally, to read these basic rights. Squinting against the sun, one of them exclaimed, "They're hard to read!" Her partner answered, "One must stand like *this*. From *here* one can read it." (So *muss man stehen. Von* hier *kann man's lesen*.) A third passerby noticed their interest in the texts and slowed to read the first sentence of the article guaranteeing the inviolability of domestic space: "*Die Wohnung ist unverletzlich*." He walked off snorting, "Then why do the cops always kick down the door?" (*Wieso treten die Bullen dann immer die Tür ein?*)

The Visible Public Sphere

Textuality and transparency complement each other; each provides the rationale for the other. During my time as a participant observer at the Enquiry Commission on Law and Ethics in Modern Medicine (Enquete Kommission Recht und Ethik der Modernen Medizin), Germany's parliamentary ethics advisory body, I came across an issue of the official Bundestag newsletter that celebrated the opening of the Paul-Löbe-Haus. One article described the architects' extremely self-conscious design choices, revealing an underlying script:

> The decision: The new buildings should hide absolutely nothing. There was to be no wallpaper, no paint, no wall hangings, no closets, no wood, no tiles, no curtains, no finery, and no lining. Nothing that would hide the texture of the construction material. The decision to show walls and ceilings as what they are: walls and ceilings that consist of cement, gravel, sand, and

whose light gray offers projection surfaces for any kind of fantasy. The decision to rely on the materials from which houses are constructed. It poses challenges to the observers.

The eye accustomed to wallpaper glides skeptically across the raw concrete, it lingers on the rough spots, and counts the points where the crane hooks lifted the pieces, it checks the transitions to expensive wood and polished glass, and it recalls all the other places where straight lines are marked with borders, and where ornate wallpapers celebrate a comeback. The soul that is spoiled by home improvement stores cannot decide: the new white paint that is on sale there looks good—and especially the cherry wood veneer. And yet, the decision not to hide anything is not a bad idea, precisely because it is not to everyone's taste.[4]

On my walks through Berlin's public architectural spaces it was hard to forget that the structures that government inhabits are symbolic and ethical as well as material. How often has a European nation had the chance in modern times to move its capital or to redesign the architecture of its government? As the geographer David Harvey powerfully reminds us, Napoleon III commissioned Georges Haussmann to redesign Paris after the 1848 revolution to prevent subsequent uprisings.[5] Reason and rationality turned the City of Light into a place of wide boulevards that make it easy to deploy troops against possible rioters and make barricades all but impossible. In Berlin, the construction of new buildings on an unprecedented scale gives clues to the German nation's understanding of governance. Architecture expresses a form of governing. In the design of buildings, ethics and aesthetics meld into one, and the architecture of government becomes a guide to the ethical sensibilities of a nation. It is in these spaces, after all, that the nation's representatives come together to decide between right and wrong. Here they find the reasons from which to construct ethical distinctions. Reason, rationality, and morality are all at stake in the work of government and also in the design of spaces in which that work is conducted.

Transparency in particular emerges as the central motif in the architecture of German politics.[6] The contemporary understanding of democracy presupposes the transparency of government, but in reconstructed Berlin transparency is seen as so elementary, so fundamental, that it has been elevated almost to a constitutional principle. Most famously, a glass dome designed by the British architect Norman Foster now caps the seat of parliament, replacing the Reichstag cupola that was destroyed by the 1933 fire that the National Socialists used to incite hatred against

Bolsheviks and Jews. Controversial at first, Foster's glass dome is now the most beloved and visited site in Berlin.[7] Everywhere, as in the Paul-Löbe-Haus, the public's gaze is invited into the spaces of political power, and the public eye is asked to legitimate the new politics of reunited and, to some, uncomfortably expanded Germany.

Glass is used metaphorically and materially to signify a commitment to the transparency of political processes, as if transparency were a material property of objects that, when properly installed, in itself suffices to reveal the inner workings of political institutions.[8] The very self-conscious attempt at making visible the insides of political buildings may be read as a recognition of the citizens' *right to know* the inner workings of the democratic institutions that they house; their conspicuous ostentation even implies the citizens' *duty to know* what happens inside. In these transparent spaces, the elected representatives of the people act as the people's conscience. Here they open to public inspection their reasoning as sovereign representatives.

The architecture of government was designed explicitly and purposefully to connect the former East and West Berlin. Viewed from above, or on a map, the Kanzleramt and its walled garden join the stretch of buildings that house the parliamentarians' offices to form what is called the Ribbon of the Federation (*Band des Bundes*). But this ribbon equally resembles a safety pin that twice crosses the river Spree. This river used to mark the border between the halves of the divided city, and on its western side several white crosses commemorate those who unsuccessfully tried to cross its once deadly waters. Today the safety pin ties together not only the two formerly severed parts of the German capital, but also the executive and the legislative branches.

Farther up the river I crossed Wilhelmstrasse, a major north–south artery connecting the central districts of Mitte and Kreuzberg, along which still stand some of the monumental buildings once home to the Nazi government, including the Air Force Ministry, which now houses the Ministry of Finance, and the SS headquarters. Today one finds there a Stasi Museum, which displays not only the sometimes ingenious instruments with which the East German government kept tabs on its citizens, but also the equally ingenious ways that East Germans invented to escape from state surveillance and from the territory of the state itself.

Berlin is also proud of its illustrious scientific heritage and its world renown for innovation and creativity, which continued into the 1940s. On the outside of the ochre-colored Berlin studio of the ARD, Germany's association of public broadcasters, I saw a silver plaque commemorating the former location of the Helmholtz Institute, which was built in the 1870s and destroyed in 1945. The bottom line of the inscription explains that it was here a century ago, in December 1900, that the physicist Max Planck delivered a talk in which he founded quantum theory. This is an understated way to describe a scientific revolution. In that year, Planck shattered conventional wisdom by postulating that the amount of energy that could be absorbed and emitted by atoms and molecules was not arbitrary, as classical physics had assumed, but came in discrete amounts called quanta.

As I continued past the ARD studio, I saw rows of seemingly identical buildings in the typical East German high-rise architecture lining both sides of the river. These buildings, the notorious *Plattenbauten* made up of prefabricated concrete rectangles, were cheap and fast to build; their homogeneity expressed the state ideology of a proletarian and non-hierarchical citizenry. For the inhabitants I spoke with, this homogeneity also meant anonymity and melancholy, and when the Wall fell, those who could afford it abandoned these apartments for others more amenable to individual self-expression.

On the north side of the river Spree, I saw two outwardly dissimilar buildings standing side by side. Both would become important to me as fieldwork sites. To the right, on Schiffbauerdamm, was a five-story block in the gray East German style, containing the offices of the five Enquete Kommissions formed to advise parliament in its fourteenth session. On the top floor sat the office of the Enquete Kommission on Law and Ethics in Modern Medicine, where I conducted my internship. This commission of parliamentarians and experts had the mandate to evaluate the legal and ethical consequences and implications of developments in modern medicine and to recommend appropriate legislative responses. To its left, an older, beautifully renovated building assembled under its roof a collection of internationally oriented academic institutions, including Humboldt University's Institute for European Ethnology, with which I was affiliated during my stay in Berlin and which has since moved to Gendarmenmarkt.

My passage into German bioethics as a research topic was also a passage into Berlin. The city is a palimpsest, and it was impossible for me to live and work there without becoming a reader of its layered texts. In the process I began to realize that the "ethical" in Germany is about much more than abstract moral principles—it is also to be found in the very materials from which the city, the society, and the nation are rebuilding themselves.[9] Berlin, perhaps more than other cities, wears visibly, even self-consciously, the destruction of World War II and the dilapidation of years of socialist rule. Gray buildings line wide boulevards, and sad-looking prefabricated high-rises surround Alexanderplatz. On one of several tiled facades, a youthful Erich Honecker inspires workers to put their labor into the service of the then-hopeful East Germany. Occasionally there are bullet holes on the sides of a building. There are gaps in rows of townhouses where bombs landed. And there are ruins purposefully left as reminders of the devastation.[10] Sometimes plaques next to such gaps commemorate the fallen by listing their names and dates of birth, along with the date of the particular event that ended their lives.

Even more powerful than the reminders of ruination were the city's unexpected juxtapositions. New glass facades of transparent post-Wall Berlin sit next to buildings from the Bismarck era. Monumental Nazi architecture comes right up to the now bland and dated-looking concrete architecture of antifascist East Germany. Sometimes road signs at an intersection show arrows pointing in opposite directions, toward the landmarks of each half of formerly divided Berlin. One sign I saw near Wilhelmstrasse, for example, had arrows pointing east to the district of Friedrichshain (historically home to Berlin's factory workers), to Alexanderplatz (the East Berlin central plaza and train stop), and to the East Berlin airport, Schönefeld, while symmetrical arrows pointed in the opposite direction to the district of Charlottenburg (historically a center of shopping and nightlife), to Zoo (the West Berlin central plaza and train stop), and to the West Berlin airport, Tegel. These ubiquitous juxtapositions practically forced me to think historically, and materially, about the ongoing reconstruction of Germany, its politics, and its ethics.

Berlin does not hide its scars or try to put on masks. It is honest in its fragility, and it invites people to be honest and self-revealing in turn. Berliners, I often felt during my two years there, are more intent on querying people and getting to the root of their life plans, their motivations, and their characters than were friends I had made in America. The openness of Berlin's urban archaeology matched the openness of its residents' personal archaeologies. Today this scarred, sacred, reconstituted assemblage of buildings in the heart of Berlin is the site of German attempts to work out the nation's new political identity. A part of those attempts is the constitution of a space for public moral discourse—including bioethics, the subject of this book.

Creating Readers

Even the casual visitor to Berlin almost involuntarily becomes a reader. The city's public architecture and monuments practically ask for interpretation, turning literacy from a basic right into a national duty.

Everywhere, plaques inform imagined readers about the history of particular buildings or places. They describe in detail the historical events that took place there, and many recount moments of resistance. Some of these plaques are barely noticeable, and one discovers them only accidentally, or after spending some time in the city, while others are more prominent. In front of the Reichstag there are ninety-six irregular black sheets of slate sunk into the ground spaced a couple of inches apart. They symbolize the ninety-six Reichstag representatives of the Weimar Republic whom the National Socialists murdered between 1933 and 1945. The edges of the sheets are inscribed with their names, their dates of birth, and their dates and places of death. Some have empty spaces in case further research reveals new information.[11]

Public reading turns people into citizens, and the signs and posters around the city convey all kinds of political information. All over the city's public billboards, and in public places like train stations, one can find paragraphs of the Basic Law. In May 2005, in preparation for the vote on the constitution of the European Union (EU), I saw government-sponsored posters advertising the EU in bus stops throughout Berlin-Mitte. The posters quoted one or another article of the EU constitution, typically a basic right, and in small print below argued at length that a strong EU was a good thing for Germany. In a bookstore, I even saw a legal commentary on the bestseller shelf.

Illiteracy, in turn, is nearly inconceivable in Germany. Take the novel *The Reader* (*Der Vorleser*), by Berlin law professor and occasional detective story author Bernhard Schlink. The book tells of the sentimental education of a fifteen-year-old student, Michael, who falls in love with Hanna, a woman twenty-one years his senior. Hanna continually asks Michael to read books to her, and their relationship comes to be expressed largely through the ritualized act of reading. When Hanna suddenly disappears, Michael initially blames himself for her desertion. Years later, while training to be a lawyer, Michael witnesses a trial of Nazi concentration camp workers, and in the courtroom he recognizes Hanna. She stands accused of having written a protocol that led to the murder of innocent inmates in the concentration camp where she had worked as a supervisor during the war. At this point Michael realizes that Hanna is illiterate, but for reasons of conscience he keeps this realization to himself. The shame of being illiterate in Germany prevents Hanna from exonerating herself, and Michael protects her secret by not intervening. Illiteracy dooms Hanna in her own eyes and in the eyes of her former lover, who has since then dedicated his life to reading the letter of the law. Hanna is sentenced to life in prison for crimes she seemingly cannot have committed. There she learns to read and write, and eighteen years later, shortly before being released, she kills herself.

What interests me about the novel is the peculiar way in which Hanna's illiteracy becomes the guiding metaphor for Germany's incomprehension of its own history. Hanna takes some responsibility for her actions by not claiming ignorance publicly and thereby exonerating herself. Her refusal to claim ignorance can be read as a refusal to look the other way. When she does learn to read, in effect, she recovers the past, and by extension, learns to decode her position in it fully. Then, she has no recourse but to kill herself. Michael, on the other hand, becomes a proper citizen of the postwar *Rechtsstaat*. Whereas he learns to read the rights of the law, Hanna learns to read the wrongs of the past—with results that prove intolerable.

This bestseller, which has been translated into twenty-seven languages and won numerous literary prizes, has been praised by German reviewers for its skillful building of suspense and its concluding revelation. Clearly Schlink knew his audience well enough to know that the mystery of Hanna's silence would work in Germany. I myself, also *gebildet* in German ways, remember being predictably surprised when I discovered the solution to Hanna's inexplicable behavior. German readers have told me they reacted similarly to the climactic disclosure. But an American friend to whom I gave the book deduced from the early sections that Hanna was illiterate and found the rest of the book a disappointing buildup to a foregone conclusion.

Normativity—Look It Up!

In Berlin I myself experienced the inconceivability of illiteracy in all kinds of everyday interactions. When I first arrived in the city, I occasionally asked for directions. My requests were simple. Sometimes I just wanted to go to the nearest train station; other times, I wanted information on the participants in a workshop, or I wanted to know an expert's opinion on a certain topic. The answer I repeatedly received was, "Look it up!" For directions, I was told to look at the maps that I discovered were hung all over the city. For other information, I was told that I should do research on the Internet, or I should check out the relevant book from the library and read about it for myself. I began to realize that in almost all interactions in Germany one is dealing with highly literate people, most of whom know where the information they want is to be found. The people who point one to websites or city maps reveal a belief that everything in the world has its place, and a concurrent belief that it is one's obligation to know those places. When people answered my queries, they were not so much bewildered that I did not know some elementary fact as they were surprised, and a little censorious, that I did not know where to look it up.

This dense intertextuality has profound consequences for being at home in Germany, as the reliance on texts effectively excludes those not in a position to read. Germany draws sharp lines between those who belong and those who don't, and it makes few concessions to outsiders. Even in a city with a history as cosmopolitan and heterogeneous as Berlin's, the outsider who cannot read German, and Germany, is quickly found out and put in his or her place. People who are recognized as Germans, on the other hand, no longer need to prove that they know their way around; they can be lost without losing their place in the community. Once I was on my way to some friends' apartment in an unfamiliar part of town. Unable to find the street I was looking for, I found myself stuck in a confusing square with no idea which way to go. Eventually I asked a man with a Dalmatian sitting on a bench for directions. He turned out to be drunk and answered with philosophical candor, "I don't even know where I am myself. . . ." The next person I asked, a blue-collar worker apparently on his way home from work, answered, "I don't know . . . I've lived here only for one year . . . at the most." I finally found the street about two hundred feet farther down. What stayed with me was the sense that these persons, by all appearances German, were unable to give directions around their immediate neighborhood. Perhaps knowing that they could look it up in theory meant that they no longer had to do so in practice? Yet Germans expect obvious foreigners to know the society and the culture intimately before they can gain admission.[12]

In learning to find my way around by reading the city, rather than by asking its inhabitants, I came to understand the paramount importance of written information, broadly understood, in German culture. I learned to see the material and conceptual landscapes of government as mutually constitutive, and the concepts that I acquired in Germany came to feel as real as the very buildings their users inhabited; the transparent buildings, in turn, seemed to express, and even shape, how "natives" think—about ethics in particular. In the following chapters I will take up the invitation, and expectation, to be a reader of German culture as I read the bioethical debates that I had come to Germany to study, in part as exercises in producing authoritative, public, moral texts.

Grappling with Bioethics

This book is concerned with bioethics in contemporary Germany. Rather than looking through the eyes of a bioethicist, however, I want to approach the subject at an angle, or along a curved line, circling around bioethics while looking at it sideways.[13] After working for several months in the parliamentary Enquete Kommission on Law and Ethics in Modern Medicine, and after numerous interviews with ethicists, stem cell researchers, physicians, and politicians, I realized that bioethics in Germany[14] is a much more complex phenomenon than a description of institutional practices alone could hope to grasp or convey. In reflecting on my observations in the two ethics commissions and on my experiences in Germany over two years, the particular performances of bioethics that I witnessed in the commissions seemed to merge into the larger transformation of post-reunification Germany—a nation becoming surer of itself while still struggling to master the darkest chapter in its history. Bioethics itself came to look like a formation at once old and new, one that was at every step shaped and reshaped by preexisting and reemerging historical and cultural constraints. It was, moreover, a product of a very particular kind of state: Germany, as I show throughout, is a pedagogical state, a reasoning state, and an ordering state.

The debate over human embryonic stem cell research had just begun when I arrived in Germany, and this debate became the lens that focused my observations and analysis. What particularly struck me in following these debates was that the issue in perpetual question was *not* the moral status of one or another biological construct per se, but that of humanity and personhood as German ideals. As I followed the discussions explicitly concerned with the humanity of unclaimed entities of a status yet to be determined, I found that these debates were tacitly centered on questions of institutional and parliamentary legitimacy. At stake, in other words, were questions about German nature and the nature of Germany.

In debating these questions, the German parliament was caught up in a conscious process of self-renewal and self-redefinition; bioethics was only one of its vehicles. My initial interest in why some ethical positions were seen as right and others as wrong receded as I became interested in the deeper cultural forms that shape bioethics, in the history of the pedagogical forms that transport values, and in the material environments that shape ethical consciousness. I wanted to understand, and make visible, not only what it meant to make statements about right and wrong in the life sciences in Germany at this moment in time, but also how this moment relates to Germany's past moments. In my attempts to make bioethics visible to the reader, I will of course develop, or presume, a kind of transparency of my own. All vision is partial, and making anything visible involves emphasizing some things and de-emphasizing others. The lines between and around both text and context will necessarily reflect this particular writer's capacity and preference to draw such lines and make such distinctions.

German bioethics is part of Germany's continuing project of postwar sense-making—of separating the lessons of the Holocaust into meaningful texts and contexts. The implicit, overriding questions that underlie German bioethics are the questions that have pervaded all of German public life for several decades: How could the Holocaust have happened? And how can we make sure that it will never happen again? In this book I show some of the ways in which the German state selectively sees and remembers this chapter of its history and thereby constructs meaningful continuities and discontinuities between the past, the present, and in relation to scientific research, the future.

To craft a morally sustainable continuity for German history, some things must be always kept in view, others made invisible, and still others perceived as aberrant discontinuities. This effort involves, I will argue, making the state visible to the citizens (producing *transparency*), making sure the citizens internalize the state's principles and support them (training *conscience*), and making sure that the appropriate lines are drawn around German territory and history (configuring *Germany* itself). To produce a working model of transparency, the viewers' eyes need to be trained to perceive it in just the right ways. To act as autonomous individuals, citizens need to show that they have reasons that the collective deems appropriate. For a new *Rechtsstaat*, and an abandoned *Unrechtsstaat*, to emerge in clear outline and with precise demarcations, history must be (re)written in a way that accentuates the contrast between present democracy and preceding dictatorships.

In the following chapters I hope to show what Germans mean when they speak of bioethics, when they call for *transparency*, and when they demand the right to follow their *conscience*, as *Germans*, in morally opaque or problematic circumstances. I analyze in ethnographic and historical detail how these three categories—transparency, conscience, and Germany—work in contemporary Germany. These categories are, to me, constitutive of German (bio)ethics; yet they feel open-ended and abstract enough to convey the sense that German thought with respect to these categories will remain in motion long after I am done writing.

In developing these themes, I have organized the book into six chapters, each of which combines ethnographic observation with explorations of Germany as a site of memory and of history in the making. In chapter 1 I introduce the institutional contexts of German bioethics as it existed a decade ago. I describe the separate and interactive workings of the two German institutions that were officially charged with "doing bioethics," the parliamentary Enquete Kommission on Law and Ethics in Modern Medicine (EK) and the executive's Nationaler Ethikrat (NER) (which later combined to form the Deutscher Ethikrat[15]). Both commissions were initially appointed to recommend a course of action on regulating human embryonic stem cell research,[16] particularly the importation of cell lines, and these were the most immediate topics under discussion when I began my ethnographic fieldwork with one of these commissions. By tracing each institution's genesis, projected image, and operations vis-à-vis the other, I show that the content of bioethics—in this case ethical stem cell research policy—cannot be separated from the workings of these institutions. I want to emphasize here that my argument does not hinge on properties particular to human embryonic stem cells or on details of stem cell research; rather, my book focuses on the German stem cell debates mainly because they were widely considered to mark a defining moment in the nation's moral maturation.

In chapter 2 I first describe my own entry into these institutional contexts in 2001. I describe the cultural norms that had to be followed to gain admission to the bioethical debates and the efforts that were made to show, and hide from view, the right things at the right time. Drawing on several months of fieldwork as a participant observer in the Enquete Kommission on Law and Ethics in Modern Medicine, I then demonstrate the many contingencies that are involved in producing a consensus that in the end hardens into a law that all must obey. I show how, in the process of grappling with developments in biomedicine, the state attempted to craft a persuasive discourse of public morality.

In chapter 3 I take up the first of my analytic categories—namely, *transparency*. As we have already seen, talk of transparency not only pervades politics but is built self-consciously into the material structure of governmental workplaces. I show that transparency was part of each ethics commission's mandate and that these official bodies felt a duty to make themselves transparent. I ask how transparency was accomplished in the practice of bioethics. I show that the production of transparency required a selective training of vision and an equally selective blocking out or highlighting of what gets seen. Transparency, in other words, emerges here as a problem for social analysis in its own right, as a part of what needs to be explained and taken into account in thinking about how power masks its operations, even when it declares its intention to be self-revealing.

Chapter 4 addresses the category of *conscience*. Talk of conscience is also an integral part of German public morality, and I found that scientists and politicians alike frame doing science ethically as a matter of conscience. I probe the meaning of these kinds of statements by asking how a question becomes a matter of conscience. Using examples from conscientious objection to abortion, I argue that the state in effect enters into a contract with its citizens under which citizens reciprocate the state's transparency by laying open their reasons for refusing to go along with important state interests. When German citizens claim to act on their constitutionally autonomous conscience, I suggest that they are in effect providing reasons in a manner that the collective deems appropriate.

The third category, *Germany*, forms the subject of chapter 5. While many Western nation-states have had relatively stable borders throughout most of the twentieth century, Germany has undergone enormous shifts in both national territory and national self-understanding. The most recent such shift was the reunification of the democratic West with its socialist Eastern counterpart. I show that in the context of reunification, German identity has shifted as West German understandings of transparency and conscience have been overlaid on preexisting East German understandings, with repercussions for bioethics in the reunified nation. I show in this way that contemporary German (bio)ethics is premised on a particular idea of German reunification and the resulting identity of the German *Volk*.

In the sixth chapter I demonstrate how each of these three concepts flowed into the Stem Cell Law (*Stammzellgesetz*), which constituted a formal national solution to the ethical dilemma of research on human embryos. I show how the codification of these concepts in the 2002 law created the framework for conducting ethical research in a Germany that is determined to serve as a model for Europe and the world. In practice, however, stem cell researchers continued to experience the sense of paralysis brought about by Germany's continuing failure to resolve the paradoxes of postwar nation building.

: : :

Many people have stereotyped Germany as a place of order, and many authors have described German inflexibility in bending rules. Less attention has been paid to the process by which this order comes about in the first place. My project is a search for the sources of order in Germany—including the order we call bioethics. In this book I describe some of the ways in which German institutions order minds and bodies. I show how persons are taught to see and act and think in certain ways that are specifically German. I am interested in how this war-scarred and traumatized nation struggles to craft a persuasive discourse of public morality—one that will, as one of its by-products, produce ethically acceptable research.

Throughout my research on bioethics, it was difficult for me to be a "neutral" anthropologist and not to have an opinion on the topic I was investigating. Engaged Germans kept asking me which side of the stem cell issue I came down on. Despite its "life-and-death" rhetoric, however, the debate seemed essentially nonjudgmental. People were curious about what I thought, but no one ever tried to convert me one way or the other. It was important only that I had an opinion, not what the opinion was. It was as if definite answers to ethical questions did not matter so long as I could show that I had struggled with the question. To this extent, then, my book is German: I grapple with the questions that preoccupied Germany for a time, and the grappling itself seems more important, to me as to many Germans, than the specific solution that the society arrived at, together.

1

A Tale of Two Commissions

When I arrived in Berlin in June 2001 to begin my ethnographic fieldwork, it seemed that the entire nation was caught up in a controversy over the moral status of human embryonic stem cells. Human embryonic stem cells, first isolated in 1998 at the University of Wisconsin at Madison, are versatile precursor cells that have the potential to develop into many other kinds of cells in the body of an adult organism. They therefore hold therapeutic promise in areas of regenerative medicine such as organ and tissue replacement. The production or harvesting of embryonic stem cells usually requires the destruction of human embryos, which makes their use controversial.

The nature of the embryo had become a matter of public debate in part because new technological capabilities had made the embryo visible and manipulable, and in so doing had brought it into society. Embryos could now be imagined as society's weakest members—indeed, those most in need of protection. The question at hand was whether these tiny precursors of tissues, organs, and even human beings deserved any legal protection. Were early embryos mere lumps of cells, usable for research, or did their potential personhood entitle them to basic rights to life and dignity? I began following the debate closely, and over the course of about a year, the positions in the debate became firmer. During that time parliament debated and passed a law that regulated embryonic stem cell research. For the German parliament, as we will see, protecting embryos became a way of showing that even the smallest and weakest were treated with the respect that every human being deserves.

After the Stem Cell Law was enacted, people described the controversy as the *Stammzelldebatte* and referred to it in the past tense. To say "we have had the *Stammzelldebatte*" was to check off the topic as being over and done with: everyone had said what they had to say, and the legislation that issued from parliament was now binding on all. The term linked stem cells to the process of public deliberation and at the same time confined the controversy to a particular moment in time. A topically specific debate had taken place, the nation had participated, a consensus had been reached, and the issue was now closed.

But was it really so cut and dried? Just a few months ago the opponents of stem cell research had claimed that research on the human embryo was akin to murder, while proponents had claimed that their constitutional right to research was being infringed. To one side stem cells had been precursors to full human beings, while to the other side they had been mere matter. How could these radically contradictory existential positions have been reconciled to the extent that no open questions remained?

Two Ethical Visions

Before we can answer this question, we need to ask how the legal and moral status of embryos emerged as a problem in Germany in the first place, for the state as well as for citizens. How did the questions of what defines human-ness and what counts as a human being—possibly the deepest ontological questions arising from the life sciences—arrive for resolution in parliament?

The relations between parliament and the chancellor—in theory cooperative, as the chancellor represents the parliamentary majority—turned antagonistic in a struggle over the definition of ethical stem cell research. The rivalry became inscribed in two ethics commissions attached to the two branches of government. While the parliamentary commission had earlier worked away in relative obscurity, the emergence in 2001 of a second commission attached to the executive made it clear that ethical acts and judgments are not objectively discernible to just any informed and knowledgeable observer. The presence of *two* commissions was partly responsible for reframing the ethical stakes; the struggle between them made visible to outsiders like me the selectivity of visions that had to construct themselves as all-encompassing. The stem cell problem was the starting point for both commissions. But in both venues the question of the moral status of the embryo, and the stem cells derived from it, was transformed into the question of who can legitimately decide matters of national moral significance. The question "Who belongs to the moral community of beings whose human dignity is inviolable?" fused with the question "Who belongs to the moral community that can legitimately define the boundaries of being human?"

Max Weber's distinctions among types of legitimate domination[1] help make sense of the dynamics that developed between the two commissions: the parliamentary Enquete Kommission on Law and Ethics in Modern Medicine (EK) and the cabinet-appointed Nationaler Ethikrat (NER). One could say that the former commission relied on bureaucratic authority to legitimate its claims, while the latter relied on charismatic authority. The EK drew on precedents and historical memories, while the NER crafted new rules for the game. Yet the forms of the two commissions began to converge in some respects over time: the bureaucratic commission gained charisma in its righteous struggle against the "illegitimately" conceived newcomer, while the charismatic commission fell into routines of its own that began to approximate bureaucratic rationality. As we will see throughout this chapter, the two institutions remained apart in their ideas about how public morality ought to be created.

Building an Ethical Imperative—The Ethics Lag

An observer of German political culture quickly finds that law, as both the expression and the guarantee of social order, is regarded as the immutable foundation of German political identity. Billboards I saw in Berlin displayed articles of Germany's Basic Law and referenced constitutional principles that protect individual liberties against government intrusion. In everyday conversations and in media reporting, even the smallest infractions led to calls for quick sanctions and warnings that intolerable conditions would result if the deviation was not addressed immediately.

Science, in contrast, is portrayed as constantly progressing as it refines the tools with which it inquires into the natural world. Nature, and in this case human nature, thus becomes mutable, putting the social order at risk from scientific advances.[2] Popular media quickly picked up stories of potentially life-changing scientific improvements, and billboard campaigns I saw all over Germany sought to temper the inquisitive impulse by raising questions and providing possible answers intended to give people pause.³

This wide portrayal, and perception, of law as static and science as mobile generates a sense that the law is lagging behind the sciences—what critical legal scholars have called the perception of a "law lag."⁴ Law, many think, arrives too late to regulate effectively. This lag results in a paradoxical situation: on the one hand Germans call for ethical debates in advance of scientific developments, while on the other hand they want to benefit from those scientific developments that by their very nature cannot be predicted and so brought under ethical supervision.⁵

In conversation Germans use the words "legal" and "ethical" almost synonymously, and the "law lag" has also been perceived as what I will call an "ethics lag." As we will see, the precise relation between these terms is more complicated. In discussing bioethics, Germans often assume that science will proceed in uncontrollable ways unless one imposes tight restrictions on its progress. Science and medicine, Germans frequently remind themselves, can easily spin out of control unless a whole series of safeguards is in effect. One needs strong institutions, comprehensive representation (transparency), strong individuals (conscience), and a modern state (Germany) that draws careful boundaries around itself, distinguishing itself chronologically and ideologically from Hitler's National Socialism and East Germany's state socialism.

In part as a response to this perceived "ethics lag," parliament appointed the Enquete Kommission on Law and Ethics in Modern Medicine and charged it with clarifying ethical concepts, structuring ethical debate, facilitating exchanges between law and science, and anticipating future ethical dilemmas. In 2001 Chancellor Gerhard Schröder decided to get advice on bioethical questions from a commission independent of parliament, and he set up the National Ethics Council (Nationaler Ethikrat, or NER).

Both national ethics commissions, the parliamentary Enquete Kommission and the chancellor's National Ethics Council, were set up in part to address the pressing question of how to regulate human embryonic stem cell research. This novel form of research had arrived in Germany and appeared to threaten the nation's moral fabric. The commissions would restore order by preparing the law to react to these developments. The fact that the two bodies addressed the law lag (or the ethics lag) differently shows that the stem cell debates were not simply about stem cells, but more importantly, about the need to regulate them "in a legitimate way." In other words, the two commissions were as much a response to perceived disorder as an expression of it.

Despite all the talk of lagging, ethical "damage control" does not always follow developments in medicine, science, and technology. A congress on cloning that took place in Berlin in May 2003, for example, offered quite another picture. The Massachusetts-based company Advanced Cell Technology (ACT) had claimed in 2001 that it had created a human embryo for the sole purpose of deriving embryonic stem cells from it. A conference speaker announced that this claim was "false." Ludger Honnefelder, an influential German ethicist, nevertheless asked the pathologist Eckhard Wolf how ACT's claim should be evaluated ethically. Wolf replied that he did not know, but added that "one ought to act as though the clone already existed, and then judge it on ethical grounds." In other words, it was the role of ethics to foresee and steer science's advances—in this case, to assume "worst-case scenarios" and regulate them even before they come into existence.

In the work of the two ethics commissions as well, practice often contradicted the idea of an ethics lag. Both constantly searched for "good topics," defined not as topics that had captured the public imagination, but as those that were not (yet) in the spotlight and yet were sufficiently important to need regulation. Ideally, the commissions' practices implied, one regulates *before* regulation becomes necessary, not after. As the ethicist and EK member Honnefelder said at the 2003 Berlin congress, "The goal is a dialogue that occurs not after the fact but *before the fact.*" And yet, once topics on the frontiers of science and medicine were identified, the commissions felt hesitant about their ability to evaluate these rapidly evolving issues. As we will see in this chapter and the one that follows, the effort to create ethical order was messy business.

Veilings and Unveilings

In 1991 the Bundestag voted to transfer its seat from Bonn to Berlin. The move to the restored and refurbished Reichstag was accomplished relatively quickly, but not before a spectacular art installation prepared the way. The artist Christo, best known for wrapping objects and buildings, and his wife and artistic collaborator Jeanne-Claude had spent twenty-four years struggling to win permission to veil the Berlin Reichstag, long imagined as the future seat of the German parliament. In 1995 parliament debated the aesthetic and symbolic merits of the veiling for a full seventy minutes and finally voted to allow the artists to proceed. Construction of the governmental quarters in the new capital had not yet begun, and the Reichstag building stood monolithically at the very edge of West Berlin, within meters of where the Wall had been. Before the Wall was built in 1961, only the black market had flourished on that barren land. In the decades after, the damaged building had been marginally used as a conference center and as a home for the poorly attended exhibition *Questions to German History* (*Fragen an die deutsche Geschichte*). West Berliners used the vast grassy areas surrounding the building for informal soccer games.

The installation was described, and very possibly experienced, as cathartic. In a remarkable engineering feat, a hundred thousand square meters of cloth, weighing more than sixty-one metric tons, were unfolded to cover every square centimeter of the Reichstag building. The veiling in reflective cloth, resembling silk in texture and appearance, with folds a meter deep, became a magnet for visitors. For two summer weeks the lawns around the building were filled with people, some equipped with blankets and pillows for extended contemplation, others selling drinks and snacks to sustain the democratic imagination.[6] The atmosphere was lightly pensive, and newspapers reported that one could almost sense a collective exhalation.[7] With repressed memories thus brought back into consciousness, but elegantly draped and safely reflected upon, Germany's political spirit and soul were symbolically purified, making it possible once again to move forward in history.

The material reconstruction of the Reichstag began as soon as the newly invested building showed its stone face again. Plans to modernize the building were driven by concerns over transparency, and the most visible exterior change was the addition of a glass dome, signaling the transparency of German democracy to European neighbors who had expressed concerns over the powerful nation once again growing in their midst. Major renovations also occurred inside the building, where immense glass walls continued the theme of transparency. The building was formally opened in April 1999, and parliament began holding its plenary sessions there in September of that year.

New Kanzler, New Kanzleramt

While the Reichstag was veiled and unveiled, construction also began on the chancellor's residence, the Kanzleramt, only a short distance away. Then-chancellor Helmut Kohl had dismissed the architects' plans for a building that would be aesthetically and spatially linked to the parliamentary buildings. He wanted a freestanding monolith, he told the architects, twice as tall as they had planned and clearly demarcated from the parliamentary constructions.[8] At eight times the size of the White House in Washington, D.C., the Kanzleramt became one of the biggest government buildings in the world.⁹ Commentators quickly nicknamed the emerging monumental cube-shaped structure, with its large circular glass panes, "the washing machine." The building would end up being thirty-six meters tall—taller by far than the Reichstag, were it not for the latter's glass dome. As we will see throughout this chapter, if parliament stands taller than the executive in legitimacy or authority, it is because it has transparency, both rhetorically and materially, on its side.

The Kanzleramt took a long time to complete, and Chancellor Schröder was temporarily forced to govern from the Staatsratsgebäude at the Schlossplatz, which was known as Marx-Engels-Platz when Walter Ulbricht and then Erich Honecker ruled East Germany from the very same site. The Staatsratsgebäude, built with parts taken from the facade of the demolished Berlin Schloss, was conveniently empty in that period and, like other official buildings in the former East Berlin, was awaiting a decision on its fate. From there, the chancellor and his staff moved into their imposing quarters on April 30, 2001. Confronting each other with their glass walls, the buildings of the chancellor and the parliament seem to represent two fighters in the federal ring: one rescued from a dark past and projecting a sense of duty to make itself transparent; the other, thoroughly modern, looking forward to a bright future for Germany, premised on architectural and technological progress.

The Kanzleramt, when viewed from the Reichstag across the lawn, looks monumental and futuristic. It projects a bold vision, and one can easily imagine it as home to not only a new Germany but also a new ethics. The Reichstag, on the other hand, when viewed from the Kanzleramt, looks almost playful in its elaborate ornateness. The plenary hall, half below ground, with curtains drawn during sessions, is more like a theater in which visions of a new kind are projected against the historic walls. The building itself, restored and conserved as a palimpsest of history, contains within it many traces of Germany's "unmasterable past."¹⁰ The inner walls preserve Cyrillic graffiti by Russian soldiers who stormed the building in 1945, and (as noted above) plaques commemorate the parliamentarians killed by the Nazis.

From the observation deck of the Reichstag, the fragmented panorama of German history is open to view. One can see the Quadriga on top of the Brandenburg Gate, which Napoleon captured in 1806 on his march through Berlin and which was returned to its place in 1814. To the west is the Siegessäule, whose gold-plated angel marks the victory of Germany over France in the war of 1870–1871. To the east is the Rotes Rathaus, which has housed the city government since 1869. To its left is the Berliner Dom, built by Kaiser Wilhelm II in 1905 as a Protestant counterweight to St. Peter's Basilica in Rome. To the right is the famous Hotel Adlon, built in 1907, partially burnt in 1945, demolished in 1984, and rebuilt in 1997 as a home for Berlin's most distinguished and discriminating guests. There is the golden roof of the Synagogue, burnt down in 1938 and restored to its original splendor fifty years later. There is the Soviet Memorial, dedicated to the Russian soldiers who died storming Berlin in April 1945. In the distance is the Fernsehturm, a towering symbol of East Germany's optimistic faith in building a better society through science, technology, and the media. Closer by is the S-Bahn Station Friedrichstrasse, which used to be the final train stop for East Berliners denied access to the free West. To the north is the Charité, the world-famous hospital where I was born and which still occupies the location where it was founded in 1710, then far outside the Berlin city walls. One can also see Potsdamer Platz and the masses of cranes that continue to construct and reconstruct the heart of the capital city. There is the Brandenburg Gate, through which East Germans streamed after the Berlin Wall fell on the night of November 9, 1989. Behind the gate one can see the new Memorial to the Murdered Jews of Europe, opened in May 2005 on the sixtieth anniversary of the end of World War II. Even its completion has not ended the controversy about its appropriateness. There is the newly built seven-story, glass-clad Lehrter Bahnhof, which is now Europe's largest train station. These structures offer not only discrepant architectural styles but also competing visions of the reconstructed nation.

Having come full circle looking *outward* at the past and the future from the roof of the Reichstag, we will now turn *inward* to the work of government. Looking down into one of the inner courtyards of the Reichstag, one can see an overgrown dedication "To the Population." The letters *Der Bevölkerung* are barely visible now, just as the debate over the use of that word—as distinguished from the dedication to the German *Volk* (the people) that marks the facade of the Reichstag—has also faded from memory.[11] From the Reichstag one can peer into the meeting rooms of the Paul-Löbe-Haus, where parliamentary commissions, including the Enquete Kommission, hold their meetings. But how had this commission come into existence?

Parliamentary Ethics—The Enquete Kommission

In the summer of 1999, after almost a year of parliamentary debate, members of the governing coalition of Social Democrats and Greens proposed forming an Enquete Kommission on "Human Rights and Ethics for a Medicine of the Future." It was to consist of fifteen members of parliament and fifteen technical experts. Its advocates offered several reasons. In the United States the National Bioethics Advisory Council (NBAC) had just recommended federal support for research on human embryos. American researchers working for a private biotech company were making efforts to produce a human embryo to derive stem cells from it. There was concern that Germany's strict Embryo Protection Law would become meaningless if it was not adapted and strengthened against such external pressures. Further, the Council of Europe's Convention on Human Rights and Biomedicine (better known in Germany as the "Bioethics Convention") had been hotly debated. Because the convention permits research prohibited in Germany, there had been strong public opposition to signing it. For example, the convention's cloning protocol merely prohibits cloning "human beings," but does not define that term. Because the convention leaves key definitions up to individual nations, protesting Germans claimed, states could ratify it but still clone "human beings" by simply defining human life as beginning at a later point in development.

The governing coalition, however, was internally divided and could not initially agree on a viable legislative proposal. After the first attempt to form a parliamentary ethics commission failed, there was renewed debate over the purpose of such a body. Numerous advocacy groups for patients and people with disabilities, as well as anti-bioethics groups and individuals, demanded a commission addressing the ethical questions posed by biomedical research. The Party of German Socialism (Partei des Deutschen Sozialismus, or PDS), the successor party to the East German Socialist Unity Party (Sozialistische Einheitspartei Deutschlands, or SED), then submitted a new proposal to form a commission, which was backed by many outside parliament. The strongest support came from disability rights groups, who had also vocally opposed the Bioethics Convention for its omissions and its lack of clearly defined legal concepts. They worried that the convention gave implicit permission to conduct research on subjects incapable of giving consent and to pass on the results of genetic tests to employers and insurance companies. They also noted the absence of a provision that would allow individuals to bring lawsuits before the European Court of Human Rights.

In December 1999, in an attempt to save face, the governing parties took the initiative by coming up with their own proposal and rejecting that of the PDS.[12] The proposed commission's mandate was to develop recommendations for dealing with the Bioethics Convention. The commission was to work to include the public and to generate a culture of openness, transparency, and participation in the ethics debate. Now supported by all parties, the proposal was accepted in March 2000, almost two years after it was first suggested. The commission began its work in the following month, shortly before the announcement that the human genome had been completely sequenced. Formally titled Enquete Kommission on Law and Ethics in Modern Medicine, the commission comprised thirteen members of parliament and thirteen experts. It had three firm mandates: first, to review the social, legal, and political aspects of current medical research; second, to investigate actual research practice and determine which aspects of it were incompletely regulated (*unvollständig geregelt*); and third, to develop criteria for maintaining the inviolability of human dignity in medical research. The commission had a purely advisory role; it was to develop recommendations that would prepare the way for parliamentary action.

Wolfgang Wodarg, a physician and a member of the Social Democratic Party (SPD), had supported the commission from the beginning and had worked hard to bring it about. Many expected him to chair the commission. At the last moment, however, his party voted to make Margot von Renesse head of the commission. Renesse, a former family judge with Parkinson's disease, was generally supportive of medical research.

Enquete Kommissions are a standard tool at parliament's disposal. They are explicitly intended as an informational counterweight to the advisory bodies that the chancellor may appoint, and they thus ensure the "balance of power" between the legislature and the executive. Typically composed of equal numbers of members of parliament and outside experts, these commissions are formed to lay the conceptual groundwork for important parliamentary decisions. Some Enquete Kommissions, such as the Enquiry Commission on Demographic Transformation (a topic of particular concern in Germany), are active for a decade or more, while others, like the one on coming to terms with East German "dictatorship," produce lengthy reports in a few years and are then dissolved. The Enquete Kommission on Law and Ethics in Modern Medicine was scheduled to work until the current legislative term ended in 2002.

New Ethics—The Nationaler Ethikrat

On December 20, 2000, Chancellor Schröder published a long article entitled "The New Human—Contribution Regarding Gene Technology" ("Der Neue Mensch—Beitrag zur Gentechnik") in the weekly newspaper *Die Woche*. Schröder wrote the article in response to more liberal regulations in other EU states, and he warned against Germany deviating from an emerging European consensus. He argued that the nation needed to remain economically and scientifically competitive with its closest neighbors. By speaking out "against a politics of ideological blinkers and fundamental prohibitions," Schröder signaled a change in the government's position and, for the first time since the late 1980s, started a nationwide debate on the future of gene technology.

One month later, on January 18, 2001, Schröder appointed Ulla Schmidt as Minister of Health. Schmidt supported Schröder's program and suspended work on the restrictive Law on Reproductive Medicine (*Fortpflanzungsmedizingesetz*) that her predecessor, Green Party member Andrea Fischer, had proposed. This law would have prohibited preimplantation genetic diagnosis (PGD), a form of genetic testing that allows one to select against certain genetic diseases in embryos produced through in vitro fertilization (IVF). This law was intended to give wide protection to the diversity of human lives, and its suspension aligned the Ministry of Health more closely with Schröder's broadly research-friendly agenda.

In February 2001, while still residing in East Berlin, the chancellor began talking about forming a national ethics council that would be answerable to the executive. His intention was ostensibly to broaden the debate on gene technologies and to represent German interests more effectively in international bioethics debates. One close observer of what would be the Nationaler Ethikrat (NER) told me that Schröder's appointment of the council was a legitimate attempt to oppose the minority opinion that the EK represented with a body that would more closely reflect the ruling majority opinion in parliament. The idea for such a council was not new, and there had been several calls for a federal ethics commission modeled on French or American national ethics councils.

It became clear that the chancellor held a more optimistic view of scientific progress than the EK and that he considered the EK to be working too slowly and obstructing scientific progress. In the spring of 2001, debates took place within the chancellor's circles about whether a separate national ethics council should be formed and, if so, where it should be located. Some suggested that it could be attached to the German president, traditionally a neutral elder statesman with largely symbolic power,[13] or to the Academy of Sciences in nearby Potsdam, also an institution dedicated to politically neutral pursuits.

The chancellor's proposal drew widespread criticism. Germans seemed to trust democratically elected representatives more than the chancellor with ethical decision making. To many people I spoke with, Schröder's initiative looked like a step designed to override parliament's supreme legislative and ethical authority. By forming a "national" ethics council of his own, they said, the chancellor seemed to imply that his commission could express the ethical will of the German people, a role that rightfully belonged to the sovereign legislator.[14] If the chancellor's commission was formed at all, it was clear that it would have to fulfill certain requirements and follow certain rules to be considered legitimate. As we will see in this chapter, the struggle for legitimacy between Germany's two national ethics commissions would make visible divergent visions not only of what it means to make ethical decisions in Germany today, but also of what the nation had been in the past and might be in the future.

Schröder, a pragmatist, was not given to hesitation. On Monday, April 30, 2001, the chancellor moved into the colossal Kanzleramt—the "washing machine"—to begin his work. Two days later, on Wednesday, his cabinet met for the first time in its new rooms and discussed the formation of the ethics commission. The other important topic on the agenda that day was the regulation of the *Dosenpfand* (deposits for beverage cans). That same afternoon a press secretary for the administration announced that the cabinet had decided unanimously to go ahead with creating the Nationaler Ethikrat.

Schröder invited twenty-three representatives from medicine and theology, industry and unions, science and sociology, to serve on the commission for a term of four years. The newspapers pointed out that several had already declined. Of those chosen, only seven did not have the rank of professor, and a mere three did not have a doctorate. The press secretary emphasized that this council of experts would be independent in choosing its topics and in organizing and administering itself, and that it would contribute to making the discussions over gene technologies "much more intensive and transparent than had been the case." Both the parliamentary and the executive commissions had a mandate to be transparent, but as we will see, this word was construed differently in each case.

The meeting room of the NER, and its administrative offices, are located in the Berlin-Brandenburgische Akademie der Wissenschaften (BBAW) on Gendarmenmarkt. This historic plaza in the former East Berlin boasts the beautifully restored Schauspielhaus and two identical eighteenth-century structures: the German Cathedral, which now houses the exhibit *Fragen an die deutsche Geschichte*, and the French Cathedral, originally built for use by the Huguenots living in Berlin and now a museum of Huguenot history and a conference center for the Evangelical Academy.

The BBAW itself is the successor of the Leibniz Academy, founded in 1700 by Gottfried Wilhelm von Leibniz, the philosopher, scientist, diplomat, and lawyer who, with Sir Isaac Newton, is credited with inventing calculus. Over three centuries the academy has gone through many mutations, but it retains its reputation as a place of scholarship and academic integrity. Even during the forty years of East Germany's existence, the academy, then called the Akademie der Wissenschaften der DDR, maintained a degree of independence from the state. It invited new members regardless of national origin, thereby earning international respect. Locating the NER at the BBAW, which was engaged in several big projects to strengthen German science (the most ambitious perhaps being a 122-volume edition of the writings of Marx and Engels, about half of which has been published at this writing), helped confer on the chancellor's ethics council the authority and legitimacy of an academy with a long history of illustrious scholarship and ethical probity.

Looking Back—The Enquete Kommission in History

While the NER had no immediate predecessors, the EK carried the burden and the benefit of the earlier Enquete Kommission on Opportunities and Risks of Gene Technology (Chancen und Risiken der Gentechnologie), whose popularity was still fresh in the public's memory. That older commission cast a shadow over the newer commission's attempts to define itself. One day in winter 2002, the director of the EK secretariat, where I was an intern, confessed to her staff during one of our regular biweekly morning meetings that the EK had missed opportunities to foster collaboration between the staff and the commission members. The staff could have let their strengths and competencies flow more actively into the discussion the way the earlier staff had done in the 1980s. The members of that commission had not been quite so familiar with the topics under discussion, and they had been grateful to the staff for substantive advice and help.

As if in response to such wistful memories, Klaus Schmölling came to visit the secretariat one day in early March 2002. A dignified elderly man, he radiated confidence and wisdom, and he was deeply familiar with the operations of parliamentary commissions in Germany. Schmölling had been the director of the secretariat of the commission on gene technology from the middle of 1984 until the end of 1986. That commission had been one of the best-known and most influential in postwar Germany, and its final report, considered the most comprehensive ever written on the future of gene technology, became an instant bestseller. In interacting with the former director, the current secretariat showed great respect for the past and an admiring deference toward this executor of the very model of the responsible dialogue between science, the state, and the public that our commission hoped to generate.

Schmölling reminisced with the satisfaction of someone showing off a sometimes creaky engine that was running smoothly for a change. His commission had been appointed not long after the 1975 Asilomar conference, and "everyone was still impressed with regulation that came from within science itself." Normally, he added, "an EK's recommendations are never accepted in that way, and with such speed." The legislative outcome of the debates informed by the commission's report was the Gene Technology Law (*Gentechnikgesetz*) of 1990, an act whose foresight and thoroughness were highly praised at the time, but whose shortcomings have since come to light. The law, he told us, is now widely considered to be too complicated, too ambitious in its efforts to regulate contingencies, and too technically uninformed (*fachfremd*) in that it leaves too many decisions to bureaucrats who are unfamiliar with the needs of science. Moreover, the law's technical definitions were now considered out of date and in need of adaptation.

Asked to make recommendations to "our" commission, Schmölling said that the official mandate should be as concrete as possible and that the commission should stick to it closely, rather than letting the topic spill over (*ausufern lassen*) and inventing new questions—all of which were merely excuses for making itself so indispensable that it could then recommend its own reinstitution (*Wiedereinsetzung*). A staff member later echoed this sentiment, telling me that parliamentary commissions by their very nature have a mandate so vague that it requires significant creativity, initiative, and self-discipline to keep their task within manageable limits. In his observation, as in Schmölling's use of the word *ausufern*, I heard a fear of overflowing, of being overwhelmed, that I was to hear throughout my fieldwork in Germany.

Back in the 1980s, Schmölling remembered, there had been enormous time pressure and great passion for the topic, and the goal had been to write a report that could be read by the average person on the street: "One also wants to convince the baker." A good test, he told us, was to have the "second secretary" read it; if she understood it, then it was simple enough. For Schmölling, the baker down the street and the presumptive pink-collar worker were the ideal audience because they represented a common enough denominator to stand in for the broader public. The actual second secretary of the secretariat, who was sitting at the table, was visibly embarrassed. Schmölling added that he had considered hiring a journalist to go over the report, but that this would not be necessary in our case because our director herself had a journalistic background. The problem, according to Schmölling, was that German researchers "don't yet understand how to express themselves intelligibly. It is different with the Americans. The OTA (Office of Technology Assessment) reports were a delight (*Genuss*) to read." Perhaps he did not know that those reports too had been produced not by experts, but by committees and staffs much like the one he himself had supervised.[15]

Schmölling's earlier commission left some other important legacies beyond the Gene Technology Law itself. Several of its members had risen to prominence in the field of science policy. Most visibly, the commission's president and the report's coauthor, Wolf-Michael Catenhusen, was now in high demand as a science policy expert and a significant player in regulating science and technology. Wolfgang van den Daele, then a professor at Bielefeld and an expert member, was now a member of Schröder's NER and a professor at the Wissenschaftszentrum Berlin (WZB), a major center for policy-related research in the social sciences. Former staff member Regine Kollek also served on the NER. Ernst-Ludwig Winnacker, an expert member, became head of the German Research Society (Deutsche Forschungsgemeinschaft, or DFG—Germany's equivalent of the National Science Foundation) and for some time was also a member of the NER. Clearly the earlier commission had produced a powerful cadre of science policy experts along with viable recommendations for regulating research.

Another accidental continuity involved Schmölling himself. After the commission had completed its work, both he (the head of secretariat) and Catenhusen (the commission head) were transferred to the Committee for Education and Research (Ausschuss für Bildung und Forschung), which was supposed to turn the commission's recommendations into concrete policy. In effect, authors became interpreters of their own work as Schmölling and Catenhusen handed the commission's report and its recommendations over to themselves for implementation.

Looking Around—The EK and the NER

Even more than the earlier Enquete Kommission on gene technology, the NER was present in our EK's consciousness throughout my time there. The rivalry between the two ethics commissions was brought to my attention from day one. Comparisons with the NER would become a dominant theme in the EK's self-definition, and they would resurface at regular intervals. Stories of the NER's blunders were chuckled at with more than a hint of schadenfreude.

The rivalry between the EK and the NER played out in issues of content as well. There was almost a competition to come up with the more ingenious, or more socially robust and sustainable, regulation. For a while the NER toyed with the idea of suggesting a variant of a cutoff date for embryonic stem cell imports (allowing the importation of only stem cell lines produced before a certain date so as not to encourage the destruction of new embryos to establish new lines). Rather than specifying a single date for all future imports, their idea was to allow the importation of only those stem cell lines that had been established at least six months prior to the German researcher's application. As the EK quickly pointed out, however, such a restriction could be easily circumvented by ordering particular lines to be produced in advance and then, six months later, applying to import them, thereby thwarting the goal of preventing the destruction of further embryos for German research.

When the NER was first appointed, it needed an expert staff. As a new kind of advisory body, it did not have immediate access to an administrative apparatus of the kind that the Bundestag provided to the EK. When the NER set out to hire a few well-qualified people for its staff, several members of the EK secretariat applied. The NER ended up hiring one of them, a woman who had left around Christmas, about two weeks before I arrived for my several months of participant observation. She was half-jokingly referred to as a "defector"—someone who had sold out—since she had traded the legitimacy of the EK for the much more dubious NER, where her income had risen several notches on the national pay scale.

The first director of the NER secretariat, Simon Golin, a trained theoretical physicist, had chosen for himself the title of "General Secretary" (*Generalsekretär*), much to the amusement of the EK staff. The military-sounding title also designated a position of power in large political parties, including the Socialist Unity Party of the former East Germany. It was rumored that this title revealed the holder's desire for hierarchy and discipline. His salary class of B3, people said, was normally reserved for someone who commanded a military unit of about 900.[16] The implication was that he was extravagantly overpaid for running a secretariat with a small number of subordinates, who themselves were very highly paid by national civil service standards. There was some bewilderment at a professorial salary level for someone who came with no obvious prior exposure to bioethics.

Making a new institution function smoothly without clear precedents is a difficult business. Golin tried to accomplish this by creating an atmosphere of pure professionalism. In order to produce an "esprit de corps" among the NER staff, as he himself once put it, he at one point opened letters addressed to individuals who worked in the secretariat. After several reputed run-ins with staff members who resented this behavior, and who found offensive his unilateral redrawing of the line between public and private, the general secretary was eventually let go. The philosopher who replaced him as head of the secretariat changed his position title from "general secretary" to "office director."

I myself had one brief exposure to the general secretary. When I once inquired about an internship at the NER to compare it with the EK, Golin encouraged me to submit my résumé, but warned that the application process was rather competitive. When I told him why I was interested in the position, he seemed favorably disposed, but informed me that I should not expect to work on any interesting bioethical issues. Rather, he said, "we basically need someone who puts out the breakfast rolls on the mornings of the commission meetings." Since I was interested mainly in the commission meetings themselves, which took place only once a month, I ended up not applying. While this choice may have made my perspective more partial, my exclusive affiliation with the EK gave me different, and possibly deeper, insights into the two institutions' relations by allowing for less guarded discussions with my co-workers.

Golin also visited the EK secretariat once. Having been in office for only six weeks at the time, he came to get advice from our secretariat director, Cornelia Beek, on how to administer his own secretariat. Cornelia was unable to tell him much, I later heard, because the mandate and structure of the NER were so different from the EK's and because the professional environment Golin wanted was so different from the more relaxed atmosphere that she had cultivated. According to one of my co-workers, who had not met Golin before, he looked as if he had come from another planet—as if he was not quite there, or utterly preoccupied with other things.

Cornelia thought that the EK had been appointed too late and that its topic was too urgent to work on effectively. With the public looking over one's shoulder, it was hard to concentrate. She was critical of her own publicity work for the EK, but said that the NER's was even worse. That would not be the case for much longer, however. One day in February, one of the EK's scientific advisors said he had learned that the NER had now hired a press secretary, someone who had previously coordinated public relations for the BBAW. That person's sole job would be to communicate with and maintain friendly relations with the media. The EK staffer conceded that the NER had picked the right "label" and noted its savvy in marketing itself. Other EK members openly wondered whether the EK would survive the current legislature. Less pessimistically, some thought that a new EK would be constituted because the Bundestag could not afford to be without an expert advisory body now that the executive had appointed its own bioethics commission.

The EK maintained its boundaries in relation to the NER by always referring to itself as legitimated through the democratic process. This legitimacy was not sufficiently obvious to outsiders that it did not need to be reaffirmed through repeated claims. One reason for the EK's distancing move was that the media also continually juxtaposed the EK and the NER as two equivalent voices in the debate. The NER, on the other hand, welcomed the parallel treatment, since it brought that commission into the public consciousness and positioned it as a legitimate counterpart to the EK. Moreover, whereas the EK was concerned over whether it would be reinstated after the upcoming elections, the NER was appointed for at least four years, and was thus in a way more stable. It seemed that the democratic process, rather than stabilizing the EK, only added to its fragility.

If questions about the EK's legitimacy did not arise in public, they did arise within the commission itself. One day I witnessed a debate over the meaning of "democratic legitimation." It started when the commission was formulating a section of its final report outlining recommendations for future action. The discussion turned toward specifying what kinds of commissions should be appointed in order to help focus future debates. Democratic legitimacy emerged as a key criterion. One expert insisted that the EK was legitimate because elected representatives of the *Volk* had instituted it. He contrasted the EK with the NER, which to him was an expertocracy and not democratically legitimated. Another expert countered that the EK was not legitimate by that criterion either, since all those present (there were only expert members in the room at this meeting) were there only because they were recognized as experts. The first speaker replied that this might be true for the individual members, but as a whole the commission had been democratically appointed. In the end, the sentence containing the contentious phrase "democratically legitimated" was struck from the report, but disagreement remained over whether this bold cutting of the Gordian knot had solved the problem.

In informal conversations with insiders, it was hard to maintain that the NER was somehow less legitimate because the executive had appointed it. The chancellor, some pointed out, is also elected, and having the chancellor pick experts for an executive commission is no less democratic than having political parties pick experts to sit on a parliamentary commission. In the end, it often seemed that the experts did most of the work in the EK anyway, while its politician members served mainly to grant formal legitimacy to the commission—and to draw media attention to it.

While seeking models for future deliberative forums, the EK staff produced a list of ethics councils in Europe. During the subsequent commission meeting, the atmosphere became tense when people tried to fit the German case, with its two rival commissions, into the list. One EK member, who was *also* a member of the NER, insisted that the NER be included in the list "for completeness's sake." After a brief debate, the commission decided that the NER could be included, at which point the doubly appointed expert began reading a text on the NER that she had prepared in advance. Renesse agreed that the presence of two rival commissions was not novel in Germany. After all, the Enquete Kommission on gene technology had worked for the Bundestag at the same time as the Benda Commission appointed by the executive.[17] After some more discussion over whether Germany should be given a place among the other nations or a special section unto itself, the commission decided to compromise by adding a special chart on "German Ethics Commissions."

Can the Nationaler Ethikrat Be Ethical?

The NER attracted more public criticism than the EK. From the beginning, detractors questioned the council's supposed independence from a chancellor who had handpicked its members and who was also providing it with generous funding of 4.2 million DM per year (later 2.14 million euros). The chancellor, as opposed to parliament, represents only one political orientation and therefore possesses in the public eye less capacity to form a collective consensus. Thus, while independence from parliament never became an issue for the EK, the NER's independence from the chancellor remained a great concern. Critics wanted the chancellor to clarify the NER's relation to the EK. Members of parliament, in particular, were concerned that the chancellor's initiative would render obsolete parliament's own bioethical advisory commission, thereby intruding on a domain that they claimed ought to be managed by elected representatives. Significantly, during the parliamentary stem cell debate on January 30, 2002, Schröder spoke not as chancellor, but as a representative, thus putting himself on a par with any other representative.

While the NER was financed at the chancellor's discretion and could spend its money as it pleased, the EK was financed very modestly according to public guidelines, and it was accountable for its expenditures to a degree that prevented extravagance or display. The NER's greater financial resources allowed it to host numerous highly visible events and to commission a great number of the expert reports that are part of such an institution's currency—at least in the EK's eyes. During my internship with the Enquete Kommission, one commission member remarked to me during an interview that the NER's influence was directly proportional to its resources. Resignedly, she told me that "the NER is working with power and great financial backing, buying itself endless expertise (*kauft sich unendlichen Sachverstand*) in an attempt to overthrow the existing legal situation in Germany . . . and perhaps they will succeed."

Some critics disapproved of the NER's composition, particularly the absence of persons with disabilities, and were concerned that the voices of the weak would go unheard. The EK had among its political appointees Ilja Seifert, a member of parliament who used a wheelchair. Seifert had been active in mobilizing persons with disabilities in Berlin and all over Germany and had given this constituency a national platform. As I will show later, the mere presence of a disabled member gave the EK a special kind of legitimacy, at least in the eyes of the media, although this particular politician's PDS membership represented a handicap of a different kind. The NER, by contrast, had no disabled members, but only a proxy: Christiane Lohkamp, the speaker for the national patient advocacy group for Huntington's disease. The NER would correct this oversight the following year, when it replaced a departing member with the Munich author, actor, and playwright Peter Radtke, another wheelchair user. Radtke has *Glasknochenkrankheit* (osteogenesis imperfecta), a rare genetic disease that affects the production of bone-strengthening collagen and thus renders bones extraordinarily fragile. Showing remarkable outer strength despite his inner fragility, Radtke has written extensively on the social consequences of new technologies. His 2001 autobiography is entitled *Karriere mit 99 Brüchen* (*Career with 99 Fractures*).

The NER's function was evidently conceived as handing down ethics to the people. The seven-paragraph summary statement that the chancellor presented to the cabinet on April 25, 2001, asserted, first, that the NER would be a national forum for dialogue on ethical questions in the life sciences. The second paragraph explained the commission's composition and stated that the NER would join together (*bündeln*) the interdisciplinary discourse of the natural sciences, medicine, theology, philosophy, and the social and juridical sciences. The statement went on to say that the NER would organize the societal and political debate through the incorporation of different interest groups. It would offer information and opportunities for discussion to citizens—but would do so only after its own internal deliberations were concluded. In what some saw as an attempt to dominate ethical debates, the NER would also have the mandate to produce ethical position papers requested not only by the administration, but also by the Bundestag. Spreading its influence still further, the NER would collaborate with other national ethics councils, mainly those of European states. Finally, the NER would be financed by the administration.

Some of the chancellor's nominees refused to go along with what they perceived as an uncharacteristically opaque process. The prominent sociologist Elisabeth Beck-Gernsheim, for example, said in an interview with *Die Tageszeitung* that she turned down the call because "the nominating process was not transparent."[18] Interestingly, there were few such concerns over the transparency of the EK's nominating process. There, as we have seen, leadership questions were decided at the last moment by party fiat, and each party in the commission then chose experts to its own liking. Transparency, it seems, is presumed to be built into the structure of the parliament, as Foster's dome is built onto the Reichstag, while the chancellor has to go out of his way to demonstrate it. In any event, Beck-Gernsheim told the newspaper that she had received a phone call from the Kanzleramt late in the previous week inquiring about her willingness to participate, but that she could not get an answer to her question of what, precisely, the NER would be deciding. Nor could they address her concerns about the potential for competitive interference with a third national ethics commission, previously located at the Department of Health.[19] That commission had, over time, become increasingly critical of the chancellor's easy acceptance of gene technologies, and Beck-Gernsheim had the impression that the NER was an attempt to manufacture ethical recommendations more in line with the chancellor's views. Furthermore, in her opinion, the natural sciences were overrepresented on the NER, while there were not enough voices critical of new technologies. Rather than allowing herself to be tainted by joining this dubious outfit, she refused to participate in its operations altogether.[20]

Other critical voices nevertheless joined the NER because it meant a chance to codetermine future directions of research. Regine Kollek, for example, is a onetime biologist who left the laboratory bench to become a professor of sociology. As she told me and others following a talk at the Institute for European Ethnology, she had read so much sociology of science that she took for granted the idea that "we are constructing reality." For a while this thinking "outside of certain boxes" allowed her to be a more creative scientist, and she was able to think up novel experiments. Eventually, however, she could no longer continue doing science in good conscience (*guten Gewissens*) and so switched fields. Although she had left science when its constructedness became apparent, her subsequent critique did not turn against science as a whole, but against science as an elitist and unimaginative endeavor. She saw it as part of her role to restore democracy and creativity to the study of nature. When ethnology students asked her why she had agreed to join the NER, she recounted that she had received the call from the chancellor's office just before going away for a long weekend to her second residence. When the caller pressured her to decide immediately, she refused and spent the weekend calling all her friends on her cell phone. When each of them recommended that she do it, she agreed, giving as her reason the need to "push the boundaries a little in our direction." When a student asked why she, in particular, had been chosen, Kollek surmised that Chancellor Schröder had to include some critical voices, such as hers, in order to legitimate the NER.[21] The balance, she noted, was still tilted toward relatively unchecked scientific progress.

When the list of nominees was published, it quickly confirmed people's suspicions. Although several well-known theologians and even two bishops were among the members, there was a sense that well-known names from the circles of scientific and medical research outweighed the cautionary voices. These names included Detlev Ganten, then head of the Max Delbrück Center for Molecular Medicine; Peter Propping, a human geneticist; Christiane Nüsslein-Volhard, a Nobel Prize–winning developmental biologist; and Ernst-Ludwig Winnacker, head of the German Research Society (DFG), the main provider of research funding in Germany.

On May 3, 2001, the day after the NER's official formation, the DFG, which had previously claimed that research on human embryonic stem cells was not vitally necessary so long as experiments on human adult stem cells could proceed, reversed its opinion and recommended that research be conducted on both types of stem cells. The DFG now concluded that importing embryonic stem cells for research purposes was legitimate. Since the DFG president, Winnacker, had been asked to serve on the NER just a few days before, critics speculated that this new constellation (i.e., that he was now in Schröder's service) had influenced the research advocate's revised judgment, a speculation Winnacker denied. Soon afterward, on May 31, 2001, parliament held a plenary debate to discuss the place and the future of gene technologies, such as preimplantation genetic diagnosis (PGD) and stem cell research, in Germany. Parliament reached no definitive conclusion, but it became clear that all political parties were internally divided on these issues.

During that month, the prime minister of the state of North Rhine–Westphalia, Wolfgang Clement, announced that he would support the work of neurologist Oliver Brüstle, Germany's most prominent stem cell researcher, on stem cell lines imported from Israel. Shortly after parliament had agreed on the need for more debate, it turned out that Clement had used the time of collective deliberation to travel with Brüstle to the Rambam Medical Center in Haifa, Israel. Their object was to negotiate an agreement for the importation of Israeli stem cell lines into Germany. Many Germans were upset when they learned that a politician and a scientist had tried to generate facts on the ground, and to all appearances might have presented the public with a fait accompli while the ethical status of those very facts was still under debate. Clement and Brüstle had violated the precept that an ethical dialogue had to occur *before* the fact, not after, and fueled fears that regulation of research might once again arrive too late.

The churches, in particular, protested against any loosening of the stringent 1991 Embryo Protection Law. Gene technology was one of the main themes of the twenty-ninth annual Day of the German Protestant Church, held in Frankfurt that June. The head of the Evangelical Church demanded that all embryo research and PGD be prohibited. But other important individuals and institutions also took a stance against the chancellor's vision. On May 18, 2001, the German president Johannes Rau, in his annual Berliner Rede,[22] spoke out against embryo research. Rau argued for setting and maintaining ethical limits on research. He indirectly criticized Schröder's December 2000 article in *Die Woche* by saying that "where human dignity is at stake, economic arguments no longer hold." Rau argued that ethical research requires the setting of firm boundaries and that there must be no crossing of the Rubicon in the name of research. In June 2001, shortly after the inaugural session of Schröder's Nationaler Ethikrat, the president of the Max Planck Society, Hubert Markl, in his annual speech to the society, rebutted President Rau and argued that neither commissions nor councils could make decisions that require each individual's conscience to be activated. Markl argued against Rau's view of a research Rubicon as a static ethical limit. Instead, he described the ancient river as flowing in a bed that conscientious researchers had to craft and recraft continuously. The metaphor of the Rubicon as a limit for ethically acceptable research would become one of several images defining and even controlling the German bioethics debate.

At its first meeting, on July 6, 2001, the NER voted on rules of order (*Geschäftsordnung*) for its future meetings. In the first sentence the NER claimed ethical sovereignty by reiterating its independence from external institutions. Members would represent only their personal convictions and be bound only by their own conscience (*nur ihrem Gewissen unterworfen*). They obligated themselves to reveal conflicts of interest and to treat their meetings and documents as confidential. Since the NER made its decision by majority vote, it had a working quorum (*beschlussfähig*) when more than half of its twenty-four or twenty-five members were present.

Over the summer of 2001, the debate on the ethical limits on research became more heated, and the question of *what* those limits were became deeply intertwined with the question of *who* should define them. Both questions acquired urgency through international developments. While Germany debated whether to ban the importation of stem cells derived from spare embryos from IVF, thus avoiding the instrumentalization of human life for research, the Massachusetts biotechnology company Advanced Cell Technology (ACT), as mentioned earlier, had announced that July that it had for the first time accomplished that very instrumentalization with no apparent ethical misgivings. According to the news, ACT said it had produced human embryos in vitro for the sole purpose of deriving embryonic stem cells from them, a practice that the German Embryo Protection Law prohibits. In the same month the German coalition government decided that it would wait for the stem cell reports of both the Enquete Kommission and the Nationaler Ethikrat before calling for a parliamentary vote on the permissibility of embryonic stem cell research. The DFG, in turn, agreed to postpone its decision on Brüstle's application for research funding until after the parliamentary debate. On July 5, 2001, parliament voted against a prohibition on stem cell imports proposed by the conservative Christian Democratic Union (CDU). The liberal Free Democratic Party (FDP), on the other hand, wanted to permit both research with human embryonic stem cells and PGD. In the United States, in the meantime, President George W. Bush sent a clear signal in his speech of August 9, 2001. He announced that public funding would be available for research only on those embryonic stem cell lines that already existed on the day of his speech. There were, in America, no legal limits to deriving new cell lines using other funding sources.

The German parliamentary debate on the importation of embryonic stem cells was postponed at least once because of the emergency following the terrorist attacks of September 11, 2001, but a date was finally set for January 30, 2002. In November 2001 ACT claimed to have cloned a human embryo for the first time and developed it to the four- to six-cell stage. Not only had they produced embryos for research, but they were now producing identical copies of those embryos. Their announcement intensified the sense that the work of ethics always comes too late, after science and technology have created realities on the ground.

On November 12, 2001, the Enquete Kommission published its report on stem cell research. The first hundred densely printed pages cover, with textbook thoroughness, the scientific, historical, legal, and ethical background on human embryonic stem cell research that is presumably needed before one can render a fully informed and legitimate ethical judgment. They are followed by a single page summarizing two strands of argument. The first strand, signed by all twenty-six commission members, demands that Germany take all necessary steps to prevent the importation of human embryonic stem cells. It states that the goals of basic research can be met by work on stem cell lines that are not derived from human embryos, such as stem cells from primates, from cord blood, or from adults. The second strand, signed by twelve of the twenty-six commission members, takes a more realistic stance. Recognizing that complete prohibition of importation may be unconstitutional or unenforceable, it recommends tolerating importation in exceptional circumstances under the strictest conditions and charging a "transparently working, officially legitimated control authority" (*transparent arbeitende staatlich legitimierte Kontrollbehörde*) with supervising the process. The remaining fifty pages of the report cover alternative and noncontroversial kinds of stem cells and give an overview of research regulation in other nations.

Just two and a half weeks later, on November 28, 2001, the Nationaler Ethikrat announced the result of *its* deliberations in a slender and highly readable publication of just twenty-four pages. After a brief discussion of the general legal, ethical, and social arguments for and against human embryonic stem cell research (including a section on the symbolic importance of protecting embryos), the report enumerates the pros and cons of importing embryonic stem cells. It then presents four options, labeled A through D in increasing restrictiveness, for addressing the question of importation. Option A argues that deriving stem cells from spare human embryos is ethically permissible both in Germany and abroad and that therefore the importation of stem cells, under certain conditions, is permissible as well. Option B permits the importation of human embryonic stem cells provided that they were not created for the purpose of research alone. Fifteen of the twenty-five council members voted for Option B, with nine of those also voting for Option A. Option C places a moratorium on the importation of human embryonic stem cells, during which alternatives should be explored, possible consequences of importation evaluated, and supervisory institutions created. Option D argues that the instrumentalization of human embryos that is inseparable from the derivation of human embryonic stem cells makes their importation ethically impermissible in principle. Ten of the twenty-five council members voted for Option C, with four of those also voting for Option D.

Although the two votes seem almost equivalent in their formal results, the logics underlying them were significantly different. The majority of the EK, in effect, voted *against* importation in principle, but allowed exceptions, while the majority of the NER voted *for* importation in principle, but imposed tight controls. The difference was significant in a nation that considers the principles and precise reasoning that underlie a political decision paramount.

Ethics Commissions as Saalordner

How did the ethics commissions justify and conceive their own roles? How did they order themselves and the discourse around them? Both commissions claimed to function as *Saalordner*; that is, as neutral providers of order. In this role, both represented themselves as sources of clarity and principle, but in different senses. An EK member found fault with the fact that everyone was talking when only serious experts, speaking for everyone, should have a voice. NER members, by contrast, located the problem in the poor framing of the issue by the public. The EK's solution, therefore, was to act as a commission in which all interests were represented and all voices could be systematically and professionally heard, while the more openly elitist NER solved the problem by reframing the issue so that everyone could see the real dilemmas clearly. Put differently, the EK saw its task as aiding in the clearer articulation of public opinion, but the NER sought to lead or guide the public through its own more informed opinions. The struggle between a commission "of the people" and another "for the people" exposed different visions of democratic legitimation.

At a podium discussion at the Urania on November 14, 2001 (a forum dedicated to Humboldtian ideals of publicizing science, founded in 1888, and the oldest such forum in the world), I heard a view articulated that I have since heard from many others. The event started fifteen minutes late because a room built for an audience of six hundred seemed deserted, with only about fifty persons in attendance. Unfortunately the event was competing with the qualifying game of the national soccer league. The stage was carefully set: the speakers were sitting on the proscenium of a fancy movie theater, with a red velvet curtain behind them. During the discussion, a physician and technology assessment expert expressed his view of the role of ethics commissions. Ethics, he said, is not the "conscience of science" (*wissenschaftliches Gewissen*), but a *Saalordner*. He claimed that people often conflated morality and ethics. Morality was something that everyone had and that everyone must have. Different moralities were often incompatible. This led to conflicts. Solving such conflicts was the main task of ethics. This meant that the ethicist must *not* be the ultimate decision maker. Instead, the ethicist must bring the different parties together so that they can solve the conflict.

The speaker gave an example from embryo research: Everyone acknowledges that the embryo is a form of life. Some take this to mean that the embryo is part of the biological domain, with no or few special rights. Others take it to mean that the embryo is a bearer of special rights. The ethicist's task is to order the options for acting and point out their consequences. In general, this speaker claimed, the different parties in a debate were very willing to let themselves be ordered (*sich ordnen zu lassen*). The speaker gave another example from the "wild" United States: to some, abortion means exercising a woman's rights, while to others abortion is murder and justifies murder—of abortion doctors—in return. In such a case, the speaker conceded, what was needed was not an ethicist, but the police. The ethicist's capacity for ordering docs have limits. Later in the discussion the same individual spoke again: he said that ethicists did not want to speak for anyone else. One had to distinguish between ethicists and lobbyists. Science and ethics merely inform politics. And as if to underscore the efficacy of his profession, he claimed that "the debate [on stem cells] is becoming more and more ordered. . . . The *Saalordner* can sit back."

This speaker saw *Saalordner* as people who place others in the appropriate groups by pointing out the implications of their arguments and then let those groups debate the respective merits of those arguments rationally. But *Saalordner* appear in the press in a somewhat different guise. They are the hired hands who enforce and maintain order at tumultuous events like rock concerts or demonstrations by removing disturbing influences from the grounds. *Saalordner* are there to generate order, then, but in common usage they often resort to force in doing so. References to *Saalordner* in the everyday world typically include descriptions of struggle and physical violence. When there is no sign of trouble, the *Saalordner* are invisible. They become visible only when there is a problem.

On January 24, 2002, less than one week before the long-awaited parliamentary debate, the NER held its first public meeting. Many had waited for this event, given that the NER had promised transparency. After the NER put all the arguments together, it had come up with four options. As described above, one categorically rejected importation, one permitted importation with some caveats, and two steered in between. Spiros Simitis, a data privacy lawyer and the NER president, later described how inquisitive journalists had practically coerced the council members into voting on the four options so that parliament would know which one the NER was favoring. In the future, Simitis said, the NER would avoid votes. After members of the NER had presented the council's report on the importation of stem cells, the public had numerous questions for the council. One citizen asked about the significance of the vote at the citizen conference on bioethics in Dresden two months earlier (which I discuss in chapter 3), where all the women had voted *against* legalizing preimplantation genetic diagnosis. Council member Wolfgang van den Daele, professor of norm formation and environment (*Normbildung and Umwelt*) at the Berlin WZB, responded that the citizen opinion was a *Gutachten*, normally an evaluation by an expert, not a decision. He said that "it was a *Gutachten* among others, exactly as ours is. And it is also important that *all the men* were *for* PGD." In an interview van den Daele specified that "they [the ethics commissions] are not supposed to decide the conflicts, but to present them as clearly as possible, and to preclude demagoguery, false conclusions, and quick ascriptions of blame. Nothing more. In ethics councils there are no specialists for moral questions. Such experts don't exist. Everyone must find his or her own answer—just as in the case of abortion."[23] Van den Daele's views were consistent with the NER's framing role, but members of the audience were not so easily appeased. They still wanted to know precisely how this council of professors and specialists had reached its decision—and even *what* they had decided.

As orderers of discourse, both ethics commissions functioned just so—to quell the disturbance of public passions. These ethical *Saalordner* brought their competence, and the reasoned objectivity of facts, to a debate otherwise dominated by the public's perceived incompetence and by the turbulent subjectivity of emotions. In 2002, after the Enquete Kommission concluded its work, a commission member recounted to me the landscape into which his commission was fitted:

> Before this Enquete [Kommission] existed, there were the most adventurous and, excuse me, the stupidest discussions on this topic. In all of these afternoon television shows, where everyone who had ever whistled a tune could call himself a star, there people now had to speak their mind on preimplantation genetic diagnosis, even if they could not even pronounce the word. But they had to say whether they found it great or awful. . . . It was simply a ridiculous, stupid discussion, and marked by utter ignorance of the facts [*Sachunkenntnis*]. . . . From lifestyle shows to afternoon shows to teenager shows, it was awful, *just awful*, what was being done. And then there were of course people who had grappled with this bioethics in a serious manner, and there of course with the appropriate effects on biomedicine, and they wanted to give all of this a serious foundation. And everyone believed that the Enquete Kommission could be effective in that respect. Surprisingly it has accomplished this goal. I myself could hardly believe it; that it could have *such* an effect.

The Nationaler Ethikrat was similarly convinced of its ability to bring order—and indeed, of its having brought order—into a chaotic world, and of its mandate to do so. When on January 23, 2003, the NER presented its second position paper, on preimplantation genetic diagnosis, the document was nearly eighty pages long, almost four times the length of its first, twenty-four-page position paper on the importation of human embryonic stem cells. The introductory comments by NER members were telling. In the report, they said, they had compiled whatever knowledge was accepted by the scientific community (*wissenschaftlich weitgehend gesichert*). Thus the first part of the report recapitulated the development of the human being before birth, then discussed different diagnostic methods, what could be learned from them, and subsequent options for action. The second part, as long as the first, described and explained the council's ethical judgment. The presenters said they had attempted to make the language as comprehensible as possible. The appendix provided images so that everyone could "*noch einmal nachvollziehen*" (recapitulate conceptually) the development of the embryo inside the mother. One NER member, referring to the report's first part, said that "one cannot break laws of nature. . . . It would be good if the first part of the position paper would have the result that people will start discussing *the real problems* instead of merely apparent ones [*Scheinprobleme*]. The real problems are numerous and difficult enough." The speaker implied that, all this time, the disordered and unaided public had been debating the "wrong problems." Presumably, the content of the public debate would henceforth be significant only to the extent that it fit the frame built by the NER.

The NER's self-understanding became even more evident in September 2002, when its president, Spiros Simitis, was invited to present Germany's institutional response to biotechnology before the U.S. President's Council on Bioethics. The Bundestag debate was over, and the question of stem cell importation was resolved. The EK had ceased to exist in August 2002 and had, for the time being, literally been erased from the landscape of German bioethics. The NER had emerged as the more visible ethics commission, and Simitis was in Washington as the NER's ambassador.

Simitis described the mission of the NER not as substituting for the parliamentary decision-making process, but as "providing parliament and government with arguments." He emphasized the NER's positive influence as authoritative *Saalordner*. The NER, he implied, had succeeded in putting clamoring voices in perspective, clarified the positions, brought the right reasons into the discussion, and calmed the previously turbulent debate. According to this picture, everyone's hopes had been pinned on the NER. The scientists were waiting for a positive ethical evaluation of their research programs, and the NER had "the press and the mass media continuously outside our doors." Moreover, Simitis represented "the general public" as anticipating that "at last there's someone who's going to *say something*." The NER, in his account, rose to the challenge and produced its report, "as we had promised," more than a month before the parliamentary decision.

Simitis was careful, but dismissive, in how he talked about the EK. Not surprisingly, he said nothing of the widespread indignation and concern over legitimacy that had first confronted the NER. Instead, he hinted that it was the EK whose judgment had lacked the neutrality and objectivity required for ethical legitimacy, while the NER had been given, and had accepted, the role of mediating between opposed subjective positions. Simitis said that the newly appointed NER was immediately under pressure to address the question of importing stem cells. The issue was pressing because the first projects for the importation of stem cells were already under way and because the university institutes and professors who were planning these projects, as well as the German scientific community at large, were very favorably inclined toward importation. The EK, "a parliamentary commission dealing with these problems," he noted, was already overly critical of these projects and was preparing a negative report for parliament, which would have to decide on how to regulate potential imports in just a few months. Simitis did not add that the EK's report turned out to be neither extreme or even unusual. On the contrary, the parliamentary vote was congruent with the majority opinion of the EK (*no, but*), while it rejected the position held by the majority of the NER (*yes, but*). As Simitis told it, however, it was the NER vote—a conditional yes to importation—that was the opinion that prevailed in parliament.

Soon after Simitis returned from his visit to the President's Council on Bioethics, which held its meetings in public in accordance with American law and practice, the NER made its own meetings public as well. Unlike the EK, whose procedures were fixed in rulebooks, the NER was free to change its procedures as it went along. So, while the EK went abroad mainly in search of wider networks and stronger arguments, the NER, true to its more charismatic style of leadership, brought home a more skillful style of political self-presentation.[24]

"This Is Not Bioethics"—"Bioethics Is a Dirty Word"

On January 7, 2002, the day I joined the Enquete Kommission secretariat as an intern, the two scientific staff members who were in the office came up to me and introduced themselves. Johann Ach was from Bavaria, but he now lived in the northeast German city of Rostock, from where he commuted to work in the commission every week. He had begun to study theology, but two well-meaning philosophy professors in Augsburg had dissuaded him. He switched to the University of Münster, where he spent several years working on ethical issues in science and medicine and wrote a dissertation on the ethics of organ transplantation.

Matthias Wolfschmidt was in his mid-thirties, had studied philosophy, history, and literature, and was a trained veterinarian. After two years in veterinary practice, he came to the Bundestag to work for a parliamentary representative and immersed himself in health politics, consumer protection politics, and biomedical ethics. In the mid-1990s, when public and parliamentary discussions were centered on the organ transplantation law and on the Council of Europe's Bioethics Convention, he switched to a different member of parliament. In that position, he worked on environmental and gene technology policy. When the EK was formed in 2000, he applied to work as a staff member and was chosen from several dozen applicants. Both quickly switched from the formal *Sie* to the informal *Du* in talking to me.

They asked why I had chosen to come to the EK, and I told them that I was interested in the ethics commission as a form of institutionalized morality. They then both insisted that the EK was not an ethics commission: "A national ethics commission does not exist in Germany." Even the formation of the NER, to all appearances a deliberative body concerned with bioethics, was what they called a "political decision." I asked how the bioethical is separated from the political, and Johann, a trained ethicist with utilitarian inclinations, explained that the problem with the NER was that no one on it had at any point thought seriously about bioethical questions. He added that he found his own work in the EK, and especially in the secretariat, frustrating as well because it too had nothing to do with bioethics, or even with ethics more generally, which he defined as resting on academic freedom.

A pessimist by temperament, Johann wanted to destroy any illusion that I had now arrived at an important bioethical node. I knew from others, however, that feelings of insignificance and irrelevance were widespread among those working in the field that others elsewhere call bioethics. Throughout my time in Germany, I hardly met anyone who referred to themselves as bioethicists or who accepted the label uncritically. Whenever I asked, in interviews or in other contexts, how my interlocutors had come to the field of bioethics, they typically responded that they were not bioethicists, but rather philosophers, or ethicists, or physicians. "Bioethics" seemed to be a dirty word to many Germans, smacking of the applied, the hybrid, the foreign, and the unscientific. The word itself was marked as foreign or alien in the German language: on a follow-up trip to Germany in 2005, I found a relatively recent edition of a *Fremdwörterbuch* (a dictionary of foreign-derived words) that boasted on its back cover that it had now added such new (and fashionable?) terms as *Bioethik*.

An accidental encounter shed light on possible reasons for the collective denigration of bioethics. In the middle of August 2002, I met

Matthias Bormuth, an ethicist and prolific writer on ethics in medicine and psychiatry, through a mutual friend. Bormuth had no connection to the official bioethics forums in Berlin, but my friend introduced him to me as a possible source of insight. Bormuth had studied medicine in the 1980s and worked for five years as an *Assistent* in the field of psychiatry before leaving medical practice in the mid-1990s. He told me that there was too much practical routine for his taste and that he had found himself reading in other fields every evening. Eventually he had decided to cross disciplines in pursuit of more theoretical questions. He had been awarded a fellowship to study for a year at the German Literature Archive in Marbach, then became a fellow in the second year of Tübingen University's graduate program (*Graduiertenkolleg*) in *Ethik in den Wissenschaften*. There he had written a dissertation on Karl Jaspers and the history of psychiatry, then received a half-time position in Tübingen that would continue for another five years. I asked what he would do after those five years. With the resignation typical of aspiring German academics, he said, "By then I will either have made it, or I am never going to make it."

The distinction between theory and praxis reverberated throughout our conversation, emphasizing the ambiguous place of bioethics in Germany. Bormuth had left psychiatry because he found its practice messy and wanted to immerse himself in clean theory. Ethics to him was a theoretical pursuit, not an applied one, and bioethics, because it defined itself in relation to messy problems, could not be a pure science like ethics.

To Bormuth, one of the greatest problems in bioethics was *Etikettenschwindel*, or false labeling. Institutes were being formed that had pompous and overblown names and titles that had nothing to do with their actual activities. With mild disdain, he proceeded to list a number of so-called inter- and transdisciplinary institutes that, according to him, had nothing at all to do with transdisciplinarity. In a sense, to Bormuth, all sciences are one, and they are unified by the scientific method, which begins with a question and develops hypotheses that can be tested using transparent methods. Any other kind of inquiry—especially one straddling theory and practice—does not deserve the label science.

When I asked whether he as an ethicist had any contact with scientists, he indignantly replied, "But we ourselves *are* scientists!" While in the United States the term "scientist" connotes someone who works in the natural sciences, the German word *Wissenschaftler* is broader in its meaning: it encompasses the social and human sciences (*Geisteswissenschaften*) as well as the natural sciences. I then explained that I meant to ask whether he was involved in the application of ethics in the natural or medical sciences, to which he responded that, to him, the important thing was research, not its eventual application. He told me that all he wanted was to do research and to point out the various positions, their consequences, and the logical connections between them. Application, he said, is no longer research. In applying ethics, the pure science suffers. He therefore preferred to leave the messy parts to others who felt a desire to become involved in politics.

He was amused by the idea, held by many bioethicists, that bioethics is a kind of value-free (*wertneutral*) *Saalordner*. In his conception, any form of pure research or science becomes impure through its application to problems in the world. It was the same with bioethics. In the transition from the library to the hospital, Bormuth implied, clean ethics turned into unclean *bio*ethics when it mixed with the messy practice of the medical and life sciences. It was not surprising, then, that I could not find a single person in Germany who self-identified as a *Bioethiker*.

In an interview a year later, the former president of the Enquete Kommission, Margot von Renesse, gave a more concretely, even viscerally, historical explanation for the taint attaching to "bioethics" in Germany. She too explained that the rejection of the term could be traced back to the Convention on Human Rights in Biomedicine. This convention, she said, was initially worked out following a German initiative, and it was altered after Germany saw the need for changes in what the convention had defined as permissible research. Opponents of the convention began calling it the "Bioethics Convention," and the word bioethics, Renesse told me, became synonymous with a branch of philosophy that serves to legitimate the unbridled march of biological and medical science. Bioethics became an insult (*Schimpfwort*), even a fighting word (*Kampfwort*), to the point where the word itself caused its opponents to cringe in pain. When the EK began its meetings, Renesse remembered, the word was used to mark one's enemies. She suggested that a better term would be *Medizinethik* or even simply *Ethik*, without further qualifications.

The Bundestag Comes to Life—Sternstunde des Parlaments

Thoroughly ordered, institutionalized, and narrativized by the two commissions, bioethical discourse was eventually ready to move into the parliamentary plenary hall. On Wednesday, January 30, 2002, parliament debated the merits of three proposals for a law to regulate the importation of human embryonic stem cells, each of which had numerous cosponsors across party lines. One bill urged a categorical ban on the importation of these cells, while another argued for controlled importation. The third bill sought a compromise by prohibiting importation as a matter of principle, but allowing for certain exceptions. For several months conflict over this issue had been raging in the media and among interested sections of the public. According to the newspapers, Oliver Brüstle, whose application for federal research funding had sparked the stem cell debate in Germany, was under police protection. Experts of all stripes had filled the media with interviews, statements, and prognoses, and both sides had marshaled arguments and images with great conviction. Embryos had been dismissively called lumps of cells or projected to be potential persons. After postponing its plenary debate several times, parliament had come together at last so that the voices representing the people could be heard.

On the afternoon of this momentous day, the members of the EK secretariat and I walked from our office on Schiffbauerdamm over to the Reichstag. Three weeks before the plenary debate, the secretariat had received official confirmation that up to thirty-nine seats (for thirteen members of the secretariat, thirteen commission experts, and thirteen guests) would be reserved for our group. A sentence centrally located on the page specified that we would be paid for the afternoon visit to the debate as though it were work, since it counted as an occasion for education in government. (*"Ihr Plenarbesuch beim DBT ist als staatspolitische Bildungsveranstaltung förderungwürdig im Sinne der Urlaubsbestimmungen des öffentlichen Dienstes."*) On the back of the same sheet we found rules for visiting the Bundestag. We learned that we would need to hand over all coats, file folders, newspapers, cameras, and binoculars. In spite of the ubiquitous emphasis on transparency, it was only when citizens were stripped of all recording devices that they were safely admissible to the site of *staatspolitische Bildung*. Nor would visitors be permitted to openly express agreement or disagreement with the speakers. Moreover, to ensure order, the leader of the visitor group would have to be able to identify each of the group's members.

Every few minutes attendants with badges and walkie-talkies corralled a group of people in a small anteroom. Extending the theme of transparency that permeates the Reichstag, this oval room, as well as its ceiling, was made of glass. After going through metal detectors and being herded through the glass cage, we rode the mirrored elevators upstairs to the visitor gallery, where those without ID cards had to check all coats and bags, while those with ID cards like the one I was carrying at that time were not checked at all. When I asked for leniency on behalf of one of my guests, I was told with a shrug, "You know how rules are." In the gallery I ended up sitting about fifty feet behind and above Chancellor Schröder, who had taken a seat among his party in order to be able to speak as a regular member of parliament. I was permitted to come this close to the head of the German state unchecked because I was carrying a yellow plastic ID card with, as it turned out, a misprinted expiration date on it.

Although the debates were broadcast on the Phoenix channel, the official television station devoted to parliamentary proceedings and other governmental actions, a number of visitors had come to witness what was billed as a historic moment, when the basic rights of human dignity and freedom of research would join in battle. Today's decision would place the embryo either on the side of human nature or on the side of nature pure and simple.

Throughout the five-hour debate there was suppressed commotion as school classes and members of the military entered the central visitor gallery to watch and left again after their allotted thirty minutes. As an intern at the EK, I was a temporary member of the parliamentary administration. In that capacity I was able, along with my guests, to observe the five-hour debate in its full length—a luxury I could never have enjoyed as a mere student of the proceedings. The Protestant bishop of Berlin, a member of the NER, sat on the *Ehrentribüne*, while Golin, the general secretary of the NER, sat with the secretariat of the EK. On the galleries dozens of television cameras were pointed at the speakers' podium below. Through the skylight above the plenary hall, reflected in countless mirrors, I could make out against the sky the silhouettes of tourists circling up the spiral staircases of the building's glass dome. As I later ascertained, individuals in the plenary hall, due perhaps to the lighting, were not visible from above.

The debate itself progressed in a highly orderly fashion. Each of the three bills under consideration had a specific amount of time allotted to it, which was calculated according to the number of cosponsors and their party affiliations. The members of parliament recapitulated many arguments that I had heard in the commission meetings. Instead of novel arguments, there emerged novel weightings. The first speaker for the compromise bill was the president of the EK, Margot von Renesse. She emphatically argued that even though the ethical dilemmas are labyrinthine, there must be no vampire medicine in a world where some live at the expense of others. She argued for a compromise in which there would be no embryo-expending research (*embryonenverbrauchende Forschung*) in Germany or, if possible, anywhere else. The "deadline"

(*Stichtag*) for stem cell production built into her bill, she argued, meant that "no embryo will be damaged through our draft law." She claimed the support of the scientific community: "Scientists have told us in large numbers that they can live with our proposal." Strictly demarcating the authority of science from that of parliament, however, she was adamant that "here the legislator decides, and not scientists." Her fellow parliamentarians applauded her assertion. Her wish for a compromise seemed to rest on two insights: First, she knew that "a law is strong enough only if it is followed *voluntarily*," which was why only the law with the greatest support could be successfully implemented. Second, taking a constructivist turn, she refused arguments for the human status of stem cells based on simple observation: "Nature is no signpost either, since humans define nature too." Parliament would decide this question on grounds of conscience, without recourse to any presumed truths of science.

Ulrike Flach, a member of the liberal party (FDP) and the main author of the most permissive of the three bills, spoke first for that bill. She also emphasized the sovereignty of parliament, but in a slightly different way. She began by reminding her colleagues that "the EK and the NER have made their recommendations, but the decision lies *here*." She dismissed Renesse's compromise position as inconsistent. Claiming to speak for the collective, Flach said that "people expect an *unequivocal* decision on values." If Renesse were consistent, she argued, then "stem cell research would not only be permitted, it would also be morally necessary." She drew attention to those who were suffering from currently incurable diseases and regretted the fact that "the voice of patients in Germany is unfortunately not as strong as that of some organizations; and it is for *those* patients we want to speak." Like Renesse before her, she based the moral strength of her argument on the claim that she was speaking for the weakest voices in society—but hers were different voices. In her opinion, Germany ought to be more permissive not only because voiceless patients needed the cures promised by stem cell research, but also because that was how other Western countries had decided. Looking beyond German borders, she sought moral support from other nations: "*Other* culturally accomplished and moral countries *do* permit it."

Monika Knoche, the Green Party representative in the EK and the first speaker for the most restrictive bill, the one that would prohibit the importation of human stem cells under any circumstances, made the most high-sounding claim, giving parliament the greatest power of all—power over the nature of human-ness. In dramatic tones, she reminded her listeners that "we decide today whether the embryo belongs to the human species." Another proponent of the bill soon supported her, adding that in such a decision "other countries cannot be a standard for us." He argued that the German parliament had to cast its vote of conscience autonomously, without reference to the actions of others. There could be no external guidelines in a decision of such high consequence. In matters of ethics, he implied, each nation had to set its own standards and be true to its own internal principles. Only by prohibiting stem cell research altogether, he said, could "progress made in Germany" receive the label "*ethisch unbedenklich*" (ethically unproblematic).

Ethics was linked to national sovereignty in several ways. Some liberally minded representatives argued that if Germany did not permit the research, it would migrate to countries that could not afford regulation. The assumption underlying this argument was that ethics was a matter of having sufficient resources and that only rich countries could afford to implement ethical regulations, which by implication meant that poorer countries, like those formerly belonging to the Eastern Bloc, were doomed to an unethical existence. In other words, while other countries could not set a moral standard for Germany, Germany could afford to set a moral standard for them. The liberals, however, seemed not to draw the obvious conclusion that before Germany could become a moral beacon to the world, it would have to stake out its own autonomous position. The connection between wealth and ethics further implied that one's ethics were good only if they had been "bought" at a high price.

Angela Merkel, the head of the conservative Christian Democrats (who in 2005 was elected Germany's first woman chancellor), argued that good science could not be separated from ethics and that the interconnectedness of our social worlds made it imperative for Germans to exert influence internationally if they wanted to maintain *their* values and standards at home. She argued that Germany had to take a strong stance against embryonic stem cell research and persuade its closest neighbors to join in so as to preserve a realm of ethical probity. Gerhard Schröder, by contrast, cited the more permissive regimes of the United States, Israel, and Australia and said that he saw proof of science's awareness of its responsibility in the fact that science and the DFG had not yet gone ahead with the research, even though it had been legally permissible all along.

As the time for the vote drew nearer, the low hum in the plenary hall grew louder as the decided helped the still undecided make up their minds. Several people made the rounds through the legislative chamber, whispering with representatives from different parties, lobbying for last-minute votes. Merkel seemed especially active in gathering her colleagues' support for the most restrictive bill. While the most permissive of the three bills was widely seen as not having much of a chance, there was still uncertainty over whether importation would be prohibited altogether or whether exceptions would be made in accordance with the compromise recommended by the majority of the EK.

When the time came to vote, each member of parliament marked one of the colored pieces of paper provided and stuck it into the ballot box. Given Germany's emphasis on transparency, it was perhaps not surprising that even the ballot box was made of Plexiglas, revealing the colorful ballots inside. More surprisingly, two men then carried this very transparent box into an adjoining room, where the votes were counted out of everyone's sight. Someone in my hearing muttered, "Of course corruption sits elsewhere!" Casting the votes was a transparent process, but counting the votes was not; participation was a public process, but the mechanics were perhaps inevitably black-boxed. The votes were later attached to names, however, and made available to the public, so that in the end transparency was restored.

Twenty minutes later the results were announced. In the first count, 617 votes were cast, of which 18 were not valid because members of parliament had not signed their names legibly. Legitimacy once again was tied to making one's position readable. With two No votes and two abstentions, the bill to prohibit all imports received the most votes (263), the bill recommending conditional prohibition was a close second (226), and the bill recommending free importation came in third (106), thereby dropping out of the next cycle. In the second vote, the compromise bill won a clear majority, since it had gained the support of most liberals. The importation of human embryonic stem cells would be prohibited in principle, but permitted under exceptional circumstances.

When the Bundestag cast its vote after nearly five hours of reasoned debate, newspapers and commentators called the process "parliament's starry hour" (*Sternstunde des Parlaments*). This phrase alludes to Stefan Zweig's book *Sternstunden der Menschheit*, a collection of historical miniatures recounting moments in world history that determined the further course of humankind. The phrase works on several levels. Not only had parliament reasserted its relevance by showing itself capable of debating complex topics and coming to reasoned conclusions, but in the process it had also decided on the course of humankind, defining what stages of human development are available for scientific research. Parliament had in effect answered the metaphysical questions of when life begins and what life is human.

After parliament had voted on the principle that would become the basis of the future law, standing political committees went to work to draft the precise text of the law in all its technical detail. On April 25, 2002, the Bundestag passed the Stem Cell Law by a wide margin. The law prohibits the importation of embryonic stem cell lines in principle, but makes exceptions for "research aims of high priority" (*hochrangige Forschungsziele*), even if those aims are interesting only for basic research. This redefined notion of "high priority" stood in stark contrast to the justification that researchers had used in the preceding debates. There they had spoken of developing therapies, and many people who ended up supporting stem cell research had done so because they assumed that therapies based on that research would soon become available.[25]

Conclusions

The January debate in the Bundestag, the *Sternstunde* of parliament, conformed to German ideals of rational debate: it was balanced, appropriately weighted according to party representation, and devoid of emotional arguments. What role did the two ethics commissions play in allowing parliament to perform its function of reasoned deliberation? In enabling the debate, the commissions evidently helped to achieve a preordained order. Yet looking at the commissions side by side reveals just how contingent this order was. As commission reports were translated into party positions and then into law, much of this contingency was erased.

I asked myself by what means this order was made to appear as such. What was it about the kind of order imposed by the commissions that made the earlier public discourse appear "disorderly"? Discipline is part of the answer. Ethics commissions claim to be value-neutral *Saalordner*. According to their own rhetoric, these commissions review the factual circumstances that exist in advance of an ethical judgment. They summarize the scientific status quo. They then point out positions and delineate ethical arguments for and against certain actions, given particular prior commitments. Then they draw out the conclusions that follow from these commitments without presuming to impose any conclusions on the legislature.

Yet we have also seen that the two commissions held different views of what it means to act as *Saalordner*. The EK member had complained that prior to the EK, everyone talked as they pleased, regardless of their qualifications or knowledge, while the NER member had hoped that the NER report would enable people to articulate and debate the real issues. The EK rationalized the public's disorderly thoughts, whereas the NER offered a rational framework within which all choices had to be formulated. The EK drew its authority from representation; the NER, from expertise and media visibility—and direct representation of competing values in society. In its work on stem cells, as in its work on other issues, the EK tried to incorporate all reasonable viewpoints into a comprehensive report; the NER sketched out four possible bases for action. The EK, on several fact-gathering trips to other nations that also grappled with the ethics of human embryonic stem cell research, turned to the outside world to collect missing information and expert knowledge; the NER used its U.S. visit to consolidate its right to represent German bioethics. In sum, the commissions differed in their interpretation of democratic legitimation, they worked with different definitions of "bioethics," and they articulated different views of Germany's place in the world.

Of course, insiders in both commissions retained an extremely clear sense of the relative positions and values of the two bodies and their members, but the distinctions were less clear to outsiders. In part, this was the result of conscious and unconscious strategic choices. The NER had greater resources than the EK, but it also managed those resources well, and it succeeded in surrounding itself with an aura of objectivity and legitimacy while the EK largely dropped off the radar screen. Informal conversations I had with friends and acquaintances indicated that the NER's strategy of showy self-presentation had on the whole succeeded. Many of my interlocutors were hardly aware that there was a second national ethics commission; most had only ever heard of the NER. Unlike the EK, the NER had managed to build up its charismatic authority. Its grand style made it the more interesting commission. Even its name, which linked the concepts of "nation" and "ethics," practically guaranteed it greater relevance than the EK, which was charged "merely" with inquiring into certain subjects in order to prepare for parliamentary debates.[26]

The enactment of a law did not, of course, replace contingency with complete certitude. As I show in chapter 3, even after ethics had been institutionalized, the need still arose for including the voices of ordinary citizens. Citizen engagement is part of the German political imagination, but the citizens, too, needed to be organized and institutionalized in a "citizen conference." Only when the citizens are explicitly construed as such do their voices become legitimate inputs to the state. As long as they freely speak their minds and feelings, their voices are taken as illegitimate chatter. They need to be *taught* how to speak as citizens. Thus the production of a German bioethics is at the same time a production of a particular German social order, that of deliberative politics. Ethical order is tied to social order, and the bioethics debate and its multiple institutionalizations made the constructedness of that social order transparent. Before widening the lens to view the citizen conference, I will therefore take a more closely focused look at the contingency of order by describing more specifically how the EK made ethics by disciplining disorder.

2

Disciplining Disorder

Learning to See the Right Things

After concluding my work in the Enquete Kommission, I co-taught a seminar in Berlin. A student in that seminar, when asked how ethics commissions should decide on a particular issue, sidestepped the problem by saying, "I don't care what they do, as long as they decide what is *right*." To her it was the outcome that mattered, not the process of deciding. But her (unintended) play on words also points to another ambiguity: to "decide what is right" can mean *choosing* correctly from several pre-existing options, or it can mean *creating* a sense of "rightness" in the first place.

That second sense of rightness pervades the German discourse on bioethics. Once a member of parliament said in my presence that the outcome of the debate was unimportant; it was the debate itself that mattered. The important thing was that "in Germany we have these good debates." On another occasion another member of parliament said, with a hint of self-satisfaction, "The rest of the world is envious of our level and breadth of debate." After the EK had published reports on stem cell research, genetic testing, and biological materials as intellectual property, commission president Margot von Renesse told members during a meeting, "The public must see that *ev-*

ery argument was weighed and taken seriously. We have advanced all three topics." In other words, the commission's precise recommendations were less important than the demonstration that it had grappled with the problems.

In this chapter I show how the state develops its own capacity to reason morally by crafting a place from which to speak, how it codifies and routinizes this capacity in textual production, and how it represents the results to outsiders. How did the EK make sense of the ethical challenges brought on by modern medicine, and how, in that process, did it order the biomedical world to make it possible to act ethically within it? How did it create ethical subjects and subjectivities? How did it produce order materially by allocating space, structuring discourse, and coauthoring reports to the legislature? How, in other words, did a group of functionaries and experts author a series of documents that then served as the basis for decisions by the German parliament?

In answering these questions, I draw on several months of participant-observation fieldwork with the parliamentary ethics commission.[1] But even though I was inside this process and observing it on a daily basis, I never felt completely part of it. My understandings and those of the commission staff never completely aligned. It was as if, despite all their efforts to make the commission's workings transparent to me, I kept running into unwritten rules like invisible glass walls. Many times I did not see quite what I was supposed to see, but instead noticed what no one else thought it important to notice.

Becoming an Ethical Insider

I first contacted the office of the parliamentary Enquete Kommission on Law and Ethics in Modern Medicine (EK) on Friday, November 16, 2001. I had seen an announcement for a public hearing on the "European Discourse on Ethical Questions of Modern Medicine" arranged by the commission, and I wanted to know whether visitors could attend for the entire time, what the order of the invited speakers would be, and what place the Enquete Kommission inhabited in the larger political landscape of German bioethics. I was referred to Ingo Härtel, the person who had organized the event. After answering my questions, he told me with genuine regret that he had to go out of town that day to attend another official hearing in Bonn and was sorry to miss the fun part of a meeting in which he had invested so much work.

When I attended the hearing the following Monday, I had to give up my passport and go through a metal detector upon entering the Reichstag building. Attendants brought people to the meeting room in groups. Coffee and sandwiches were sold at the door, but for "security reasons" people were asked to eat outside the meeting room. Next to the door I saw neatly sorted piles of all the invited experts' statements in several languages. Inside, the spoken statements were simultaneously translated into German and English, but the number of available headsets was limited. The first sentence I heard upon entering the room was that international cooperation among law-abiding states (*Rechtsstaaten*) was a necessity. It could no longer be left up to individual nations to develop ethical guidelines for regulating biomedical research; in an increasingly interconnected world, one needed to harmonize ethics internationally as well. This rationale explained the multiple translations of the statements. It also brought home to me that in speaking bioethics, the German state was also articulating its place in the world.

Among the people I met that November day was a German political science student who was an intern with the Enquete Kommission. When I mentioned my interest in bioethics, he told me of another available intern position. The commission would be interested in my work, he said, and working for it would give me access to relevant people and let me inquire into issues related to stem cell research. He suggested that I call one of the staff members, and I reached Ingo Härtel again two days later. After some friendly small talk, he agreed to support my application for an internship, and it was quickly approved.

In those days, the topic of human embryonic stem cell research dominated the public discourse. On the radio news that evening, I heard that the Nationaler Ethikrat (NER), a rival ethics body, had postponed its statement on the importation of stem cells until the end of the following week because its internal debate was not yet over. The news also quoted Oliver Brüstle, the neurologist whose application to import stem cells into Germany had prompted the national *Stammzelldebatte* in the first place, as warning that German researchers would lose touch with scientific communities around the world if the nation hesitated any longer. Like speaking bioethics, doing science seemed to be an instrument for (re)positioning Germany respectably within the world.

In December, three weeks before beginning my work with the commission, I had an anxiety dream, thus replicating a national trauma on a personal scale. In my dream, people barraged me with all kinds of ethical problems and dilemmas, which I was unable to solve or properly address. I woke up feeling quite unprepared for my new job and unsure about what I was expected to do. This sense of instability was a precursor of confusion to come.

The First Day

January 7, 2002, was my first day at the Enquete Kommission. I had been told beforehand that the working day began at eight in the morning, and I arrived on time, managing to muster the willpower to get up a good hour before sunrise in the cold, dark Berlin winter. The commission's secretariat was located on the top floor of an East German office building from which one had a clear view of the Reichstag, the seat of parliament. The building also happened to be conveniently next door to the Institute for European Ethnology, which I was associated with as a research fellow. This fortunate coincidence allowed me to meet anthropologist friends for lunchtime conversation when I wanted other ethnographically trained eyes to look at what I was seeing. I was assigned a north-facing office. For the next two months, until spring, I would not see daylight again except during lunch hours and on weekends. I spent my time at the EK in a physical darkness that seemed to match my dream state of ethical bewilderment as well as my condition of ethnographic innocence.

The commission building stood next to the river Spree, which runs through the city. I could often hear the tourist boats that went up and down the river, showing the landmarks along Berlin's waterways to an interested audience. The boats typically played a tape with explanations of the buildings along the route. As each went by, bystanders at fixed locations could hear the same phrases repeated every time. One particular boat that passed by the EK building on its way from the former East Berlin to the former West Berlin frequently broadcast the words *"Aber das ist jetzt alles Vergangenheit"* (But all that is now a thing of the past). This constant commemoration of the vanished past took on special meaning for me in a place where a wide-open future was ostensibly taking shape. It reinforced my sense that each forward step taken by the German state—even in the realm of bioethics—was still oriented by a look at the past.

On my first official errand, I acquired the pass I needed to enter the government buildings. The person who handled the paperwork, Herr Mützel, was a friendly man who had himself been a student at nearby Humboldt University. He had spent a lot of time on the street I was living on, Linienstrasse, which in his college days had been home to many student cafes and the East Berlin red-light district. A dedicated civil servant, he was very concerned to follow rules to the letter. He made sure that I read the documents I had to sign and practically quizzed me on the rules I would need to obey. The function of the quizzing was as much to make sure I understood the rules I had agreed to as it was to hold me accountable in the event of a transgression. On his door hung a quote from Rousseau: "The freedom of man does not consist in doing what he wants to do, but in *not* having to do what he does *not* want to do." I thought to myself that this statement of the imperative of noncoercion implied that it was not necessarily important to know what we did want as long as we knew what we didn't want. This negative sentiment would appear ever more important as both ethics commissions came to draw boundaries not only around certain scientific practices, but also around the territory in which German law (and German ethics) applied. Crafting ethics became a way of spelling out what "we Germans" did *not* want to take place "within our borders."

My trip to get keys to the building I now worked in took me to the adjoining Wilhelmstrasse. With machinelike efficiency a passport to the Bundestag was produced for me, and within three minutes I held in my hand this new identity card, complete with my picture. Although the passport's dates of validity were misprinted (it said 2001 instead of 2002), I never had any trouble passing through official doors from then on. Flashing this little piece of yellow plastic, for which I had only signed a few documents saying, among other things, that I would not hand over to third parties materials vital to national security, allowed me to bypass the metal detectors, bag checks, and other security procedures that unofficial visitors to official buildings had to undergo. With a stroke of the pen I had gone from distrusted outsider to trusted insider. My mere signature, by some accounts an illegible scrawl, had converted me to the category of "safe persons" and had rendered unnecessary the security apparatus designed to keep these institutions of democracy safe from external threats.

Ethics Made Transparent

Over the next few weeks of my internship, Cornelia Beek, the director of the secretariat, took me to various meetings that she had to attend in her official capacity. She told me that she was happy to bring me along because the point of an internship was to give me an idea of how everything related to the commission worked. She said she saw it as her responsibility, and her pleasure, to try to make everything as transparent for me as possible. On that first day I also met two of the three scientific staff members, Johann Ach and Matthias Wolfschmidt. Ingo Härtel, a trained physician and philosopher, was the third scientific staff member, but he was again in Bonn that day. All the staff members, except for the director and associate director, switched to first names within minutes, and I will refer to them by their first names throughout. Despite their disillusionment with the commission as a whole, Johann and Matthias also had some praise for it: "Not all secretariats have as open an ear for the wishes of interns as this one." They too told me they sincerely wished to give me whatever information and material I could possibly ask for. Transparency, as I found out, meant literally being shown everything and seeing everything. My internship in parliament was therefore an exercise in visual pedagogy, as I was shown how this particular part of government functioned. At the same time, in being trained how to see, I was also trained to be an observer. To aim at making the workings of something transparent presupposes that those workings are in themselves complicated and opaque. Cornelia Beek in a sense acknowledged that one does not begin with perfect insight; rather, transparency is the outcome of acquiring the background information that one needs before something can *appear* transparent.

Transparency was, as I have already noted, built into the physical architecture of the government. On our first trips out of our office I noticed that the large interior halls of many government buildings were designed to look like external facades, so that even when one was safely inside a building, one got the feeling of being still on its outside. The people looking down from their windows completed that impression. A perhaps unintended side effect of this architecture was that one constantly felt watched from the many windows that covered the inner and outer walls of the building. It is as if, even when one was inside one of the parliamentary buildings, one was constantly made aware of being in a public space—indeed, a materialized public sphere. Or as if those who worked for parliament must never forget that they were doing public work. Nonetheless, all that transparency did not by itself constitute me as a true insider. But that story comes later.

The contrast between insiders and outsiders was brought home to me a few weeks into my time at the commission, when I witnessed how the government distinguishes self from nonself. On our way to a commission meeting one afternoon, the three scientific staff members and I, all authorized to bypass security, walked into the Bundestag building. As we went into the Reichstag, I saw people take off metal objects and put them in plastic containers for scanning, then walk through a metal detector so sensitive that it registered even those visitors who were wearing a belt with a metal buckle. Being in a bit of a rush, the four of us were walking side by side, with the result that one of my co-workers, while heatedly gesticulating, accidentally walked through the metal detector carrying a backpack. Suddenly a shrill alarm went off and the red danger indicators lit up to maximum amplitude and began flashing. The guards rushed toward my co-worker, who now realized that the alarm was singling *him* out because he had forgotten to identify himself as a safe insider. He mumbled an excuse, pulled out his small photo ID card, and attached it to his jacket, whereupon the guards, appeased, sat back down, nodded, and waved him through.

It only happened once that an official challenged me inside the Reichstag. After a debate on genetic testing within one of the Bundestag parties, I was exploring the building on my own, looking more closely at the internal structures that our normal work routine left too little time to take in. I was lingering in the inner halls, absorbing these otherwise closed-off spaces, when suddenly a man in uniform came up to me and asked for my pass. I pulled it out and attached it so it was visible. Since it was color-coded, he did not notice that it had a wrong date printed on it, and he walked off again, satisfied.

Such encounters reminded me of a story about Karl Marx that I had read as a student in East Germany. Marx was strolling through a forest with his inquisitive young daughter. When they came upon an anthill, the girl disdainfully observed that some ants seemed to laze around without carrying their weight in a collective swarming with busyness. To demonstrate the function of these seemingly free-riding ants to the child, Marx poked a stick into the anthill. Suddenly the phlegmatic ants leapt into action and started to attack the stick fiercely. They were not as lazy and useless as they had seemed, but instead served the vital function of protecting and defending the anthill so that the other ants could go about their business in peace. The moral of the story was that in a perfectly functioning society, such as an anthill or a socialist state, every member had a role and knew when and how to exercise it. In turn, only when something happened to disturb this well-ordered society did it become necessary for defensive roles to be exercised. In my case, external markers of identity provided the needed assurance that all was in order and in place around the German government's normally inactive defense team.

My own role in the commission, however, remained opaque to the people I both worked with and studied. I was officially an insider, but from time to time my outsider status would resurface. Although I was there to learn the ways of parliamentary administration, my goal was never to become either a more efficient civil servant or a more knowledgeable citizen. Instead, I wanted to find patterns of practice and make explicit their embeddedness in the larger culture. Parliament was not prepared for such an insider role, however, and the persons with whom I interacted quickly put me into categories that did not quite capture my function. In an institution geared to codification, I remained troublesomely unclassifiable.

First Impressions

In the beginning of 2002, coinciding with my arrival, the newly completed Paul-Löbe-Haus was opened for parliament. An imposing rectangular structure with cylindrical insets of glass and concrete, this was the building, located right next to the Reichstag, that Germans had dubbed the "Motor of the Republic." The EK's meetings moved from an old rectangular room in the Reichstag to one of the circular rooms in a "piston" of the new building. In this room the commission fulfilled its mandate to prepare reports based on which parliament would legislate ethical constraints on modern biomedicine. Present at most meetings were a segment of the commission, several aides working for its parliamentarian members, and the secretariat. The secretariat staff members noted the changes that commission members had proposed in the texts being reviewed, but they very rarely intervened in the discussions. When I joined the secretariat and began attending commission meetings, the widely anticipated report on stem cell research had just been completed, and the report on genetic testing was in preparation.

The first commission meeting I observed took place on Monday, January 14. In EK meetings the experts were almost always present, while the parliamentarians attended much more rarely. My impression was that the experts took their work on the commission more seriously and devoted more time to it, while a number of the parliamentarians came only to those meetings at which a report was presented to the public and media coverage was assured. Sometimes members I knew only from photographs in the commission's publicity brochure came to a public meeting and gave interviews on their bioethical stances before wide-eyed television cameras. For them, the work of producing public morality was worthwhile only when it was broadcast to a larger audience. In fact, the image-making news media kept pushing everyone toward more charismatic styles. Over time, I felt that the experts too were learning how to represent their opinions more effectively as they imitated the parliamentarians' argumentative styles.

On the next day, January 15, the members of the secretariat (minus the student helpers) met in the director's office. I was given special permission to attend. Previous interns, presumably with less investment in understanding the inner workings of the commission, had not been asked to sit in on these "strategy meetings." The director's office was large and sunny, and a palm tree completed the oddly tropical ambience. Large posters of two views of Earth as seen from outer space hinted at the global aspirations, if not the actual dimensions, of German bioethics. Another poster in the hallway, a "map" of the human genome in its entirety, seemed to gesture toward the responsibility of bioethics for even the most minute and intimate features of life. The meeting ran from nine until ten in the morning, and the one-hour time limit necessitated verbal economy. After going through the agenda for the next commission meeting, we got an update on the intranet service of the Bundestag and on the deliberations of the French National Bioethics Commission on the then current Perruche case, in which the parents of a child born with severe disabilities had sued their doctor for failing to notify them of visible abnormalities in the fetus.[2] We then summarized the textual changes mandated by the commission members and distributed among the staff the responsibility for entering them into the text to be discussed at the next commission meeting. My own task was left unspecified; I was there to learn.

Next, the staff wanted to hear what impressions the previous day's commission meeting had left on my own untutored and as yet unjaded mind. I told them how impressed I had been with the well-reasoned opinions and the depth of the discussion and said that I had not seen such detailed considerations of social issues in the United States. I had also noticed that, in spite of the round table, the seating arrangement had seemed rather confrontational, and that the secretariat and its staff had seemed to be in opposition to the commission. Most specifically, I had noticed that one of the staff members had been held almost personally responsible for perceived flaws in the text, so I asked how that text had been produced. I was told that it had been written by the secretariat on the basis of discussions in smaller, topic-based groups. This explained to some extent why that staff member, after some particularly depreciating comments ([commission member]: ". . . *if* Herr [. . .] is able to do this . . .") had lapsed into the third person singular in his replies ("Herr [. . .] *can* do this . . ."). He seemed no longer able to respond as an individual, but instead relegated himself to the position of a subordinate staff member answering to the officially superior expert. While he privately complained about what he perceived as unfair treatment, in public he felt he had to obey. Staff members frequently vented their frustrations with commission members among themselves, but friction in the machinery of ethics remained invisible in official contexts.

The director of the secretariat mentioned as an aside that throughout the previous day's meeting, the commission had been technically incapable of making any decisions (*nicht beschlussfähig*) because there had been no quorum.³ If anyone had raised that question during the discussion, then the meeting would have had to be adjourned. To the great relief of the director, no one had asked that question out loud, and the commission's decisions had therefore been valid. Legitimacy therefore seemed unexpectedly contingent and unpredictable, as articulating an ethical stance successfully depended on an agreement that the commission kept tacit. After she had pointed it out, I noticed the issue of *Beschlussfähigkeit* throughout my time with the commission. At the following meeting on January 21, for example, only four of the twenty-six members were present at 11 a.m. when the meeting was scheduled to begin. At 11:30 there were thirteen, and at noon there were sixteen. Those who came in late squeezed through between the round table and the circular wall. On the way they shook everyone's hand, whispered a quick greeting, and gave a practiced smile.

Such breaches of administrative rules might have undermined the prized legitimacy that the Enquete Kommission held up so high when it questioned the legitimacy of its rival, the NER. The fact that everyone played along shows that pragmatic concerns trumped the ideal of a dialogue among all viewpoints, for which all in fact would have had to be present all the time. That pragmatism, as it happens, was vindicated. While the NER's legitimacy remained a matter of public debate, no one ever publicly questioned the legitimacy of the EK. Once legitimacy was formally established, it seemingly no longer needed to be defended unless it was challenged from within—but pragmatic concerns over finishing its tasks on time made such challenges unlikely.

A Place for Disability

When the EK moved to the circular meeting rooms of the Paul-Löbe-Haus, the seating went from something resembling a rigid horseshoe—the pattern in the old meeting room—to a perfect circle. In the old room members of the same political parties had always sat together. The circular form blurred the party lines somewhat, even if it did not altogether dissolve them. Although the circle introduced mobility (people did not sit in the same place at each meeting), it was still clear from how people arranged themselves where the affinities were. The one fixed point was the PDS representative, Ilja Seifert, whose wheelchair and occasional need to leave the room required him to sit close to the door. Seifert had been paralyzed from the waist down in a swimming accident when he was 16. The PDS representative not only sat awkwardly in the circle, but he sometimes had a difficult time inserting pieces of text that the commission would recognize as legitimate because of the ambiguous status of his party, the successor to the former East German Socialist Unity party. West Germans saw the PDS as a direct outgrowth of a thoroughly corrupt party that should have been dismantled along with the bankrupt nation it had governed. It was only in the former East German states that the PDS attracted voters, many of whom used that venue to reclaim a part of the socialist history they felt West Germans were trying to discard or erase. The party was nevertheless quite marginal in German politics, and both West and East Germans regarded it with suspicion. In the 2002 parliamentary elections later that year, the PDS did not get the five percent of the votes required for parliamentary representation, and Seifert lost his mandate.

The media treatment of Seifert, however, left no doubt that any legitimacy he lacked on political grounds was compensated for by his disability. On February 18, when the EK publicly presented its report on genetic testing, the commission members were present in greater numbers than ever before. In place of the 10 or 11 members who typically showed up, I counted as many as 18 when the time for the official announcement came nearer. Outside, camera teams were waiting, and commission members kept slipping out to give interviews on genetic testing and its consequences for society. The PDS commission member in the wheelchair was also interviewed, and the camera seemed to be directed more at the wheelchair than at the person in it. Even when his interview was over and the cameras swept the room to capture the general scene, the wheelchair again and again became a focal point that the cameras could not seem to let go of. It was as if Seifert's impairment automatically placed him at the center of the ethics debate.

Not only did the disabled seek out the EK, but the EK sought out the disabled. In 2000, the commission traveled to the West German city of Bielefeld in order to organize an event there to engage people with disabilities and assure them that their needs would be adequately represented. Bielefeld, where I spent one year as a student, contains a well-known home for the disabled, and various public displays are designed to bring their presence into collective consciousness. On the trams gliding through the city in 2002 were printed sentences meant to evoke sympathy with the disabled and to remind Germans of their equality before the law. One Bielefeld tram carried the sentence, "Not to be disabled is not an accomplishment . . ." (*Nicht behindert zu sein ist kein Verdienst* . . .). This sentence was a quotation from then president Richard von Weizsäcker, and it continued, ". . . but a gift that can be taken away from any of us, at any time."

Particularly prominent was the 1994 addition to Article 3 of the German Basic Law, which guarantees that "no one may be disadvantaged because of one's disability." (*Niemand darf aufgrund seiner Behinderung benachteiligt werden.*) The amendment to Article 3 also appeared on public billboards in Berlin announcing the exhibit *Fragen an die deutsche Geschichte* at the German Cathedral on Gendarmenmarkt. The years of National Socialism had shown how quickly a part of the population could become vulnerable, and the change to the Basic Law reflected a will not to exclude or render vulnerable any part of the population. This amendment, as websites promoting the interests of the disabled point out, is merely an antidiscrimination law (*Antidiskriminierungsgesetz*), rather than a more far-reaching equal positioning law (*Gleichstellungsgesetz*). In other words, the law prohibits discrimination against the disabled in employment and so forth, but it does not mandate that they be given all the opportunities and access that the nondisabled enjoy. A long step toward the latter goal was taken in May 2002, when the *Bundesgleichstellungsgesetz* went into effect. This law mandates the removal of all barriers that keep the disabled from fully participating in civic life. In Berlin signs marking wheelchair accessibility have been proliferating ever since.

Although the EK could hardly have neglected them, people with disabilities made certain that they were visible at crucial moments. One afternoon, after sitting through a committee meeting debating changes proposed by the liberal party to a law on genetic testing, I found perhaps a dozen people in wheelchairs outside the meeting room. They were there to take advantage of the television cameras set up to interview politicians on the results of that day's deliberations.

Yet if Germany seemed sensitized as never before to the special needs of people with disabilities, the brand new parliamentary buildings introduced massive physical obstacles to their unhindered access. In a triumph of elegance and form over function, most doors are very heavy and move extremely slowly, opening at sometimes awkward angles. They were hard for Seifert, in his wheelchair, to open, though fortunately for him he had an assistant with him at all times. For ordinary citizens with disabilities, access to the seats of government remains more a theoretical promise than a physical reality.

Yet physical disability did not easily translate into political aptitude. Ingo Härtel once told me that patient groups were relatively numerous in Germany, but they did not know how to exercise their power effectively. For example, they tried to shape legislation on patient rights by lobbying parliamentarians who sat on relevant commissions, but these elected officials often were not in a position to help because they did not control the drafting of actual recommendations. A better way, Ingo suggested, would have been to connect with the staff members who in fact wrote the reports. Perhaps no one thought to do this, he added, because it would violate the *Dienstweg*, the official channels through which politics is supposed to work. Even political lobbying, it seems, has to be depoliticized in Germany as a practical matter, through interest groups' obedience to the formal rules of the game.

Dienstweg

The *Dienstweg*, or using the proper administrative channels, kept mattering at odd times. When my six-week internship came to a close, I wanted to renew my position and continue on for a second term. Since there was no precedent in the commission for such a request, I phoned Herr Mützel, the friendly civil servant who had authorized my internship and quizzed me on the rules the first time around. He pondered for a while how my request for an extension might be fulfilled. He was generally unwilling to make any exceptions or interpret existing rules in a new way. When I kept pressing, he told me that he could not justify an extension on the basis of the reasons I had given him, but implied that another reason might convince him. His office did not authorize him to extend an already existing internship, but it did permit him to authorize new internships. He had found a loophole that might let me through, but only if *I* made the proper motions.

Although my interlocutor knew what I wanted perfectly well, he refused to simply go ahead with the authorization. Rather, he told me the words that I would have to say back to him. To show that I was *mündig* (capable), I had to say, with my own mouth, not that I was applying for a status extension, but that I wanted to do an internship in the Bundestag for such and such particular reasons. For the act to become official, it was important for me to articulate the request myself, in my own voice. My intention had to be verbalized in a prescribed way, and Herr Mützel, as the personification of the state, could help me with every step except the voicing of the request. After I had repeated his phrases, he was satisfied, marked his paperwork to show that I had applied, and saw no further obstacles to granting my request. The whole procedure (most reminiscent of a public swearing in or an exchange of vows) felt like a repeat of the first time he had quizzed me, when I had to speak back to him the words that I had on paper in front of me. A few days later I received an official letter, identical to the first one two months earlier, confirming the dates of my "new" internship and congratulating me on this professional opportunity.

Du und Sie—*More Ways of Creating Insider-ness*

Eating was a ritual that held things together at the EK.[4] The secretariat often organized potlucks—typically breakfasts—and people paid attention to who would bring what and who would sit where around the table. Similarly, it was significant who would go to lunch with whom. Typically the scientific staff members went as a group, and the secretaries and student helpers went as another group. I myself went with the scientific staff most of the time, and our discussions usually revolved around politics. On the rare occasions when I went with the secretaries and student helpers, conversation tended to be more personal.

One day, after I had been working with the EK for a month, the secretariat was meeting to celebrate one or another festive occasion. In the middle of the meal, the director stood up and gave a small speech, looking strangely awkward. At the end she said, "If you don't mind including me in your circle: my name is Cornelia." With the exception of the director and the associate director, members of the secretariat had been using first names and the informal *Du* pronoun with one another all along. The director was now offering that greater degree of intimacy to the people working below her. The associate director was absent that day, so he was not included in the offer.

The question of whether to use *Du* or *Sie* is highly nuanced. In any German bookstore one can find books offering guidance on when to choose which appellation. Some books on business etiquette, for example, offer detailed descriptions of the situations in which *Du* or *Sie* is appropriate, broken down by profession, by size of the company, and even by the context of the interaction (e.g., whether customers are present). The distinction between more and less formal ways of addressing one another (present in most Indo-European languages) gives speakers great flexibility in modulating proximity and distance between persons through speech; yet it also opens the door to perceived slights and signs of disrespect. To give a personal example, my grandmother had written a letter to some younger acquaintances with whom she communicated regularly, and in it she had "offered them" the *Du*.[5] When her correspondents did not reply for a while and then, perhaps through oversight, continued to use the *Sie*, my grandmother took offense and from then on insisted on using *Sie* herself, even after the corresponding couple switched to *Du* as my grandmother had suggested. In the end, no formal guide seems to substitute for one's own subjective assessment. As one online columnist put it, "It is best to wait until the Du-feeling (*das Du-Gefühl*) develops."

After "Frau Beek" had transformed herself into "Cornelia," the members of the secretariat were silent for a moment, and then kept talking, carefully avoiding addressing the director directly. I remember feeling as if a line among us had been subtly redrawn; some intimate and previously hidden part of the director had now been revealed to this circle and was waiting to be included in it. It was not until about fifteen minutes later that the first person addressed her by her first name, a move everyone noticed. Others soon followed, and later during lunch, the secretariat members were placing bets among themselves on how long it would take for the more formal associate director to follow the director's lead. To my knowledge, he never did, and he remained the only person who was addressed formally, and who so addressed others, throughout the life of the EK.

Weeks later Cornelia told me that she now saw the problem of combining informality with hierarchy. First names were so intimate, she said, that it became difficult to give orders. Instead, the orders began to sound like requests for favors, but like favors that one's subordinates could not refuse. They were still orders, of course, but now masquerading as favors.

A member of the NER secretariat told me that slips of informality had occurred there as well when hierarchies were subverted because of an accidentally uttered informal *Du* instead of the more formal *Sie*. In general, she too believed, the informal *Du* was a mistake in the office (*Geschäftsstelle*) because it made giving orders more difficult and refusing to obey orders easier. The intimacy of the address destroyed the hierarchy on which the functioning of the office depended. When the *Du* slipped out accidentally, it was typically ignored, and people went back to using *Sie* as if nothing had happened. Public and private relations remained for the most part strictly separated.[6]

Writing Bioethics

Making bioethics at the Enquete Kommission seemed synonymous at times with making paper. The commission's entire function consisted, it seemed, of producing, destroying, or conserving paper. During my first days at the commission I heard the noisy copier next door to my office running almost all the time, interrupted only by the even louder sounds of the magnificent industrial-strength shredder that stood next to it. Paper was incessantly produced and just as incessantly destroyed. Sometimes there was an error in the page layout; sometimes the copier had been programmed to produce more copies than needed. Often thick packets of copies were thrown out without having been read or even looked at. A few weeks before the official closing down of the EK, a mountain of paper—it must have been at least one cubic meter— was delivered to the EK and stored in one corner of the copy room. I watched with disbelief as people copied their way through the mountain and it quickly dwindled to nothing. As a compulsive reader I initially found this turnover of paper deeply frustrating. Over time, after I had accepted the fact that I would never read everything that passed through the commission, the sound of the copier became more soothing—almost a reassurance of our own productivity.

Commission members were equally overwhelmed by the amount of paper that the commission generated. Two persons told me in interviews that they received more papers than they could possibly go through in time for the commission meetings. Ulrike Riedel, for example, an expert member chosen by the Green Party, apologized to me for the mess of paper in her office. I asked her how she managed to keep up with all the material being published in the field of bioethics at the moment. She shook her head and said, "It's terrible. Especially the Enquete Kommission." Pointing to a stack of papers about a foot high, she added, "This is what I have received from the commission over the last fourteen days. . . . It is unbelievable. They . . . they . . . every three days I get a fat envelope. I would prefer not even to open them at all anymore. One simply doesn't get through it all." I asked her how she kept all the different arguments in her head, and she said, "I don't know . . . I also get more and more confused, and I also forget. . . . And this ethical debate has become so professionalized by now that . . . I would need to spend half of every day reading, and not forget any of it. And that's something I cannot do, and don't want to do."

Other commissioners found novel ways of mastering the paper flow. One in particular was notorious for asking his staff members, much to their frustration, to summarize all the documents that came across his desk on no more than a single page, in large font, and to include a recommendation on a course of action. Apparently this was standard practice among the politician members; only the expert commissioners— and, to begin with, the visiting anthropologist—tried to read them all.

The staff members, for whom handling and producing these texts was a full-time job, were slightly more successful at controlling chaos. Their offices were generally cluttered with papers that were in the process of being read, written, or commented on. The office furniture served to hide still more papers neatly arranged in clearly marked folders. Despite the stacks of paper that sometimes accumulated on individual desks, the closed file cabinets always conveyed the impression of order. Papers that were referenced in conversation could typically be retrieved immediately from a folder or from the computer. If they could not, then the staffer would promise to deliver the paper later or would dismiss it as unimportant.

The "archive" of the EK was also filled with mountains of paper, although staff members claimed that the collection was incomplete. While numerous official series of documents and journals were stored in the archive, much of it consisted of documents produced by the EK itself. Nearly every document produced by the commission was placed in the archive, and when I went through it one day, it seemed to be as much a collection of papers produced by the commission as a collection of papers acquired by it. I often observed that incoming papers whose status or usefulness was not easily discernible to the secretariat were "archived." The archive therefore became at once a place of order from which documents could be retrieved and a liminal place where documents that defied the existing categories of order could be stored.

Going through the archive that day, I discovered a sealed box labeled "signatures." There was no indication of the purpose for which these signatures had been collected or of how they had been put to use. Even my co-workers could not tell me any details. Given the confused history of the EK's establishment, it seemed most plausible that this sealed box contained the signatures that had helped bring the commission into existence. Perhaps it was the archive of forgotten activism.

Paper, of course, is materialized information,[7] but where did this overwhelming flow of information come from? One primary source was the Bundestag ticker service, which fed a constant stream of information to each employee's computer, updated every minute. During my time in the secretariat, news items related to bioethics would arrive about every ten minutes. This news would quickly travel through the halls of the secretariat and would often provide topics for discussion during lunch. Reading "news as it happened" became addictive for a week or two, at which point I learned to tune it out for most of the day and scanned the postings only before lunch or prior to a meeting. It was an odd feeling, however, to have access to the very same official sources and resources from which newspapers constructed their accounts for tomorrow's print version. Although I stopped printing out interesting press releases after about a week, I still had a sizeable stack of printouts when I left the secretariat.

A member of the Nationaler Ethikrat once complained to me of being overworked. She too seemed to be suffering from information overload. She could not understand why the NER seemed to be so constantly under pressure while other ethics commissions, in the United States or in France, seemed able to work without constant scrutiny from the public and the media. Why, she asked, did the American President's Council on Bioethics seem to set its own agenda, while there was always a line of people waiting with topics to be discussed by the NER? Why, in other words, did ethical issues in Germany seem to have a greater urgency, and demand immediate institutional responses, while at the same time there were such high expectations of reasoned argument from its forums of semipublic deliberation? The Kanzleramt, she said, expected regular reports from the NER, which she also deemed "burdensome."

As I sat through meetings of the EK, and also occasionally of the NER, I noticed over and over how much their members oriented their topics of analysis to the news of the day. It happened repeatedly that a commission member would distribute a scientific paper or a social scientific study referred to in a newspaper article, which would decisively shape discussion that day. Ethics, in other words, was the deliberative echo of transient political events. Even during the parliamentary debate on January 30, 2002, a representative quoted the results of a medical study from the latest edition of a physicians' weekly. A member of the NER confirmed my sense of the fluidity of the ethical agenda. When I asked her where topics for discussion came from, she said that often a topic emerged when someone carried an idea or a concern from one commission to the other, where it introduced a productive difference and was then amplified. Only when an issue traveled across contexts—from science to the media, from newsroom to commission, from one commission to another—did it become ethically interesting. The crossing of borders activated ethical sensibilities.

During breaks in commission meetings, members would sometimes distribute recently published scientific articles that they believed might change the terms of the debate or have an effect on regulation.[8] Watching parliamentarians or experts rushing around the table to distribute yet more reading material, it was hard for me not to get the sense of ethics trying to keep up with science and thus buy into the discourse of the ethics lag that the commission used to legitimate its interventions into scientific practice. The latest scientific findings continually flowed into the meetings and into our texts. During a meeting on the ethics of organ transplantation, an EK member passed around the most recent press releases of the British biotechnology firm PPL Therapeutics, which claimed that pig organs had been made compatible with humans. One expert argued that the EK's report on organ transplants needed to be more precise in stating what was being researched at the moment, but another energetically countered that the group simply did not know enough and that the report ought to be kept more general for that very reason. The commission continually tried to fix the state of the art in science in order to give its ethical judgments greater weight and validity. Although consistent with the opinion voiced by many scientists that biological concepts precede ethical ones, this effort seemed like a futile attempt to capture the rapid progress of scientific research within a cage of moral reasoning based on stable ethical principles.

The number of texts the commission processed often amazed me. It was not always clear where those texts had come from. I wondered how a final document became official; it was certainly not through the "purity" of its sources or the orderliness of its genesis. One time an expert member wanted to strike all references to two particularly well-known philosophers—Kurt Bayertz of Münster and Dieter Birnbacher of Düsseldorf—from the final document. She said that the quotes attributed to Bayertz were so general that one needed no citation at all and that those attributed to Birnbacher were useless since the philosopher stood for a particularly liberal and utilitarian position. After a brief debate on the principles of citation with Johann, the staff member who had studied with Bayertz, the expert still insisted that all references to Birnbacher be struck from the text. I remember my surprise at what appeared to me like a form of censorship.

Although the secretariat was largely responsible for producing the texts that would then be modified to become the commission's reports, members of the scientific staff did not generally discuss the content of bioethics. When I tried to initiate such discussions, one of two things happened: either I got a seemingly comprehensive rundown of all the relevant issues and positions, or the other person waited for me to ask more specific questions that could then be answered. It was apparent that the scientific staff members had already considered all possible positions and had mentally either adopted or refuted them, so that emotionally engaged debate, like that supposedly raging outside the doors of the building, was no longer necessary. Everyone's position was too well known to everyone else. The staff had, in effect, been socialized into being good *Saalordner*. The only places where discussion occurred were the places where debates were staged for an outside audience.

One staff member even told me that he had lost all passion for bioethics debates. Daily engagement had so inured him to life-and-death issues that he could now recognize a particular chain of argument after hearing a single sentence of it. He knew where it would eventually go and what its strengths and its weaknesses were. Ethics, for him, had become thoroughly routinized.

This familiarity extended not only to the ethical arguments, but also to the people making them. After I had spent an afternoon sitting through a meeting of a different commission, one of our scientific staff members took me aside to explain what I had just observed. He rattled off the participants' names and their ethical positions, ending with an evaluation of the "performance" of this particular discussion. I was reminded of the way Germans sometimes evaluate participants in a soccer game, contextualizing each person's performance in a larger frame. He then had a long discussion with the associate director about "what it all meant." Together, they sounded like two sports commentators disagreeing on the performance of opposing teams rather than two experts struggling to find a solution to a moral problem.

Without the commission's staff there could have been no text, but as in many legislative settings, for the text to be authoritative, the extent of the authors' involvement had to be kept invisible. During one of our postmortem morning meetings in the secretariat, the staff members realized that there were mistakes in an agreed-upon section of a text that was going to be the basis of the commission's discussion the following week. One paragraph in particular could not possibly work as it stood, since it contradicted not only itself but also other existing legislation. An interesting dilemma ensued as the staff members discussed whether they should distribute a version that they had cleaned up: "No one will remember anyway." But the director had moral qualms: "It is not possible to change a text that has been voted on."[9] Johann, a trained philosopher, offered a way out: "There is a difference between cleaning up the form and cleaning up the content. . . . What we would do here is only a formality." Once the contradictions were redefined as a *formal* element of the text, the director agreed with the philosopher's judgment: "So we are only going to smooth it out. That *is* our mandate, after all." The philosopher supplemented logic with common sense: "It makes no sense to work on a text that we know from the beginning can't stand the way it is. It is also bad form to prepare and bring to the meeting two text versions, since it will confuse the commission members." In resigned tones, the director asked, "What do we have to lose?"

At other times, too, form and content were held apart. When during a commission meeting one parliamentarian member suggested changes in the language of a text in order to make it more readable, another expert prevailed, saying that linguistic changes did not need to be discussed here and now; that could be done during the breaks. Readability, for this expert, was a marginal phenomenon and could be separated from the textual content.

In sum, then, bioethics seemed to proceed in stages. At first there was confusion about a new disciplinary formation. At that stage no one was certain of the arguments, and confusion reigned. Next, a group of insiders came to define and learn by heart all the positions and were enabled to negotiate them among themselves. Then these ordered positions were converted, or materialized, into texts. Finally the debate among those positions was staged for the public as parliamentarians took up the structured and filtered arguments produced by their ethical *Saalordner*.

Grammar of Democracy

But how was text actually produced in the EK? Its meeting of January 21, 2002, began with a cigarette break outside the meeting room, in the large transparent hall, while those present waited for some of the more vocal members to arrive. Without them nothing could be decided. Margot von Renesse, the commission's president, was herself a smoker, and the universal ritual of smoking together brought out her personal side. On this day she revealed to the small group around her that she got her own information about stem cells (as well as about the attacks on Afghanistan then going on) through her family members, some of whom were physicians and researchers. She joked that she and her family members were like stem cells themselves in the way that they were in constant communication with one another.

When the meeting began, Renesse easily slipped into the mode of commission president. The meetings she ran had an organic feel to them. There were no jarring interruptions, and through her ability to make discussions cohere she exuded the sense that everything was proceeding in order, as it should. Renesse herself later revealed in an interview with me that she was acutely aware of her power to shape discussions by selectively invoking rules of order. Sometimes, for example, she would cut short a lively discussion or prevent a vote because the issue had not been on the meeting's agenda.

When the discussion turned to the commission's report, there was some confusion in the room over which synopsis was under discussion. Last week's and this week's had similar title pages, and some commission members had not been able to print out the most current version. Meanwhile, a philosophy professor was getting impatient: "Either we must talk about this now in an *orderly way* and *in detail*, or . . ." Only when something had been talked about in an orderly and exhaustive manner could the discussion be closed, not before. Not only the outcome of the deliberations but also the process itself had to be orderly.

Throughout the discussion that day (as on many other days), people were designated to take dictation on parts of the text in order to record the proposed textual changes. This collaboration in producing a robust text went down to the sentence level. In some instances, different members of the commission had suggested different revisions to parts of the same sentence, and only when these parts came together, when individual perspectives were placed side by side, did a grammatically correct result emerge. Establishing order involved producing grammar in a democratic way—each person contributed fragments to the whole. One representative dialogue went in part:

> WUNDER: "What is happening now to my second sentence?"
> HONNEFELDER (*quoting in shorthand*): ". . . whether . . . whether . . ."
> Renesse asked to please have the entire sentence repeated (*den ganzen Satz, bitte*).
> HONNEFELDER: "Mr. Wodarg has the first part." (*den ersten Teil hat Herr Wodarg.*)
> RENESSE: "Mr. Wodarg: the first part, please." (*Herr Wodarg: Bitte den ersten Teil.*)

After Wodarg read out his part, the discussion proceeded. A short while later there was a vote on punctuation marks: "Who is for a period? . . . And who is for a comma?" Democracy dictated even the choice of punctuation.

In the afternoon there was a discussion over using the words *erzeugt* ("created" in the sense of "produced") and *gezeugt* ("created" in the sense of "generated"). Some wanted to speak of "*erzeugen* children" in order to mark the technological background of making a baby using IVF or other "artificial" methods. Renesse, along with others, was opposed to the idea of calling any child "*erzeugt*," regardless of where or how it had had been created (*gezeugt*). She argued that in "normal IVF" the child was still "*gezeugt*" because the natural process had simply been brought into the laboratory (*weil ja nur der natürliche Prozess ins Labor verlegt wurde*). The term *erzeugt*, she said, should be reserved for cloning and similar methods, which were "unnatural processes." The matter was then put to a vote. There were five votes for *erzeugt*; more than five for *gezeugt*. The word *gezeugt* was therefore left in the text unchanged. IVF thus became "normalized," or "naturalized," democratically, through a vote.

Having a textual resolution is not the same as resolving all the forms of behavior the text seeks to discipline, as the recent rise of wrongful birth lawsuits in Germany shows. The novelty of IVF is not only that the process of procreation now takes place in the laboratory, but also that this process is mediated by physicians and laboratory technicians who can be held liable by the biological parents for defects in their child. When the child is born with a defect that might conceivably have been detected, the parents can claim that the physician, who is involved in the "generation" of the child, has neglected his or her duty. The same parents who wish to see the child as *gezeugt* (created, as through reproduction) from their own perspective may wish under other conditions to claim it was *erzeugt* (produced) by the physician or the supporting laboratory.

There followed a similar discussion on the concept of "cryoconservation"—maintaining embryos in a viable state by freezing them to extremely low temperatures. Someone questioned whether, by calling embryos "cryoconserved" in the law, one was illegitimately attaching a criterion of being human to a technical procedure or a technological artifact. Only cryoconserved embryos can be supernumerary by definition, someone said, not "fresh" ones. What would happen, another person asked, if one could invalidate the technical criterion of cryoconservation by discovering a novel way of conserving embryos—one that did not require the extremely low temperatures (colder than minus 130 degrees Celsius) that defined cryoconservation? In this discussion the ever-changing science refused to be written into everlasting law. It was feared that law itself would lose its immutability if its applicability were made to hinge on the scientific processes that constitute the very idea of progress. At one point Renesse joked that referring to embryos as "frozen" would make it sound as if Germans were importing "strawberries from Israel."

Listening to the debates in the commission, I often felt as though I was observing engineers taking apart a piece of machinery (a word, a phrase, or a concept) in order to analyze precisely how it works. Once its workings were understood, the item under scrutiny was closed to further inspection and debate because it had been agreed upon and accepted by all; in Bruno Latour's term, it could be black-boxed again. In this process of deconstructing and reconstructing a foundation for bioethics, grammar, vocabulary, and technology played equally constitutive roles.

After following this and similar discussions intensely throughout the afternoon, I became aware of the time again only around 6 p.m. It was now dark outside, and I could see into the parliamentarians' offices to the right and left of the meeting room. Their offices were illuminated, and glass "windows" stretched from floor to ceiling. None of the blinds were drawn; a co-worker whispered that closing oneself off from public inspection would attract suspicion. Transparency, Berlin's recurrent architectural trope, was enforced by (imagined) seeing eyes. In a panoptic universe, as Foucault intuitively understood, no one could possibly police the officials as well as they policed themselves. Meanwhile, many commission members had left the room. Of the full complement of twenty-six, only seven now remained, but the decisions and changes, the cuttings and additions of text fragments, continued. In the end, this "minority report" would nevertheless count as the commission's voice.

Textual changes were often made for the sake of consistency and aesthetic harmony with other parts of the document. The authors chose words and constructed phrases in order to fit the wording of other sections; texts referred to texts referring to texts. In the end, texts very similar to this one would acquire the force of law, and citizens would understand those laws and obey them because they would find them to be coherent. The lawful texts would have the right form, and they would be seen as internally consistent. I myself later became witness to the self-authorization of such texts when a law student and I discussed the Stem Cell Law, whose less-than-orderly production I had witnessed in part. She, however, unerringly located order in it. Going through almost every sentence, she found one kind of logic or another in both the form and content of this text. As a budding lawyer, she knew how to look for order, and she found what she was looking for. She was aware of other legal texts and decisions in neighboring fields and could see an order in this text that its producers may not themselves have been aware of.

When the texts were produced, they resembled *Flickenteppiche*, or patchwork carpets, lacking a unified design.[10] Heterogeneous questions, comments, and concerns were woven together into a single document, and it seemed unlikely to the beholder of the process that any common will would ever be found behind them. *Gesetzgebung* (lawgiving, or legislation) seemed to be the epitome of a compromise between conflicting points of view. And yet the Stem Cell Law would become common will in the sense that (almost) every perspective would have been included, or at least heard, or at least have had a forum where it *might* have been heard. Perhaps this is what Bismarck famously alluded to when he said that citizens ought not to observe laws (or sausages) in the making, for then they would not be able to sleep peacefully at night. At the turn of the twenty-first century, he might have said the same for German bioethics.

"What Are *the Ethical Aspects of Organ Transplantation?"*

In order to treat certain topics in detail, the commission had divided into three smaller groups that went to work on specific subjects. I attended a number of meetings of the group that prepared a report on the ethical aspects of organ transplantation. On Monday, February 4, 2002, our subgroup met in the offices of the secretariat to discuss the text of that report. There were five expert members in the room, but none of the parliamentarians; the politicians were in their hometowns campaigning for an upcoming election.

The meeting started with a statement that the organ transplantation report needed to be clear on why the subject was being treated here and how. Justification was felt to be necessary. "Why here and why now? These are the questions we need to explain so that laypeople can understand them." Answers were immediately given: The state of medical research develops and evolves, and the law must be adapted. Transplantation was the framework within which the (current) stem cell debate was unfolding. Others added their own opinions: The problem was that organs were scarce and had to be distributed justly. There was no need to raise false hopes for stem cell therapies. The issues were autonomy and heteronomy (*Selbstbestimmtheit und Fremdbestimmtheit*). The point of the report was to educate and inform the public. The first speaker objected to this last opinion, saying that federal reports did not in general lead to an educated public. He suggested taking medical language out of the text under discussion, and indeed, out of the entire report, to increase its accessibility. Significantly, no one ever suggested taking out the legal language, which was presumed to be unremarkable and transparent.

Whenever parts of the transplantation report were unclear, the problem was attributed to the staff member who had left the EK for the NER, who had been in charge of the transplantation report until her departure. In part this was perhaps a strategy of simple convenience in that it exonerated those present in the room; on the other hand, it was perhaps more than mere coincidence that the person who had left for the questionably legitimate NER became the scapegoat for flaws in a text that wanted to claim order and legitimacy.

After lunch one commission member asked bluntly, "What precisely *do* the ethical aspects of organ transplantation consist in?" Another member responded that she was no expert on ethical questions. Ernst Luther, an expert in medical ethics from the former East Germany, offered that it was a matter of societal negotiation and just distribution, which were threatened by new biomedical developments. For him, in other words, ethics was, first, a question of proper allocation and, second, something to be decided according to principles of allocation already laid down.

This expert then offered an additional piece of text on the regulation of organ transplantation in East Germany. He said that the previous version had not taken into account the East German transplantation law, which had continued to apply there even after the fall of the Wall, as had Paragraph 218 of the Penal Code regulating abortions. Two experts from West Germany immediately shouted him down: "What's written here is complete nonsense! I don't understand this at all. This has no meaning! Either this gets rephrased, or we'll strike it from the text!" I observed several times how suggestions from the East German EK members were rejected. Their ethical sensibilities, it almost seemed, were not considered to be trained in a way that would justify their inclusion in the collective report.

Two months later, during an EK meeting in mid-April, no one seemed any surer what the ethical problems of organ transplants were. As if to normalize and minimize the problems accompanying organ transplantation, Renesse subsumed them in the more general logic that underlies capitalism: "We live in a world of scarcity, on all levels. Everything that's worth anything is also finite."[11] Organ transplants, in other words, raised no new issues worth the EK's attention. Another commissioner expressed her doubts about the group's work altogether and wondered about what recommendations for future work to include in the EK's final report: "I don't know if we can tell parliament to continue doing what we have been doing here." Toward the end of that day's meeting, some grew convinced that both the actual practice of organ transplantation and the literature already available on the topic had by now superseded the state of knowledge captured in the EK's draft report. The experts in the room expressed concern that they might be publishing an outdated text. Someone knew that a well-known professor of medical law was about to publish an edited volume on the ethics of transplantation medicine, which would surely overtake and embarrass the commission.[12] All agreed that the EK had neglected to pay attention to debates occurring in other professional domains, and the group voted to cut the section on organ transplantation from the final EK report. At this point the staff member who had been in charge of formulating much of the text on organ transplants and piecing together the fragments that commission members had given him left the room. He told me later that it was too painful to watch the commission discard practically everything he had been working on over the past few months.

Translation — The Semi-Legitimate Outsider Attempts to Produce a Legitimate Text

I myself became implicated in the production of the language of ethics, and here again I ran into the glass walls of tacit knowledge.[13] Since it quickly became known that I had spent many years living in the United States, people sometimes asked me to translate short passages from scientific or other official articles and statements or to check the translations of professionals they had employed for such jobs. The first time this happened, on my second day at the EK, the associate director of the secretariat asked me to check a translation of a document that summarized the EK's own report on existing stem cell guidelines and its further recommendations. The commission would send the document to several international ethics councils and post it on the EK's website.

I struggled for some time to turn what struck me as excessively legalistic German into something that sounded even remotely English. My efforts were thwarted by endless run-on sentences (a perfectly correct construction in German, which my American college professors used to scold me for carrying over to my English), passive constructions (a German way of taking agency out of writing and making a text more objective), and neologic nouns created from verbs (also common in German, when the "just right" noun isn't available in the preexisting language). As I translated this document that spelled out, in highly abbreviated and summarized form, the perceived ethical and legal problems of stem cell research in Germany, I thought that these linguistic contortions could stand for the broader difficulty of translating ethics and values into a novel idiom. The language for it wasn't felt to be adequate to the task just yet, and one needed to stretch it here and there to cover the new territory as it was emerging. German, I came to feel, is particularly well suited to such stretching. It easily accommodates neologisms, and its flexibility in word order and sentence structure allows it to adapt easily to new objects that enter the linguistic horizon. My effort at translation reinforced this sense of the (relative) adaptive advantages of German.

Eventually I produced a text that tried to balance the demands of fidelity to the intentions of the German writer and intelligibility to the non-German reader. I took some liberties, but tried to retain the complexity of the issue and the density of the text. I tried to change, for example, the translation of *"überzählige Embryonen"* as "supernumerary embryos," which to me did not capture well enough the connotations of the German term. I suggested that the word "supernumerary" be replaced with a less dismissive term. Even "spare embryos" seemed better, because in a context that regards embryos as human forms of life, "spare" connotes at least a potential, if not eventual, purpose. The tension inherent in a potential life that could at the same time be denied potentiality seemed to me to lie at the center of the German stem cell debate. "Supernumerary," on the other hand, invokes no such tension; it denotes excess and superfluity. In a fittingly theatrical definition taken from the *Oxford English Dictionary*, a "supernumerary" is an actor "employed in addition to the regular company, who appears on the stage but does not speak." A supernumerary, then, is dispensable, of no account, possibly without value or worth.

The idea of supernumerariness was a central concept in the stem cell debates, and the commission debated it vigorously. Once commission member Wolfgang Wodarg mentioned that he was against the concept of "supernumerary embryos" (*überzählige Embryonen*) altogether. He said he was against creating new categories, or entities, that the Embryo Protection Law might then no longer cover and with which researchers could do as they pleased. He added that "what is meant [by the term] are 'embryos that will, for various reasons, no longer be implanted in the woman.' That is how one should describe it, without defining it. Because what is at issue are certain intentions."[14] Other commission members agreed that the term was unsuitable. One member added that the description should also include the *Vorkernstadien* (the fertilized eggs before the nuclei fuse) because otherwise another debate would start about when life begins. Ludger Honnefelder, the prominent philosopher, pointed out the act of constructing: "Supernumerary embryos do not, after all, come into existence in nature." Still another member noted a danger inherent in all new concepts: "We have to take care not to introduce a new concept that everyone knows stands for something else that we don't want to name, like 'new federal states' (*neue Bundesländer*) for DDR." The commission then agreed to include in its report the recommendation that "in deciding on how to treat them, the legislator must distinguish between embryos and *Vorkernstadien* for which the term 'supernumerary' has become popular, although this term should not be used in a law."

When I finally offered my translation of the stem cell report summary, the associate director seemed less than happy. He had already received an official translation from a British professional translator and had apparently only wanted me to confirm that her work had been sound. By not doing this, I had called into question an established process that was recognized as legitimate. When I showed him where I had disagreed and why, I watched responsibility for errors in the text discerned by me be shifted back and forth, until finally the associate director, who had himself commissioned the translation, conferred with the primary translator, who convinced him to leave the text as it was. "Official translations" are required in many German contexts—for example, when foreigners need to document something related to their status—and such translations are notarized, which makes them unalterable. Perhaps I was not "democratically legitimated" and therefore could not make changes to an officially translated text? Perhaps it was structurally impossible that a semiofficial co-worker would produce an official text? It is also possible that I was altogether wrong in my interpretation of the mandate to translate. Perhaps I inadvertently made the short text transparent to a (democratic) public that I imagined and identified with "out there," whereas the official translator was speaking bureaucratese from one office to another, and hence performing the sort of translation that converts texts into the smoothly working, internally coherent, black-boxed machinery of government.

Glossary—Marking Science, Unmarking Law

Although the intended audience for the EK's reports was the indeterminate general public,[15] the public they reached seemed preconceived in certain ways. On January 16, 2002, during my second week with the commission, one of the scientific staff members asked me very politely to generate a glossary for the section on genetic testing that would be part of the EK's final report. I had told him that I was in search of an institutional role, and he had found this temporary solution. He asked me to do this, he told me, because he was so familiar with the specialized medical terminology that he did not even notice it anymore. That terminology had become commonplace and perfectly comprehensible to him.

When I, on the other hand, had read through earlier reports produced by the commission, I had noticed not only the medical vocabulary, but a number of obscure and highly specialized legal and ethical terms as well. I therefore compiled a glossary of all the terms that I thought required definition for the educated lay reader—taking myself as the surrogate. When my co-worker went through my compilation, he removed all the legal and ethical terms. He told me that neither juridical nor ethical terms had ever been explained in a commission glossary, only terms from the natural and medical sciences. Consistency demanded that I follow this practice as well.

Thus the language of science was marked, while the vocabularies of law and ethics were deemed to be unmarked. This practice reflected the broader assumption under which the EK operated: that science was the innovator and agent of change, and thus needed formal introduction, while legal and ethical concepts were stable and immutable reference points that could be looked up anywhere, and therefore needed no explanation. Juridical concepts could be left undefined because, certainly from the standpoint of a legislative body, it was the law itself that spoke ethically while it grappled with medicine and science. The legal and the ethical were the common languages in which this community could communicate. The educated lay reader, it seemed, was supposed to be thoroughly educated in law and ethics, as parliamentary representatives evidently were; only ever-changing science and medicine produced language and concepts that needed to be reconstructed in ethical language.

When I inquired further about this discrepancy, I was told that most members of parliament have legal training. Matthias, who had once worked for a member of parliament, explained that not only was the juridical profession the normative profession within parliament, but "*Juristendeutsch*" was also the normative way of expressing oneself. German legal language was seen as a form that needs neither critique nor modification to be made more transparent. While one might modify or exchange medical or ethical terms and expressions, he told me, the juridical was experienced as clear and precise, both within the EK and in the Bundestag more generally. Even when it was extremely convoluted, circuitous, complicated, and removed from everyday usage, juridical language was considered as immutable as the law itself. This explanation was borne out by my own experience. I never once heard a commission member change a legal phrase in the text, regardless of how forced the prose seemed to me. On the other hand, I observed commission members repeatedly trying to reduce the amount of medical terminology in the texts, often because it was unclear to them.

And yet it seems that the intended audience for the commission's reports was larger than the parliamentarians. If a fairly well-educated reader like myself found some of the legal and ethical terminology incomprehensible, then surely other lay readers would as well. When terms like "*supererogatorisch*" (supererogatory—defined by the *Oxford English Dictionary* as "going beyond what is commanded or required" and used by the ethics professor Honnefelder in connection with living organ donation) were tossed around in drafts of the organ transplantation report, the goal could hardly have been to advance the text's general intelligibility. As it happened, this particular term was not kept in the final report.

Other nations too were concerned with finding a language in which to articulate solutions to the ethical problems that science brought to society. On the same day that I created the glossary, January 16, 2002, the news ticker declared that George W. Bush had announced the composition of his President's Council on Bioethics, which he had been planning to form since the second half of the previous year. Bush had stated that the council's mandate was to act as the "conscience of the nation," and the list of predominantly conservative experts, who in some cases had been very outspoken along the lines that Bush himself identified with, seemed well equipped for the task. The council made a conscious effort to use plain language and to call things by their names; it effectively changed the terms of the cloning debate, for example, by reframing the distinction between therapeutic and reproductive cloning as one between "cloning for research" and "cloning to produce babies."

The Beginning of Life

At another point in my internship, the same co-worker who had asked for the glossary asked me to look for images on the Internet that would detail the precise process by which the sperm fused with the egg. The new reproductive technologies allowed novel kinds of mechanical interventions in, as well as visual access to, the reproductive process, and the statement that "life begins at conception" was no longer precise enough. Common knowledge equates "conception" with the fusion of sperm and egg, while medical practitioners who perform IVF assume that the human being begins with the formation of the individual and unique genome; that is, with the fusion of the two strands of DNA from sperm and egg. My colleague, who worked on the section on preimplantation genetic diagnosis in the final EK report, needed a more detailed depiction of the process by which, over the course of several hours after sperm and egg have joined, their genomes fuse. Biomedicine had rendered the earlier definition obsolete by deconstructing the previously coherent moment of conception into discrete phases. The visual representation was intended to render these new scientific facts accessible to the report's readers. I found an image, but my colleague ended up using a different and still more detailed one for the report. When during the plenary debate of January 30, 2002, several speakers claimed that "only the beginning can be a valid definition" for determining the point when human life becomes worthy of protection, my co-workers pointed out that the beginning point is always chosen arbitrarily.

Other formulations also demonstrated that language could be confusing regardless of the speaker's professional affiliation. One expert stated that he for himself recognized brain death as the criterion for the death of the body, but he added that all persons ought to (*sollte*) decide this for themselves. Another expert asked whether one really ought to (*sollte*), and the first expert emphatically stated that one must (*soll*): "Everyone must decide for himself or herself at what point they wish to be regarded as dead." A starting point premised on the individual's autonomy of conscience thereby shifted subtly into the inescapable duty of a legal imperative. An "ought" became a "must."

Conflict of Objectivities

When I told the ethics commissions that I wanted to study their everyday workings, I at first kept running into resistance. Both commissions seemed to think of themselves as operating on a meta-level. As detailed earlier, they saw themselves as *Saalordner* who brought order to an otherwise chaotic discourse. Both commissions were new, and both considered the end products of their labors, rather than their everyday workings, worthy of study. Studying another person's point of view, and treating it as an object of investigation, is one way of elevating oneself above the other. Being an object of social scientific study implies that one is somehow partial and holds a perspective that others can fit into grander or more inclusive narratives. In other words, having an observer inside who "exposes" the subjective basis of an institution's objective judgments might undermine its central claim to legitimacy. The initial resistance to my anthropological scrutiny might therefore be explained in the following way: perhaps my wish to study the ethics commissions appeared to their members as an attempt to expose the subjective foundations on which their objectivity, and their claim to legitimacy, rested and to trump their hard-won objectivity by claiming for myself the naturalized objectivity of the disinterested researcher.

Possibly the EK gave me access to its inner workings because people in some sense misunderstood the nature of my work. This misunderstanding was not apparent to me at first, in part because I myself had no clear idea of what I was looking for. The nature of my inquiry became clearer to me only in the process of learning about the commission—and working through its resistance. My questions to the commission evolved throughout my time there, and throughout the writing process, as similar logic surfaced in seemingly disparate domains.

It seemed in any event that while the ethics commissions could create and pronounce order, they could not allow an outsider to look behind the curtains to see them producing that order. Or at least not if that outsider came to study the process of ordering rather than to participate in it in earnest. It was, under certain circumstances, acceptable to offer one's perspective as one more point of view, but it was unacceptable to give a summary of one's perspectives because such a summary implies an observer position outside and above the scene of action.

During my time at the EK, it happened once that the curtain was lifted. While the EK reports were in draft stages, they were stored in electronic form on a Bundestag hard drive that only members of the commission and the secretariat (i.e., those with the proper password) had access to. In the secretariat, the staff had divided among themselves responsibility for working on particular sections, and each knew the domain in which he could make changes to the documents. There was almost a sense of ownership; people spoke of "my part of the text."

During the EK meeting on February 18, 2002, Renesse, the commission president, was visibly angry. Without mentioning details of what had happened, she told the group emphatically that drafts of the EK reports were confidential ("*Entwürfe sind vertraulich!*"). Later it was revealed that one of the commission members had passed on drafts of an EK report to a member of the NER, and perhaps to the NER's general secretary as well. The scientific staff members were especially irate because they considered themselves the primary authors of the largest portions of the text. They asked ironically whether they should simply place the EK's internal hard drive on the Internet and make it accessible to everyone. Cornelia reminded the staff of their duties: the commission members might talk to whomever they liked, but the staff members were sworn to secrecy. Likewise, she warned the staff members not to talk to journalists when they were present: "The commission members may talk to journalists. We [in the secretariat] are probably the only ones sticking to the rules." Staff members were also frustrated with some of the commission members who gave talks at political events and complained there about the lack of cooperation from staff members.

Draft reports were confidential because they were works in progress on which consensus had not yet been reached and which had not been voted on. Secrecy had a temporal dimension: the text was secret as long as it was being debated. Once the report was completely written and collectively legitimized—that is, once it stood for the consensus of the commission—then one hoped for maximum publicity, even though in practice the commission took very few active steps to promote its reports. To study the decision-making processes of the parliamentary EK, then, was also a transgression because it deconstructed a process whose very function was to produce a non-deconstructible collective judgment. In other words, revealing the workings of the EK challenged the very foundations of its authority to make informed ethical judgments on everyone's behalf.

Paper Wars

The EK was not concerned only about being overshadowed by the highly visible and publicity-seeking NER. The EK staff members told me that the problem with the EK was that it was too much in the public spotlight, but not in a helpful way. They were especially frustrated at spending months extensively researching and heavily documenting texts that journalists would then summarize in two or three sentences. Their piles of papers would be ignored, their labor would be made invisible, and those few journalistic lines would be all that the public ever learned about the commission's laborious work. Some staffers became too disillusioned to care. One staff member said, "When one has worked on a document for this long, then in the end one doesn't care about what's inside. It's really true." At other times textual changes were accepted with equanimity and a resigned shrug: "I am completely dispassionate on this point." (*Ich bin da völlig leidenschaftslos.*) Bioethics insiders who were not attached to either commission well understood the staffers' complaints over their invisibility. One of the organizers of events for the Protestant Academy in Berlin told me that she too was disappointed that the more glamorous NER attracted so much media attention while the more industrious EK was hardly noticed by the public.

Still, there were happy moments. Matthias was excited to get positive feedback on the report on genetic testing whose factual content he had personally overseen. A pediatrician had written to say that the report was better than any textbook on the topic that he had seen. Of course there were one or two factual errors, which the writer made sure to point out, but overall, he said, it was unsurpassed.

The EK had produced a first brief report on the protection of intellectual property in biotechnology in January 2001. It issued a much more substantive second report on stem cell research in November 2001, and it produced a hefty final report on human dignity, social ethics, PGD, genetic data, and "discourse and participation" in a democracy in the summer of 2002. The NER, by contrast, produced only a twenty-four-page booklet on human embryonic stem cells—the work, I was told, of two factions within the NER, one arguing for stem cell importation and research, the other against it. The NER report is aesthetically appealing in its symmetry and absence of footnotes. The EK's stem cell report is heavily documented and encumbered by its legalistic, convoluted phrasing and consequent lack of appeal to the lay reader.

Bruno Latour argues that a (scientific) document becomes robust by virtue of enrolling allies, whose strength is in their numbers. For a scientific paper, the paradigmatic ally is the supporting reference. The more numerous the references cited in support, the harder the paper will be to disprove—and the stronger it will be.[16] In the case of the rival ethics commissions, however, it seems that the very robustness of the longer document led to its inaccessibility and subsequent neglect. Enrolling allies in the scientific workplace apparently does not guarantee the enrolling of wider publics. Indeed, too much success in producing robust claims, so necessary to professional communities, may impede the bioethical claimant's ability to reach out to lay audiences.

A Visit to the Media

Although parliament goes to great lengths to make itself transparent to citizens in other domains, the parliamentary ethics commission deliberated behind closed doors. The EK did not allow outside visitors other than those who had an official reason to be there. There did, however, come a time when the EK desired publicity.

The EK had a difficult relationship with the media, and insiders were aware of this. Many perceived the NER's self-presentation as highly professional, even if a major paper like the *Frankfurter Allgemeine Zeitung* devoted only one column to an NER event. Insiders in parliament blamed the EK for not trying to get similar media exposure. On my third day in the commission, Cornelia, the secretariat director, a somewhat reclusive person who seemed uncomfortable in the spotlight, openly blamed herself for her lack of media savvy. Although she used to work as a journalist and was familiar with the ways of the media, she had neither built nor maintained close contacts with newspaper or television reporters, even though the EK's topics would have met with great interest all around. Even to a recently arrived reader of German newspapers and casual television watcher like myself, the ethical and legal challenges brought about by innovation in biomedicine seemed omnipresent. It would not have taken much, I felt, to position the EK in the public consciousness as an important arbiter of ethical questions relating to the biosciences. Especially given the government's self-conscious attempts to make itself transparent to a public clearly interested in learning, the EK's failure to make itself more visible remained puzzling. In an attempt to excuse this failure, Cornelia claimed that the NER's publicity work was "still worse." But while the NER's public relations would dramatically improve over time, those of the EK would not. Throughout my time in Germany, it was clear that hardly anyone had heard of the EK, and many were surprised to hear that there were two commissions addressing the same problems. In fact, when I told people that I was working for the ethics commission, they frequently assumed that I was working for the NER. It was the new and immodestly self-promoting commission that left the firmer imprint on the public mind.

In one particularly embarrassing incident, a television station mistakenly claimed that the EK had changed its opinion on PGD *after* the NER had delivered a public statement on that topic. In an effort to rectify matters, Cornelia talked to the staff at the TV station. She said they had looked embarrassed (*betreten*), but had promised a correction. As everyone knew, however, later corrections of a news item typically did not receive the same kind of attention as the original item itself. No amount of counter-steering could "correct" an initial wrong impression.

The question of priority was particularly important because the EK kept pointing to its own legitimacy in relation to the NER and to its own high standards for public statements. As I mentioned earlier, the EK's report on stem cell research, for example, was a hefty document of over a hundred fifty densely printed pages, while the NER's was a slender twenty-four pages printed on flimsy paper, suggesting both lack of substance and impermanence. While the EK report contained scientific background, graphs, international comparisons, and a lengthy bibliography, the NER's statement summarized four possible positions that were worded similarly but weighted differently.

Cornelia's annoyance over the misattribution of priority pointed less to her desire for publicity than to her sense of the importance of the autonomy of the EK's ethical judgment. It was important for the EK's legitimacy for it to arrive at an ethical conclusion first and, even more importantly, without help from outside sources, especially less legitimate ones. Furthermore, once the EK had arrived at an ethical judgment, it should not alter that judgment on the basis of a competing ethics commission's evaluation. Otherwise, it would be not only the commission's autonomy that suffered, but also its standing as an original and authoritative source of ethical knowledge and provider of moral guidance. In other words, the EK had to prove that it was *mündig*—able to use its own reason without the guidance of another. In a later chapter I will show how the autonomy of ethical judgments is linked to the German notion of *Gewissen* (conscience).

Perhaps in response to the commission's perceived uneasiness with the media, one of my EK co-workers in the secretariat organized a visit to the studio of the TV station ARD (Allgemeine Rundfunkanstalten Deutschland—a cooperative of public broadcasters). If the EK had been clumsy in projecting itself onto the TV screen, then at least it could learn about why and how this communication failure had occurred.

The ARD was founded as an association of public broadcasters in the 1950s. Today it still calls itself, and is widely known as, "*Das Erste Deutsche Fernsehen*" (First German Television Station). German television has been tightly regulated from the beginning. The German constitution (Article 5, Section 1) gives all persons the right to inform themselves from publicly accessible sources and to form their own opinions. The German media therefore have the mandate to enable this process in a free and comprehensive way. The ARD main studio was just across the river from the secretariat, in a reddish building built in the 1990s. Inside, the building had the look of a prison. All the offices faced outward, and from the metal stairways that led up the building's hollow interior, one could only see closed doors. The studio was built according to ARD specifications, but we were told that many workers now found it too small. When the building was planned, the idea that Berlin might again become the capital of reunified Germany was merely in the air(waves), and the media needs of that imagined era could not be foreseen. The growing role of the media in contemporary politics similarly was not contemplated.

On our tour, Ingo Härtel asked the reporters why they always seemed to interview the same handful of politician members of the EK (who, moreover, rarely attended its meetings) while the nonparty experts never appeared on TV. The reporters replied that their mandate was political reportage. Ingo pointed out that the NER got a great deal of attention despite the complete absence of elected politicians, which seemed to contradict their statement. If the NER was even less of an official political body than the EK, then the explanation for the discrepancy in media coverage had to lie elsewhere. What Ingo did not point out is that the NER was a political body through and through. As a child of Schröder's power play, it was a challenge to the Bundestag, and hence by nature interesting to the conflict-loving news media. Even in trying to get at the reasons for the EK's lesser success, Ingo invoked just those ideas of bureaucratic legitimacy that made the EK so uncharismatic in the public eye.

Even before our ARD visit, there had been agreement all around that the NER had been much more successful at working the press and that the EK ought to have an additional press conference every now and then. There was also agreement, however, that while the NER might care about dramatic effect, the EK was "merely doing its duty, and ought not to be interested in public recognition." In other words, as an official political body, the EK was supposed to appear disinterested and to deliberate in a secluded space, while the NER, newly conceived and unambiguously self-interested, clamored for public notice.

As if to apologize for paying so little attention to the EK, the ARD journalists cited the drawbacks of their medium: they said that they had to work with segments that were ninety seconds long at the most, and if one thought about it, this was hardly more than a single printed page. They enviously pointed to the *Frankfurter Allgemeine Zeitung* (FAZ), which could hold forth (*"sich austoben"*) in its feuilleton pages in articles that would have taken up nearly an hour of television coverage. In general, they said, the FAZ kept to very high standards ("one can even read it"), whereas television always needed to proceed from the assumption of the lowest common denominator.

The NER, at any rate, seemed to be more aggressive and more innovative in courting the media. In April 2002 I heard that the general secretary of the NER had invited a group of important German journalists and reporters to what one might call a "reverse press conference." The aim was not to give the media information that they could then report on, but rather to ask the media how they would like to have information from the NER presented to them. The general secretary, in effect, wanted to know how the NER could position itself more effectively in the news. How could it get the access and recognition it deserved? The media's interest was taken for granted, and the question was how best to facilitate the reporters' function. The media would not be permitted to ask questions in turn. In other words, the general secretary was trying to fashion good relations with the media *not* by making the inner workings of the NER transparent and letting the media see in with their own eyes. Instead he wanted to let the media codetermine how the NER would portray itself. According to an insider, most of the staff members had excused themselves from what they thought might be an embarrassing act of self-denigration. Unlike their boss, they had not yet internalized an understanding of ethics that would allow them to subordinate their deliberative style to the logic of the media.

The Nationaler Ethikrat Goes Public

As we saw in chapter 1, Simitis imported the idea of public meetings from the United States to Germany in 2002. The NER amended its rules of order (*Geschäftsordnung*) in December 2002 to open its meetings from then on, to allow the publication of transcripts on the Internet,[17] and even to allow taping and filming of its meetings in individual cases, with prior permission.

When the NER held its first open meeting, on the afternoon of January 23, 2003, there was some confusion among the members of the audience about what to expect. In the morning the NER had presented its report on preimplantation genetic diagnosis (PGD), explained its structure, and defended the majority and minority positions. The event was well attended, and audience members had the opportunity to ask questions before lunch. Television cameras were moving among the rows of seats, trying to focus on the papers that people were holding in their hands. The cameras were sometimes positioned so as to look over the shoulders of unsuspecting listeners, and some audience members seemed bothered by the intrusion. (Only a few weeks earlier, TV cameras in the Bundestag press gallery had used zoom lenses to successfully decipher a handwritten note that Chancellor Schröder had passed to the minister of defense, much to both officials' embarrassment.)

The officially "open" meeting took place after lunch. There were only about ten people in the audience at this point, which did not seem entirely unexpected. The open part of the meeting had been announced on the NER web page only a few days earlier, and no invitations had been sent out through the NER mailing list, as had been done for previous NER panel discussions. The late notice meant that only people with ready Internet access had learned about the meeting. Moreover, the NER chair had emphasized throughout the morning that the public part of the meeting would be followed by a part closed to the public. Apparently many people understood him to mean that everything following lunch would be closed. Throughout the afternoon the NER debated issues concerning therapeutic cloning while a professional stenographer, hired from the Bundestag for a reputed 450 euros per meeting-hour, sat at the head of the table. The transcript would be available on the Internet a few weeks later.

The person who had introduced the report in the morning said that the NER had made every attempt to make its language as comprehensible as possible. As I explained earlier, the EK had made no similar efforts. Members of the EK had used the normative language of the law, and allowed the law to speak through the commission, on the assumption that the law needs no further translation. Then one NER member expressed the hope that the first part of the report would make people stop discussing problems that only seemed to be problems (*Scheinprobleme*) and direct attention to the real problems, for they were numerous and difficult enough. Another NER member, a lawyer, tried to resolve a widely perceived disjuncture between the abortion law and the Embryo Protection Law by explaining that the starting point was reproductive freedom: "the parents' right to have *their own* child. The disharmony . . . can thus be resolved, and there will be no *Dammbruch*." The NER was cast in these remarks as the body that educates the public to frame its ethical concerns in correct terms.

Later a member of the audience asked whether the United Kingdom's experience could serve as a good example for regulating preimplantation genetic diagnosis in Germany. In Britain, PGD could legally be used to search for an embryo with the compatibility necessary to create a suitable organ donor sibling. An NER member answered that the British had a different moral attitude and a different legal system. The United Kingdom permitted research on embryos until the fourteenth day after conception, but one also had to keep in mind that the British Human Fertilization and Embryo Authority had turned down many requests for PGD. In the NER's rendition, Britain emerged as an ambivalent "other": at once admirably restrained and out of control. It was held up as an example of sensible regulation on the one hand but denigrated, on the other, for what this regulation permitted. The United Kingdom was admirable to Germans when it exercised the kind of control and restraint that Germans would like to install in their own country; at the same time, when its authorities permitted what was unacceptable in Germany, its legal system and moral intuition were held to blame. NER members seemed to want the moral *strength* of their British neighbors, but not their moral *content*. Even though "international cooperation" was seen as necessary, what emerged in the NER report was a very German solution to problems that many states were sharing.

The NER's growing facility with the media coincided with another German cultural innovation. In Germany during the postwar decades, politics had been a modest affair, conducted in sleepy Bonn, with little of the theatrical quality of American politics. The election of Chancellor Schröder, which also marked the move of the government from Bonn to Berlin, changed many of these older preconceptions. From the beginning, Schröder broke with convention, arguing that the time had come for Germany to move in new directions. A *Spiegel* article from those years described how Schröder was driven through Berlin for the first time. Coming upon the golden victory column (*Siegessäule*) on the Strasse des 17. Juni, the chancellor asked his companion, Walter Momper, former mayor of Berlin, what it stood for. Momper replied that it was there to commemorate the Franco-Prussian War of 1870–1871. Schröder then wanted to know whether Germany had won that war. When he received an affirmative, if subdued, answer, Schröder's expression, according to *Spiegel* reporters, seemed to say, "See, we *can* do it." (*Na also, es geht doch.*)[18] The first postwar chancellor who also grew up after the war seemed no longer focused on drawing from the past the potential dangers of every possible misstep.

Karlsruhe—Merging Law and Art

There was near-universal agreement that the EK should organize an event to officially and formally conclude its work, to present its findings, and to make itself available for public questioning. The commission chose the Center for Art and Media Technology (ZKM) in Karlsruhe, a city in southwestern Germany, as the location for this event—a place nearly as inaccessible from Berlin as it could be within German borders. One reason for this choice was that Karlsruhe is home to the Bundesverfassungsgericht, the Federal Constitutional Court. The hope was that the location would underscore once more the constitutional legitimacy of the EK and its final report. Others said that the choice was an attempt to appease EK member Monika Knoche of the Green Party, whose strongly feminist proposals had been repeatedly struck from the report and who, despite her lively initial engagement, had also been denied the presidency of the commission. Knoche had introduced the most restrictive bill during the parliamentary debate of January 30,

2002. Karlsruhe was in her voting district, and holding the event there would give her an opportunity to present herself as an effective politician to her constituency.

In late January the secretariat had been charged with developing the concept for the meeting and suggesting speakers, and the topic came up in most of the morning strategy meetings. Early on someone had suggested staging a debate between the philosophers Jürgen Habermas, a staunch defender of democracy who had intervened in the cloning debates with his 2001 cautionary booklet *Auf dem Weg zur liberalen Eugenik*, and Peter Sloterdijk, a controversial figure who had achieved notoriety with his 1999 essay *Regeln für den Menschenpark*, in which he appeared to endorse genetic manipulation for breeding better human beings. The two giants of philosophy had attacked each other in a series of widely read newspaper articles. During one EK discussion, Honnefelder said that he could see no connection between art and normative politics, and that he had discussed these themes with Habermas himself, for whom he already felt sorry. Renesse suggested putting the concepts of *Humanität* and *Anthropotechniken* at the center, since they had to do with law as well as with art. While nearly all agreed that Habermas would be a great draw, there was also consensus that a democratic thread needed to be seen to run through the meeting, and some did not see Sloterdijk as an appropriate choice. The objections to Habermas came from feminists, who said that "Habermas doesn't really have much to do with the female body." (*Mit dem weiblichen Körper hat er es nicht so.*)

The secretariat got permission to invite Habermas, and Arnd Pollmann, a staff member who was a political science student of Axel Honneth (himself a former student, and now colleague, of Habermas), sent the e-mail. Habermas responded two days later. He would have time on that weekend in June in principle, but he wanted to know the concept of the event and the list of other invited speakers. Now a frantic search for a suitable counterpart began. A woman would be good, the logic ran, and she should be young, because so often it is elderly men who debate each other on panels. Someone mentioned Jutta Limbach, president of the Federal Constitutional Court in Karlsruhe. Someone else pointed out that she would no longer be president by the time of the event, thus no longer commanding the charisma of office, and her name was dropped. Next the director asked who might fit the content of Habermas's approach, but no answers were forthcoming. When no suggestions had materialized two weeks later, someone suggested that Habermas should answer questions posed by the commission members.

Apparently no adequate dialogue partner could be found for this towering theorist of public communication.

I for my part wondered why the commission, a working organ of the democratic state, wanted to put itself in the position of receiving answers to its questions from Germany's most authoritative public intellectual. Why would parliament, the embodiment and enactor of democracy, want to submit its democratic authority to a theorist, however prominent? The wish to have Habermas lead these debates not only underscored his preeminent place in Germany's political philosophy, but also showed parliament's reliance on theoretical perspectives in informing its decisions. Why was there such reverence for theory? Perhaps philosophy, with politics filtered out, was seen as a bulwark against a possible *Dammbruch*. Habermas himself seemed to feel uneasy about his potential role. Perhaps he was concerned that his instrumentalization for a political event might pollute his standing as a commentator on politics. A few days later he withdrew from the program, stating that he was "not the right person for such an event."

With all slots open again, the commission then went through a list of professors in an attempt to construct a microcosm of the German philosophical landscape. They divided the candidates into "constructivists" on the one hand and those who "believe in the natural" (*Natürlichkeit*) on the other. In between they positioned the "skeptic" Odo Marquardt. Other factors had to be balanced as well: "Now we still need a woman with a constructivist approach. Oh wait . . . we already have one. Then we may also include two more men."

Then there was a debate over whether to have Catholics or Protestants on the podium. The churches had been so involved in the bioethical debates, one member claimed, that they should now also have their say. One influential commission member, herself a Protestant, was determined to have a Catholic conduct this debate on public morality "because at least they know how to do this" (*die können das*), whereas Protestants did not have a clue. Next the discussion switched to who should moderate. Someone suggested a woman from the weekly journal *Focus* and added apologetically, "Of course she is neutral. But in this case that may not be a disadvantage."

A recurring theme in setting up the panel of EK members who would question the speakers was whether the parliamentarians or the experts on the commission should be on the panel. There was constant indecision because on the one hand, the members of parliament had certain rights that came with their office, while on the other hand, in the commission context, all members were supposed to be equal. The president had a clear preference: "On the podium there should be members of the parties, because their democratic legitimacy is weightier than that of having been invited into the commission." Another important concern was to make sure that each and every facility in Karlsruhe would be accessible by wheelchair. One commission member warned, "The disabled will cause a huge commotion if anything doesn't work properly." The process resembled the assembly of a puzzle whose final form was predetermined, but whose pieces still had to be put in their proper places. The entire discourse was to be staged from beginning to end, and it was apparently important that each constituent voice in the ideal microcosm be represented. Even a forum conceived as philosophical could not simply mimic the philosophical landscape alone; it would also have to mirror the composition of the public sphere as a whole.

On Sunday, June 2, 2002, the members of the secretariat embarked on a seven-hour train trip to Karlsruhe, a city with a microclimate that makes it 5 degrees Celsius warmer, on average, than the rest of the country. After checking into the elegant Schlosshotel, we walked through the conference center, the ZKM, and then rested in the sun before dining in the Schlosshotel's "mirror hall" at the mayor's invitation.

The event on the next day was relatively well attended, given its out-of-the-way location. More than a hundred people sat in the auditorium, and closed circuit television brought the podium discussion to those who meandered through the halls of the building. The *Dialogveranstaltung* began with an introduction that drew attention to Karlsruhe as a place of law (as home to both the Bundesverfassungsgericht and the Bundesgerichtshof [Federal Court of Justice]) and technology and of the successful fusion of these two sometimes hard to reconcile social forms. The ZKM director pointed out that the topics of the EK had been addressed there for several years—for example, in the exhibit *Der Anagrammatische Körper* and by philosophers such as Peter Sloterdijk, who was rector of an institute in this very building. Now that the EK found itself in the same space as Sloterdijk, there was some embarrassment that no one had thought to formally invite him to the event. Next, the mayor assured the audience that there were people on the panel who would engage those topics. Then Monika Knoche, the commission member who benefited most from having the event take place in Karlsruhe, recounted the resistance to having a commission on law and ethics in modern medicine in the first place: "Never has the appointment of an Enquete Kommission been such a struggle . . . all parliamentary means had to be mobilized." Pointing to the dangers of cloning, and bemoaning the loss of sexuality and sensuality (*Entsexualisierung und*

Entsinnlichung) in reproduction, she poignantly asked, "Is a human being that enters the world without the help of a woman a being with human rights? . . . When women are no longer spoken of, then something essential is in danger of getting lost."[19]

At the end of the day, the ZKM director let the members of the commission and the secretariat into the current exhibition, *Iconoclash*, which Bruno Latour had co-curated. For half an hour we wandered through the vast display hall, trying to orient ourselves in a place that was consciously intended to disorient. It seemed a strange inversion that this labyrinth should constitute the end point of months of orienting and ordering the deepest issues of human existence.

The Last Day of the Commission

August 30, 2002, a Friday, was the EK's last day in official existence. The people who had been my colleagues for the past several months had cleared their desks and were about to lock their offices for the last time before returning their keys to the authorities. When I arrived that day a little after noon, all the secretariat members were sitting around the large table in the meeting room at the end of the hallway, and the leftovers of an ample meal were spread all over the place. On a strategy board, the meal had been planned with a blue marker: Antipasti: Bread; Hauptgang: Pasta/Linguini; Dessert: Apfelstrudel. It was unclear where the money had come from. According to Cornelia, the president's "representation fund" (*Repräsentationsfond*) was available "only for hosting others" (*nur zur Bewirtung Fremder*), and she had emphasized that she would not let herself be talked into any tricks, since she would have to sign for the money.

The scientific staff members talked about their new jobs, or their lack of employment. Ingo was now working in the Department of Health and was frustrated with its insistence on routine. Some of his co-workers carefully marked their time on the job, sometimes running into the building to stamp their time cards, then coming back outside to lock up their bicycles. Although he had initially looked forward to putting his knowledge and skills to use for the government, he now found the bureaucracy paralyzing.

Johann, the philosopher who had mastered a solid body of expert knowledge on bioethics, was planning to open an institute for ethical consulting in his hometown of Rostock. He joked that before doing so he would take a few months off and collect his unemployment payments. His pride was his collection of Bundestag internal memos (*Haus-*

mitteilungen), which he had been filing away in two large folders over the past two years, as well as a heap of death notices that had been sent around during his time at the Bundestag. He joked about scaring his successor by covering his office walls with these documents. All the documents were neatly ordered and easily accessible at any moment, even if no one ever looked at them again. He was performing bureaucracy, even as he was parodying it.

Matthias, the veterinarian, would begin working at a watchdog institution ensuring the safety and transparency of food production and marketing. He too joked, however, that he would first take a long vacation and collect unemployment payments. Arnd would also collect unemployment payments and use them to finish writing his philosophy dissertation on participatory democracy. His time at the EK had been productive in the sense of giving him practical insights into problems of public participation.

All four had continued on a trajectory that followed from their involvement in the EK. They would continue to be involved in ethical advising, legislating, or educating. They had in effect become bioethicists through their work in the commission.

At the end of the meal, the secretariat members sang an old Irish folk song that reminded me of the songs sung by the pioneers in old John Wayne movies. It was a rolling melody with hopeful words evoking wide skies and the transience of all things human. After the singing, several copies of the commission's final report were passed around for signing. Matthias told me later that at the end, all the staff members thought that the final report had turned out well, and that they took pride in having produced it.

Rules, and Rules on Following Rules

Just before the commission closed down, in August 2002, I met once more with Cornelia, the director, to speak with her about some questions that had come up in my evolving understanding of the commission's role and work. I also wanted to go through the commission's archive, before it was permanently stored in an inaccessible location, to look at the documented activities of the commission from the time before I had joined it. I went to her office, pocket bulging with the recording device I had brought, and found her chatting with someone in the hallway, waiting for me to arrive. It was a hot day, and in her office we sat down across from each other at the table where I had sat so many times during the biweekly meetings of the EK staff. Although the offices were almost all cleared out, she still had a geographic map of Germany hanging on one wall and posters showing the "front and back" of Earth photographed from outer space on the opposite wall. The planet in those photographs now seemed smaller and more manageable, but also farther away and out of reach.

Cornelia and I chatted for almost two hours. I did not record our conversation, but jotted down notes as she spoke freely about different aspects of her work. Our main focus was on how parliamentary commissions operate in general. There was an odd mixture of relaxation, a product of late summer in Germany and the commission wrapping up its work, and tension, because, as she now told me, she had never quite understood what my "true purpose" had been in working for the EK. I had been, and apparently remained to the end, a boundary crosser in an institution dedicated to putting everything neatly in place.

She said that even though the commission had formally ended, she did not yet know how much longer she would be in this office or what she would be doing next. Elections were coming up, and only afterward would new parliamentary advisory commissions be called into being. Naturally she wanted a position at her present level, since the level of office was tied to her pay, so she would not know until a couple of months from now what she would be doing. In the meantime, she and her two secretaries would remain in an otherwise deserted office, pack things up, organize and move the archive, but otherwise simply sit and wait. Some of the other people working in the office had expressed bewilderment at this self-inflicted state of aestivation. They could not imagine sitting in an empty office for two months while waiting for news of further employment.

At one point in our conversation, Cornelia needed to look something up and pulled a thick folder from her shelf. Suddenly she frowned and looked annoyed: "I hate it when loose-leaf folders are not maintained properly!" (*Das kann ich ja hassen, wenn Lose-Blatt-Sammlungen nicht ordentlich gepflegt werden!*) Much as for ethics itself, the precondition of a working bureaucracy is that everyone files everything predictably and precisely, thus making it retricvable for others at the touch of a button or a glance at a label or numbering system. Only when all users of a system are in fact working according to the same rules does the machinery work. A filing error, or the failure to constantly update a loose-leaf book of rules, can render an entire reference system out of date, unreliable, even useless.

When I asked her a few formal questions about how parliamentary commissions in general are put together, Cornelia retrieved a book of guidelines to look it up. I could have done this myself, but I decided not to stop her. While she was looking for the information I had requested, she started leafing around in the book and reading other information out to me, expressing surprise at these details of public administrative policy. Even after she had read me the answer to my question, she kept reading off information far beyond what I had asked her. She seemed to take great interest, even pleasure, in informing herself further about her duties, and perhaps informing me as well. Even in her current state of temporary unemployment, she was truly a "public servant," diligently informing herself about the law. Later I would come to recognize this almost obsessive interest in written rules as part of the ethos of civil servants.

Sometimes, of course, the rules as written were not unambiguous. But there was always a solution. Whenever the regulations were unclear on some detail, or whenever I reminded Cornelia that in practice things had been a little different from what she had just read, she pulled out a companion volume of commentary where she could look for interpretations of the basic rules. There she typically found the requisite explanatory sentence: "A practice has been developed that does not completely contradict the law." (*Es hat sich eine Praxis entwickelt, die dem Recht nicht vollständig widerspricht.*) For example, there was a rule that there could not be more than nine parliamentarians in a commission, and some smaller number of experts, so that the total number could not be greater than eighteen. In practice, however, the Enquete Kommission had thirteen of each. The commentary on exceptions to the rule permitted such excesses. Another small deviation from the rule in the EK was that the SPD had given one of its allotted seats to the Greens, probably in a gesture of good will. In other words, the rules by themselves were not encompassing enough; one also needed rules on how to follow the rules.[20]

Leaving the Field—An Outsider Again

By the time my official affiliation with the EK came to a close, I had clearly worn out my welcome with Cornelia. The director, who had initially sought to make the inner workings of parliament as transparent to me as possible, was now getting impatient with my questions. We had stopped joking with one another, and she had started asking me pointed questions about what it was that an ethnographer did anyway. Perhaps my unusual interest in the minutiae of daily practice in the EK had sensitized her to the disciplinary effects of the ongoing routine and the interpretive potential that rested therein. While she had earlier categorized me as an intern, a definite insider with a desire to learn for the purpose of preparing for his future profession, she had now begun to rethink my place in the neat dichotomy of inside and outside. Although I had made it clear from the beginning that I was going to write an ethnography about the everyday practice of bioethics in Germany, my liminal status and transgressive vision now bothered her. The transparency she had sought to create for me in the beginning now seemed risky, as I had apparently not learned the right things, but instead had noticed too many of the "wrong" things. Like the general public, I had been concerned with the "wrong" problems.

After my final conversation with Cornelia that August afternoon, I went to the archive one last time and looked at the folders containing documents from trips the commission had taken. Cornelia refused to let me photocopy anything, explaining again that since this was not public information I would not be able to use any of it in my dissertation anyway. I knew from another person who worked there that the EK archive would eventually become part of the vast Bundestag archive, where it would be accessible to researchers through a complicated and lengthy application process.

In preparing for my final departure, I took several large file folders and began to organize and file my notes and printouts. I had the distinct feeling that I had acquired some bureaucratic traits myself, and when my filing was complete I realized with mild shock that my office now looked identical to those of the people I had been working with for the past few months. My notes and papers were organized according to the same categories as theirs. Since I had evidently been able to replicate their mental order, and to do so almost without thinking, I now felt confident that I had internalized it sufficiently to seek the distance necessary for analysis. I had "gone native," and it was now time to leave "the field."

3

Transparent Fictions

As I showed in the previous chapters, national ethics commissions require transparency, or *Nachvollziehbarkeit*, in order to be accepted as legitimate arbiters of nationally significant moral questions. Both the Nationaler Ethikrat and the Enquete Kommission, although in different ways and with different degrees of success, made public displays of the exercise of making the right decisions. They effectively made spectacles of (their own) transparency by opening carefully orchestrated meetings to the public or organizing more or less well-publicized and well-attended events on controversial topics such as embryonic stem cell research or genetic testing of embryos. The commissions, each in its own way, had to call attention to their internal workings and reach out to the public in order to procure the audiences that would, in turn, legitimate their work. In short, in order to do right, they needed to be seen to be doing right. In this chapter I deconstruct this use of transparency as a legitimating device. I examine the use of transparency in some of its particularly German manifestations, and I argue that transparency, as a resource for political legitimation, can take culturally specific forms.

Toward an Ethnography of Transparency

Transparency has become a key symbol in global expert discourses on political, commercial, and institutional operations. The German government, in making transparency material in the very architecture of German democracy, has virtually raised the concept to a constitutional principle.[1] The official rhetoric is that transparency is good for democracy and that the state ought to make itself as transparent as possible to its citizens. The demand for transparency came to the fore both in Germany and among its neighbors when reunification and the government's move back to Berlin raised the specter of a resurgent German superpower in the center of Europe. The shadow of National Socialism's rise to dominance through the legitimate institutions of Weimar democracy looms large in Germany, and transparency is seen as an antidote to a repetition of that dark chapter of German history. If enough light (of reason) is cast for everyone to see, the idea seems to be, then no shadows (of unreason) will be able to form. Germans almost seem to hold lack of transparency accountable for the circumstances that permitted the Nazis to take control of the German parliament, as if more "transparency" would have prevented Hitler's rise to power.

The presumed efficacy of transparency is fundamental to contemporary German democracy. That efficacy, in turn, rests on the assumption that the legitimating public gaze will be looking at the right thing at the right time and thus ensure democracy's accountable functioning.[2] By assuming that all onlookers would see the same thing if they only looked, the proponents of transparency convert it into a technology of objectivity. The emphasis on transparency thus tends to become a self-fulfilling prophecy. Democracy functions unquestioned as long as the public is convinced that everything is happening transparently; yet once democracy is seen to be transparent, it becomes irrelevant whether anyone is looking.

But how do we recognize transparency when we encounter it, if indeed it can be seen at all? The arguments *against* plain seeing are not hard to find, especially in the work of cultural anthropologists: Franz Boas wrote that we could only meaningfully *see* what we already *know*;[3] Marshall Sahlins, drawing on Walker Percy, showed that recognition is re-cognition—that precepts precede percepts;[4] and Marilyn Strathern demonstrated that the attempt to make an institution transparent can force invisible structures upon the working of the very institution whose working one is making transparent.[5] But is transparency itself, the very quality of see-through-ability, entirely accounted for by such constraints of theory? Must we not, in order to see anything, also be trained in what to look for and what to look with? How can there be transparency unless there are eyes educated to see through to what is being shown?

The connection between transparency and democracy, then, is not as self-evident as contemporary German political discourse makes it appear. In what follows I will argue that the two concepts are linked not by necessity or nature, but by culture. I will suggest that in order to perceive something as transparent, one needs to develop a set of perceptual resources, tools, and habits. How these are acquired, and what kind of training is involved in teaching us to see as if through a glass, *transparently*, will be the focus of this chapter. I argue, in effect, that if transparency ensures the functioning of German democracy by making government a legitimate subject of public scrutiny, it does so only because a preexisting understanding of democracy produces a particular idea of transparency. In other words, what we recognize as transparent depends as much on our preconceived notions of how the political system that we live in works as the working of that political system depends on its "transparency" to function effectively. It follows that the meaning of transparency, like that of democracy, is created through practice.

As an anthropologist of public morality, I am interested in how people use transparency in specific cultural contexts—how ideas of transparency are conveyed, received, and performed in the production of common codes of ethics. I take as my point of departure the insight that for transparency to function meaningfully, persons on both sides of a perceptual divide must have some shared notion of how the thing works whose workings are being made transparent. Given the impossibility of unmediated access between the domains of the seers and the seen, I focus my inquiry on the mediation itself—the *act of seeing*. I therefore ask, Who sees? With eyes conditioned by what categories? Who determines when something counts as transparent and when those conditions are met? Who, in other words, educates the knowing citizen to see transparency and thus makes government democratic?

Many social theorists and science studies scholars have addressed the role of visibility in managing relations between the state and its citizenry or between science and the public. The accounts of philosopher Michel Foucault and political theorist Yaron Ezrahi are particularly important for my story.[6] In Foucault's view the panoptic eye of power, invisibly located at the center of powerful modern institutions and constantly monitoring those in its presumed range of vision, ensures compliance. The internalization of constant surveillance has made the presence of an actual observer superfluous. Ezrahi, by contrast, drawing on the classic account of the rise of experimental science in Restoration England by Steven Shapin and Simon Schaffer,[7] argues that both science and the state rely on staged spectacles to win the assent of the governed. In Ezrahi's scheme, the dynamic of power is inverted, and the observed populace is not coerced into obedience, but rather coaxed into cooperating with, and thereby legitimating, the institutions that function as guarantors of natural and social order.

My account is situated between these two theorists of political visibility. In Germany today neither coercion nor coaxing is readily available as a strategy for ensuring harmonious relations among science, state, and public. Instead, Germany can draw on a long tradition of pedagogical thought and a similarly extensive historical and contemporary literature on producing subjects who will at once serve and express the higher will of the rational state. In other words, the state does not simply make itself visible to its citizens, nor does it discipline their bodies into obedience. Instead, it educates its citizens to see the state as it wishes to be seen. In this process, transparency is refracted through pedagogical traditions propagated by the state.

Transparency Today

Concern with reason and reasoning is paramount in modern democratic states.[8] The German state, in continual reenactments of the transparency of its decision-making processes, articulates a special obligation to make itself and its democratic functioning visible to its citizens and to the world. Having internalized the horrors of the Nazi regime, the state has taken it upon itself to cleanse and control itself, opening to public inspection all the dark corners of its heart so that no black thoughts can remain. Through transparently collective reason, the state hopes to produce an ethics and politics that will speak "for itself."

The commitment to transparency is particularly evident in times of crisis, when politicians strive to make visible the production, consumption, and working habits of the modern technological state. In the wake of the mad cow crisis, for example, calls were heeded to introduce a tracking system in Germany that would permit one to follow beef back to its point of origin. Beef packages in supermarkets now have bar codes on them that allow the seamless tracing of each steak to the herd, sometimes even to the specific cow, from which it came. This tracking technology was termed "the transparent cow" (*Gläserne Kuh*).[9] Science and industry share this commitment to transparency. It would not be Germany without a transparent car factory (*Gläserne Fabrik*) like the one Volkswagen built in Dresden to display the production of its upscale flagship model Phaeton to engineering aficionados. In the north of Berlin, on the premises of the Max Delbrück Center for Molecular Medicine, there stands a transparent laboratory (*Gläsernes Labor*) where visitors can isolate, look at, and even touch a visible chunk of DNA from a fruit salad. Here too transparency is conceived as a property of objects (DNA), metaphorically transferred to the activity (science) that produced them.

Transparency, it seems, is no mere quality whose material or symbolic use produces understanding. Rather, it expresses a carefully managed relation between the state, industry, or science on one side and assenting and legitimizing publics on the other. It is a relation between different points of view. Embedded in this difference, moreover, are assumptions of cognitive inequality: one side explains, while the other side absorbs the proffered explanation by seeing what the other puts on display. As Jean and John Comaroff put it, "Efficacy and influence, alike in rhetoric and realpolitik, lie largely in controlling the capacity to reveal and conceal, to make 'reality' appear or disappear."[10] How then should we read the architectural tropes of transparency that symbolize, all over Berlin, the government's dedication to making itself open to scrutiny and transparent to its citizens?

One late afternoon in December, when the sun had set and lights in the buildings were turned on, I took a walk through the city's government quarters. Looking inside the Paul-Löbe-Haus, the so-called motor of the republic, I watched secretaries talking on telephones, but I could not hear what they were saying. I observed parliamentarians in their offices, but I could not tell what they were working on. I looked in on commission meetings, but I could not see what the participants were debating and deciding. And although I was walking within a few meters of their windows, my eyes never met those of the people inside, as they had apparently learned not to notice the passersby. All this transparency, it seemed, was not conducive to dissolving barriers between the government and the public. What I and other passersby *did* see was a political body at work, even if we could not make out its operations in detail. In other words, what we *do* see is what we are shown: a reassuring representation of democracy that our eyes can follow. We see a peaceable image, whose mechanics are at once accessible and inaccessible to us.

Jeremy Bentham, the founding father of utilitarianism, who advocated the greatest good for the greatest number, also advocated maximum transparency in government so that the electorate could perform its essential checking function. In an attempt to ward off tyranny, the German government today similarly places a premium on complete transparency. It is only fitting, then, that a British architect, Norman Foster, should have designed the glass dome of the Reichstag that today stands for the transparency of German politics. The entire building, as I mentioned earlier, is a maze of glass and mirrors. The plenary hall of parliament, the sovereign body, is shaped like the elliptical intersection of two overlapping circles, a larger one for the representatives of the citizenry at large and a smaller one for those representing the individual states. Parliamentary sessions are open to the public by prior arrangement, and six visitor galleries seating a total of four hundred people (there are some six hundred members of parliament) are arranged in a semicircle above the parliamentarians' chairs. An official Bundestag brochure describes the intended effect on visitors: "Descending in tiers, these galleries extend so far down into the centre of the plenary chamber that everything seems close enough to touch—the plenary chamber appears to be within tangible reach, as if the spectators, too, were seated in the center of the room."[11] Indeed, when I attended the parliamentary debate on the importation of human embryonic stem cells on January 30, 2002, I ended up sitting closer to the center of the ellipse than most elected representatives. This architectural detail, according to the brochure, "reveals the far-reaching transparency of the German Bundestag . . .— a further symbolic element of its commitment, inscribed in large letters above the main portal on the west side, 'To the German people.'"

In his discussion of the production and proliferation of the panoptic control apparatuses imagined by Bentham, Foucault points out that the physical arrangement of the panoptic institution, in which individuals are segmented and segregated for easy surveillance and discipline, traps the observers just as much as the observed. In fact, Foucault, paraphrasing Bentham, continues, "The arrangement of this machine is such that its enclosed nature does not preclude a presence from the outside: we have seen that anyone may come and exercise in the central tower the function of surveillance."[12] In principle, "any member of society will have the right to come and see with his own eyes how the schools, hospitals, factories, prisons function."[13] The observers themselves are under constant observation. As Foucault notes, "The seeing machine was once a sort of dark room into which individuals spied; it has become a transparent building in which the exercise of power may be supervised by society as a whole."[14] He adds, "There is no risk, therefore, that the increase of power created by the panoptic machine may degenerate into tyranny; the disciplinary mechanism will be democratically controlled."[15] In a footnote Foucault describes Bentham's hope for a "continuous flow of visitors entering the central tower by an underground passage and then observing the circular landscape of the Panopticon," so that "the visitors occupied exactly the place of the sovereign gaze."[16] Word for word, almost, this could be a description of my own visit to the Bundestag at the time of the stem cell debate.

But what do people see from the galleries? Typically, visitors to the Bundestag debates are permitted to stay for precisely thirty minutes, and their choices as to the timing of their visit are limited. I was allowed to stay for the entire debate that January day only because the parliamentary Enquete Kommission had made special arrangements for its staff and members. For normal citizens, those thirty minutes of watching parliament from the center of the plenary hall offer only a glimpse of those proceedings that they are supposed to, and supposed to be able to, recognize as democratic. The parliamentarians themselves participate in the performance; although plenary debates are in theory a model of democratic communication and persuasion, members of parliament are under pressure to vote along party lines (I will discuss this pressure and its implications in the next chapter). This means that in practice plenary speeches are often intended less as arguments to persuade dissenting parliamentarians in order to arrive at robustly democratic solutions than as staged spectacles of dissent from opposing parties self-consciously performed for the (voting) public observing from the galleries.[17]

Even more noticeable than the visitor galleries is the glass dome, visible from afar and illuminated at night, which also makes "the people" present by ostensibly placing them in the center of democracy. The spectacle of a continuous stream of people on the double helical walkway inside the dome from early morning till late evening (the dome closes at 11 p.m.) gives outside observers the impression of lively citizen participation. But even Foster's celebrated dome does not live up to its promise of making parliamentary transactions visible to the public. While the original design included a complex system of mirrors to allow both sunlight and the eyes of the public to enter the plenary hall, a structure obstructing this view was installed later, allegedly to keep the glare of sunlight out of the parliamentarians' eyes. Now the glass dome's most transparent function is to make the visiting citizens visible.

As I walked through Berlin taking in all these ambiguous gestures at transparency, I could not help seeing the armed guards who sat behind large bulletproof windows and eyed this sightseer with suspicion. It was quite clear what *they* were doing—their intentions were all *too* transparent. Their job was to keep people out, and the glass that made us visible to one another also acted as a divider, with unequally situated eyes on each side. Similarly, the cameras that have proliferated in central

Berlin reminded me that the observers of power are often being observed by that very power, and that citizens coming to see the workings of the state are always visible to that state. Transparency *of* the state, it seems, comes at the cost of transparency *to* the state.

The legitimacy of the German state has long depended on its power to cultivate feelings of equality and inclusiveness among its citizens. Dating back to Bismarck, many services provided by the German state, such as health insurance, retirement plans, and social security, have been organized so as to express the mutual obligations of community members by instituting a mandatory system of equal access to those services in return for payments proportionate to one's income. The sense of horizontal solidarity is preserved through a lot of vertical control. Social solidarity also involves a division of labor of sorts, but one in which individual needs and contributions are not revealed or compared. Solidarity works because the precise workings of the social body are *not* transparent to any of its individual members. Today transparency is of concern to ethicists who fear that knowledge of citizens (for example, of their genetic makeup, their health records, or their personal consumer preferences) might lead to discrimination that would undermine the all-important solidarity principle if it fell into the wrong hands. The wrong kind of transparency, in the wrong hands, can be socially disruptive.

The three pairs of wrong hands that information could fall into are those of the state, of employers, and of insurance companies. Media and activists voice concern about the transparent citizen (*Gläserner Bürger*) and the transparent patient (*Gläserner Patient*). The system of solidarity, by which the state seeks to avoid discrimination against or exclusion from the social system of any individual, is seen as coming under attack from genetic tests that might introduce seemingly natural differences and hierarchies among people, with deleterious consequences if used for the wrong purposes. Genetic information in particular is considered dangerously potent because it claims to define a person's essence and his or her personal future reliably.[18] The essentializing and individualizing tendency of modern genetics counteracts the traditionally randomizing and collectivizing character of health and illness. Further, since genetic information links biological relatives to one another by definition, it is in the field of genetic testing that the right to know (for example, one's biological origins or genetic makeup) can clash with the right not to know (for example, one's genetic predispositions to incurable late-onset diseases).[19] While solidarity presumes equality within communities, genetic testing can naturalize differences.[20] In these senses, genetic knowledge produces a radical tension between Germany's historical commitment to horizontal equality and solidarity and the postwar German state's almost obsessive commitment to transparency.

Although transparency remains a positive value in politics, finding expression in Germany's political architecture, the transparency of the human body and its value to self and society has become morally ambiguous as new biomedical technologies threaten to upset established social values and political orders. Citizens and citizenries no longer want to be known quite so well by either science or the state. But while the prospect of the transparent patient, accessible to the biomedical gaze, evokes unease, the figure of the glass-domed Reichstag shows us the political body made symbolically transparent to democratic governance. The public sphere is the space in which everyone transparently participates, whose procedures are open and intelligible to everyone, and in which all the inhabitants can experience themselves as citizens.

Crafting Citizens through Bildung

The achievement of transparency, I have suggested, hinges on there being a viewer who recognizes this transparency as such. If one seeks to make transparent the response of a well-functioning democratic political body to innovations from science and technology, then the ideal observer is the autonomous citizen capable of acquiring scientific knowledge and rendering informed judgment on the means and ends of science. But such citizens may have to be made—not found.

The preoccupation with educating citizens has deep roots in German culture. Following the French Revolution of 1789, concerns over the meaning of citizenship assumed new urgency. In Germany, Count Christian Friedrich von Augustenburg solicited a series of statements and reflections from the poet Friedrich Schiller in order to help him understand the implications of the revolution for other aristocracies. In his letters to the monarch, published in 1793 as *Letters on Aesthetic Education*, Schiller endorsed Immanuel Kant's Enlightenment ideal of a polity firmly ruled by reason.[21] Schiller saw the French Revolution as a failed attempt to bring about a rational polity, and he altogether doubted its feasibility for several more centuries. Schiller writes that the liberal regime of reason in neighboring France had arrived prematurely, at a time when people had barely emancipated themselves from the hold of their animal natures. The character of citizens had to be refined, a labor of at least another century, before a constitution could be successfully implemented. In other words, Schiller writes, "one must *create* citizens for the constitution before one can *give* the citizens a constitution" (my emphases). Before one could give the people a foundation for a body politic, one needed to train their individual bodies to function in a polity. Before giving the people a text to be ruled by, one first had to turn them into proficient readers of such texts.[22]

Schiller thought that the enlightenment of reason had already progressed significantly and considered most urgent the refinement of the senses and the moral purification of the will. To this end he developed a pedagogy of aesthetics. He described the necessity of providing spaces in which to cultivate thought and understanding (*Verstand*) into holistic *Bildung*, the prerequisite for the universal rule of reason. Until citizens were thus educated to take their part in politics, they would not be capable of existing under a constitution.

Over the centuries the characteristically German concept of *Bildung* was further refined, and education and citizenship in Germany became ever more entwined. Meaning more than mere education or rote learning, the concept of *Bildung* includes ennobling one's character, training one's senses and sensibilities, and forming one's mind for critical reflection. The word (which, instructively, has no exact English or French equivalent) also carries the secondary meaning of "creation." Louis Dumont, in his *German Ideology*, has written at length about the German ideal of *Bildung*, which consists in the conscious construction and cultivation of the self while that self is at the same time insulated from the external world. Dominic Boyer, in his *Spirit and System*, has taken this concept further by showing how ideas of *Bildung*, *Geist*, and *Kultur* came to function as tropes of self-identification and became anchors of dialectical social knowledge among the *Bildungsbürgertum*, or Germany's educated elite.[23]

Let us now look at how such notions of citizenship and *Bildung* were formed, and performed, at a 2001 citizen conference (*Bürgerkonferenz*) on genetic testing and how the concept of transparency came to encode ideas of citizenship, of lay and expert knowledge, and of democracy. It is in that microcosm that we can observe how the conference organizers crafted the kinds of citizens necessary for seeing and interpreting this transparency in the "right" way, as well as a working model of the democracy of science.

Democracy Made Transparent at the German Hygiene Museum

A citizen conference held at the German Hygiene Museum in Dresden, and advertised as the first of its kind in Germany, focused on the social and ethical implications of genetic testing. The museum, a well-known transmitter of scientific knowledge to the broader public, had launched this experiment with support from the federal Ministry of Education and Research, lending its considerable resources to the cause of making genetic testing transparent to citizens. The aim of the conference, according to the museum's director, was to counter concerns about a democratic deficit by adding the voices of an informed citizenry to a scientific discourse dominated by experts.[24] Modeling their project on consensus conferences held in Scandinavia, but with methodological innovations, the conference organizers semi-randomly selected nineteen citizens to participate. Those citizens, in a series of meetings, narrowed the focus of their deliberations from genetic diagnosis in general to the three particular issues of preimplantation genetic diagnosis, prenatal genetic testing, and genetic testing in preventive medicine. In cooperation with the organizers and a scientific advisory council, the participants identified a number of experts to invite and cross-examine, and they discussed a series of statements that they had solicited in advance from authorities in relevant fields. The final round of interrogation of the experts was open to the public. At its end, the citizen jury was to pronounce its verdict.

In 2001 I received a mass-mailed invitation to attend the citizen conference as an observer. The international debates on the ethics of embryonic stem cell research and genetic testing for diseases had been keeping Berlin busy as scientists and churches, philosophers and patient advocacy groups fought over the meanings of "disease," of "life," and even of "human." As an intern and participant observer at the EK, I had heard experts testify before the parliamentary commission, and I was curious to see how their practiced presentations of the issues would compare with the amateur debate among laypersons. How would ordinary men and women discuss the social, scientific, and ethical questions raised by genetic testing?

The conference was designed to engage the lay public in a debate with experts, but first that public had to be educated. The assumption was that ordinary persons, when given the relevant facts, were capable of arriving at well-reasoned conclusions that ought to be respected. As I looked forward to attending the conference, four sets of questions immediately arose for me: First, who were these missing masses of "the public," and how could a small group of individuals be selected to represent them? Second, what would the process of educating the lay public look like? Who would the educators be, and how would those to be educated absorb the requisite knowledge? Third, what would those properly educated citizens say back to their educators? How would they integrate the knowledge claims of the experts with what they themselves held to be proper knowledge and proper ways of knowing? Would their voices be recognized as having been acceptably educated? And fourth, would they learn to speak the language of science and reason, and would their voices be heard? The only way to find out was to observe the event, and the organizers' claim that this was the first citizen conference of its kind in Germany made me even more eager to witness for myself this claimed cultural innovation.

The citizen conference was going to be the latest place for fashioning a lively version of democracy. Merely asking citizens to be passive spectators of a parliament that *"appears to be* within tangible reach," as the Bundestag brochure had put it, was clearly not what the conference organizers had in mind. They wanted to improve on the existing representative democracy by actively eliciting the voices of the citizenry. In bringing citizens from the physical center of democracy into its deliberative center, the organizers had to craft a process with citizens inside it, and in which citizens would see themselves as such. Moreover, the citizens would have to be empowered to see into the inner workings of the matter being deliberated. Most importantly, perhaps, the conference organizers would have to produce the transparency required for making expertise visible. Technical subject matter would have to become transparent enough to enable ethical debate.

Below I focus on three aspects of this complex operation. First, the location of the conference, the German Hygiene Museum, resonated with the concept of transparency in ways that added depth to the situatedness of genetic knowledge and its ethical dimensions. Historically both an extended arm of the state and an extended arm of science, the Hygiene Museum seemed almost predestined to be a place where state and science could join hands in the pedagogical mission of making "informed citizens." The museum's long-standing commitments to transparency and education shaped the process of communication between experts and citizen participants. Second, the conference participants—citizen participants, lay and professional experts, and organizers alike—were similarly engaged, however unselfconsciously, in producing working images of transparency. The organizers made transparent a version of democracy, the experts illuminated truths about genetic testing, and the citizens clarified, for themselves and their observers, the meaning of "citizenship." Third, the process of organizing the citizen conference itself turned out to produce the very objects—or subjects—(citizens, experts, laypeople, and democracy), that it claimed merely to invoke and engage. Let us look at these three aspects of the conference in turn—place, participants, and process.

Place—A Pedagogical Training Ground

A two-hour train ride through the derelict landscape of the former East Germany brought me to Dresden from Berlin. On this November morning in central Germany, as night slowly lifted its veil, the air was chilly, and traffic moved as if still sleeping. Strolling through a park, I walked slightly uphill along Lingnerstrasse until I arrived at Lingnerplatz, where the first rays of dawn illuminated the slightly elevated modernist building that houses the German Hygiene Museum.[25] A length of red carpet rose up to the museum, and a long ramp complemented the widely spaced stairs leading up to the entrance. The red carpet looked as if it was intended for guests more important than me, and the ramp felt oddly disruptive of the organic unity of the building. I found the juxtaposition jarring at first, but later realized that the combination of the carpet and the ramp stood for the recognition and respect with which citizens, especially those with disabilities, were welcomed that day. Both ramp and carpet had served to make the 2001 exhibition *Der (im)perfekte Mensch: Vom Recht auf Unvollkommenheit* accessible by wheelchair while the building was being renovated. They were then, a museum attendant told me, left in place mainly for aesthetic reasons.

The conference conveners had chosen the museum for its supposed neutrality, but what was it that made the museum neutral? One answer is that education in Germany has historically been so naturalized that even the German Hygiene Museum, with its ideologically checkered past, could appear to be a neutral backdrop for an event like a citizen conference. The idea for the Hygiene Museum, founded in 1912, came from Karl August Lingner (1861–1915), the inventor and marketer of an antibacterial mouthwash named Odol (which GlaxoSmithKline markets in its original distinctive bottle to this day) that made him famous and wealthy. A firm believer in hygiene of all kinds as well as a shrewd advertiser, he was the organizer of the first International Hygiene Exhibition in 1911, which attracted more than five million visitors and earned a profit of more than a million marks.[26] Lingner summarized his social engagement in four theses on hygiene, the last of which was that "every person must acquire unconditional trust in the science that is recognized by the state, and become convinced that this science alone brings the certainty of a cure for diseases."[27] The hygiene museum was to be the proselytizing agent through which that trust in science would become linked to citizens' awareness of their bodies' innermost workings.[28]

The Second International Hygiene Exhibition took place in 1930 (Lingner had died of oral cancer in 1915), with a life-size figure called the Transparent Human (*Gläserner Mensch*) as its main attraction. On this occasion the museum moved to Lingnerplatz, into a newly built structure that embodied the transition from the avant-garde Bauhaus architecture of the 1920s to the more monumental forms of the 1930s. While the museum's interior expresses the Bauhaus emphasis on light and transparency,[29] the exterior anticipates the monumental style of fascist architecture of the 1930s in Germany and elsewhere.

In the late 1930s the Hygiene Museum produced a model transparent factory (*Gläserne Fabrik*), in which every step of the production process was instantly accessible to the supervising eye, thereby making the whole process visible, manageable, and efficient. The clearest expression of the fusion of man and machine may have been a glass motor (*Gläserner Motor*), also manufactured in Dresden in the late 1930s and conceived perhaps as a counterpart to the Transparent Human. In one depiction of the glass motor, Adolf Hitler is shown admiring the transparent engine, surrounded by a crowd of fellow enthusiasts in suits and uniforms.[30]

In the early twentieth century antibiotics were not yet available, and public recognition that hygiene acted to prevent infections generated great enthusiasm for health education. The Hygiene Museum very quickly became one of the major organizers of exhibitions relating to the human body, its social and cultural environment, and the relations among them. Guided by the awareness that citizens still lacked the means to live lives according to the precepts of reason, the exhibitions aimed not at prescribing from above, but at educating citizens: the emphasis was on explaining connections and making them understandable so that viewers would be able to follow their own reason.[31]

Soon after its inception, the German Hygiene Museum emerged as one of the main producers of educational materials in the country, and it shipped its well-known life-size transparent figures to pedagogical institutions around the world. Another widely distributed teaching tool was a 1926 chromolithograph by the German Jewish gynecologist, artist, and popular science writer Fritz Kahn (1888–1968), entitled "The Human as a Palace of Industry" (*Der Mensch als Industriepalast*), which graphically combined native sociology, history, and philosophy. Created when the German chemical industry was the world's most advanced, the image shows a human head and torso cut open to reveal their inner workings. Pipes and wires connect the labeled parts of the schematized digestive and nervous systems like parts of a machine; homunculi operate each organ like workers at factory stations.[32]

Not only are the bodily functions made transparent to the student in this image, but the mental functions as well. The upper half of the head, traditionally seen as the locus of human consciousness, is divided into different compartments. In the front half, three people are pictured sitting around a table in a room labeled "Reason" (*Vernunft*), and in the back a single person is reading at a desk in a room called "Understanding" (*Verstand*). Reasoning is portrayed as a collective act of discussion and debate, while understanding, situated as prior to reason in German metaphysics, is shown as a solitary activity involving careful, absorbed study. In other words, the image represents both an individual body in its normal functioning state and a social body in which individual understanding and collective debate give will and direction to a collectivity of producers and consumers. The image links the idea of transparency to the concepts of labor, usefulness, and efficiency. The presentation of collaboration within one body suggests democracy. The clear visibility of physical and mental functioning normalizes this seemingly democratic division of labor by making it transparent to everyone (and thereby objective).

Dictators no less than democrats were susceptible to the lure of transparency. When the National Socialists took over the German government in 1933, they used the widespread interest in personal hygiene to propagate national ideologies of race and racial hygiene, and the Hygiene Museum seemed almost predestined to become a transmitter of knowledge of *Volksgesundheit* and the care of the *Volkskörper*. Transparency of the social body easily lent itself to nefarious uses. The aims of science education and disease prevention were reformulated to encompass eugenic aims. The museum generated numerous pedagogical materials on population politics. The Transparent Human traveled all over the world, including the United States, where the statue and many other propaganda materials published by the museum were eagerly studied out of curiosity about the "Eugenics of New Germany."[33]

Although the museum was almost completely destroyed in the 1945 bombing of Dresden, the Russians occupying eastern Germany apparently recognized the utility of an established source of authoritative health education and restored the museum in less than a year. Under Soviet military administration, the museum began operating again in September 1945, apparently cleansed of its now discredited ideology. During the four decades of socialism the museum became a central institution of state health education, fulfilling for East Germany a role similar to that of West Germany's federal Central Agency for Health Education (Bundeszentrale für gesundheitliche Aufklärung), and its exhibitions, films, and publications enjoyed wide popular appeal. It continued its production of transparent figures, eventually turning out five transparent horses, eight cows, two cells, fifty-six transparent men, and sixty-nine transparent women, one of whom was represented as pregnant. As political regimes came and went, each found the idea of transparency appealing. Since the fall of the Wall, the museum has organized a number of provocative exhibitions, such as one called *Gene Worlds* (*Genwelten*), and another called *Sex: Of Knowledge and Desire*, which was on display in late 2001 when I went to Dresden.

A visitor who enters the museum today and walks into the main exhibition hall sees one of the spectacular transparent figures that made the museum famous when the Second International Hygiene Exhibition opened in 1930. The Transparent Human (*Gläserner Mensch*), a female figure with raised arms, appears ready to reveal herself to the dissecting looks of the observer. The crowds who came to see the 1930 exhibition and its main attraction, in the shape of an educational device, testified to the magnetic power not only of an ideal of total visibility that would unveil the inner workings of the body, but also of the museum's ideological commitment to transparency.

The figure, the first three-dimensional anatomical model of the human body ever produced, used eight miles of wires to represent the nerves and blood vessels. These wires were laid over the inner organs, which were then covered with plates of clear cellon to form the shape of a life-size human body. In these works of art and of German craftsmanship, each of which took eighteen hundred hours to make, the creators were handcrafting a modern vision of the human body, melding science, rationality, and transparency. Transparency was both a teaching tool and a prelude to self-knowledge. A former assistant at the museum remembers that visitors came in from the noisy streets and found themselves in a cathedral-like hall with low, indirect lighting. There the statue stood on a pedestal, and as the lights dimmed its heart began to glow, illuminated from the inside. A melodic voice from a gramophone began to read precise explanations of the functions of the different organs as they were lit up, one after another.[34] Material transparency alone was not enough to instruct the viewer; it worked only when it was sequentially presented and pedagogically performed.

On that November morning in 2001, however, the main attraction was not in the exhibition hall, but rather in a side wing of the building, added to it like an afterthought. Here a group of ten women and nine men were participating in the first German citizen conference on bioethics. As I walked into the sterile room of glass and white walls, I had the impression of witnessing a trial, with the nineteen citizens sitting on one side like a jury and on the other, the experts taking the witness stand one after another. In the middle, a moderator and his assistant were orchestrating the order of questions and answers, and they seemed to be firmly in charge of what was said in this room. The people in the audience were told that this day was devoted to the citizen panel's questions to the experts and that the general public could ask questions only between 5 and 6 p.m. This was the first sign I got of how carefully the notion of "citizen" was being managed here. The citizens "on stage" were styled as the "real" citizens. Untrained and unprepared citizens like me were deliberately kept off the stage and excluded from pedagogical attention. In a virtual performance of Schiller's ideas, the process crafted citizens who would be competent to speak about the issues of the day. I, a mere member of the untrained public, would have to wait until the end of the day's formal proceedings to have my undisciplined concerns addressed.

Participants—Who Are the Citizens?

But how did these nineteen particular citizens end up on stage at the conference? How were they chosen to represent Germany's entire body politic? Earlier that year, the museum had put on a morally fraught and widely publicized exhibit called *The (Im)perfect Human (Der [im]perfekte Mensch: Vom Recht auf Unvollkommenheit)*. In that controversial exhibit, the museum had emerged as a champion of people with social and physical disabilities. It makes sense, therefore, to assume a degree of self-selection among the participants in the citizen conference. Those who favored the unbridled march of scientific progress are less likely to have responded to the museum's invitation to participate.

If we look more closely, we find that the making of a representative group of citizens was an elaborate process. Venturing into Germany's obscure hinterlands, the conference organizers had written to a large number of town and city governments asking for names and addresses of a certain number of citizens in their jurisdiction. Addresses could be got from the city governments because in Germany every citizen is registered with *Bürgerämter* (citizen services offices) and can be easily tracked down. The organizers then mailed out 10,000 postcards and received 292 replies, a modest response rate of 3 percent.[35] The few who responded were sent postcards inquiring about their willingness to participate. In two published accounts of the selection process, participants wrote that they had almost thrown out the postcard because they thought it was an advertisement, but then both heard the call of civic duty. That call must have struck a deep chord, as one of the accounts mentioned subsequent nightmares about participation—analogous, it seems, to my own bad dreams before joining the EK.[36]

The responses were first sorted by sex. They were then sorted again according to age (16–30, 31–60, and older than 60). Within each age group, further divisions were made according to occupation. The 16- to 30-year-olds were categorized into student, employed, and unemployed. The 31- to 60-year-olds were categorized into employed, housewife, unemployed, and other. Those over 60 were categorized into unemployed, employed, and retired. From the resulting twenty piles, two visitors to the museum drew the names of twenty persons. Nineteen ended up participating in the conference.[37]

Democratic theory holds that all citizens are equal before the law and that each voice has equal weight. These principles seemingly justify the semi-randomized selection process employed for the conference, which one participant described as similar to a lottery. But who were the nineteen citizens? Where were they from? How old were they? What were their occupations? They came from both parts of the formerly divided Germany. Of the nineteen, six were students, five were retired, two were housewives, and one was unemployed. Thus, of the nineteen, only five were earning an income. Ten of the participants were over fifty, seven were over sixty. While a cynic might read the group's makeup as mirroring Germany's aging and unemployed population, it does hint at an imbalance that has become a source of public alarm.[38]

The nineteen citizens comprised ten women and nine men. The gender distinction would be significant because all ten women would vote against PGD, then the most controversial form of genetic testing, while all but one man would vote in favor of it. Indeed, the gender division became absolute in one newspaper article, which mistakenly claimed that all eleven women had voted against the testing while all eight men had voted in favor. Opposition to genetic testing was evidently marked as feminine, to the extent of inadvertently cross-gendering the one man who had voted with the women.

The conference organizers commissioned an evaluation report, which criticizes the uneven and unrepresentative composition of the group, but explains it away as the unpredictable result of a randomized selection process.[39] As we have seen, however, this was not the case: the organizers' sorting categories had already determined that six participants would be employed, another six would be unemployed, two would be housewives, and two would be students. Given the remaining "other" categories, it is not surprising that in the end, the panel consisted of six students and six retired or unemployed persons. With another two homemakers, and one withdrawal, the panel ended up with only five employed persons in this supposed microcosm of the German citizenry.

To give legitimacy and authority to a political decision, it is standard German practice to attempt to include all possible viewpoints.[40] Only through such inclusion could the "citizens on stage" be recognized to represent the "citizenry at large." Perhaps that is why the heterogeneity of the participants was highlighted throughout the event. Thus, on the final day, the conference facilitators began by introducing each member of the group, first the women, then the men, by name, age, occupation, and place of residence. By representing the citizens in this way, the facilitators were in an important sense structuring the meaning of citizenship. Long after the conference, interviewees mentioned to me one citizen participant in particular: a nineteen-year-old female high school student of Turkish heritage. This unusual concession to Germany's Muslim "minority" had left a deep impression on everyone, and it was repeatedly held up as an example of the inclusiveness and consequent legitimacy of the group.

Critiques of deliberative bodies in Germany are almost always based on their composition rather than their functioning. All the prerequisites of proper functioning are presumed to exist once a deliberative body is constituted as such. The sine qua non is that sufficiently diverse opinions must come together at the same table and make themselves transparent to one another. When those preconditions are met, the deliberations are considered trustworthy, and the end result is almost always acceptable to all. Transparency in Germany is accomplished when others are able to follow the reasoning of the bodies that are constituted to speak for the people. In practice, however, the fullness of representation becomes a surrogate for the actual reasoning.

The conference organizers were careful to maintain a cognitive divide between citizens and experts. According to the conference brochure, they eliminated all potential citizen participants who had any vested interest in genetic diagnosis, which included representatives of insurance companies, employees of drug manufacturers, and scientists. In other words, they made an effort to create a body of citizens who would be laypersons, and through that effort they generated a highly malleable notion of citizenship. As in jury selection in the United States, the aim was to get rid of preconceived notions by erasing the possibility of prior inscriptions on participants' judgment. Unlike the experts, the citizen participants could be neither interested nor knowledgeable, but had to arrive as blank slates with respect to the issue under discussion. Pedagogical authorities could then inscribe citizenship by endowing their waiting minds with relevant knowledge.

The group of experts, as well as the body of citizen participants, was carefully crafted. After the citizen participants were selected, the organizers inquired into the specific issues they wanted to discuss. Chosen in a lottery, the participants turned to how the genetic lottery, and the threat of being selected against, may affect people's lives and society's image of the human (*Menschenbild*). As noted before, they decided to focus on three forms of genetic testing: preimplantation genetic diagnosis, prenatal testing, and preventive genetic screening of adults and of populations. Ironically, two dominant themes at the conference would be a person's right to know his or her genetic makeup and a person's right not to know what one's genes might portend, such as predisposition to genetic diseases.

The organizers then came up with a list of appropriate experts to invite, drawn from the fields of medicine, science, philosophy, theology, and patient activism. The participants picked a number of those experts as potential interlocutors, and from those who submitted statements, they selected the ones they wanted to interrogate in person. Communication between citizens and experts was always channeled through the organizers. By controlling who could speak and on what topic, the facilitators controlled the form of the citizens' dialogue, and with it the contribution of the citizens to the debate; the organizers were in effect the wall between reason and understanding in Kahn's portrayal of the body's inner workings.

The evaluation report devotes special attention to the role and function of the moderator, to his perceived shortcomings, and to potential remedies for future events of this kind. The participants' main criticism was that, although the moderator had plenty of experience in professional mediation, he was not an expert in the scientific subject under discussion. The report interprets this criticism as a call for an *Universalmoderator*, someone who could accompany the communications of the participants neutrally and also "correct" the group if necessary.[41] The citizens themselves seemed to be aware of the need for their eyes and ears to be professionally trained. According to the evaluation report, the majority of participants wanted a strong supervisor, someone who could lead them to see "right." Like the student in my Berlin seminar, the participants seemed to have come with a sense that there were prior standards of rightness.

Process—Education in Citizenship

Since the event was advertised as the first citizen conference in Germany, it was up to the nineteen citizens on stage to represent the concerns of the entire citizenry. The participants took this role very seriously, and one could almost feel the solemn responsibility that had descended on their shoulders. Yet throughout the conference the participants seemed unsure of their role in the event. At one point a female participant asked the roomful of experts in a pleading voice, "What is it that you expect from us?" A slightly aggravated male expert responded, "What do *you* expect your citizen vote to accomplish? Is it not somewhat exaggerated to see yourself as a counterpart to the Nationaler Ethikrat, or even to the Enquete Kommission?" A male participant stated his own goal, probably expressing what others were thinking: "I want the public to take note of our decision." Gaining courage, the female participant then refuted the challenge to the authority of the conference: "The Nationaler Ethikrat is composed of experts, the *Bürgerkonferenz* of laypersons. I don't find it exaggerated to see ourselves as a counterpart."

Both sides continually reinscribed the expert-citizen dichotomy, with the supposedly autonomous citizens easily slipping into a subordinate role. After one expert presentation that drew extensively on information from the Internet, the same male participant asked if the expert could "assign those websites as homework," to which the expert sarcastically replied that, as responsible adult citizens (*mündige Bürger*[42]), they could surely find those web pages on their own. At another point one of the participants sought expert confirmation for her own beliefs. She had given ethics advice to a relative and wanted to have her personal judgment validated by an ethics expert.

The participants admitted they were nervous about being put on the spot in this way and that they were happy to be there as a group, with shared responsibility. It was only toward the end of the conference that the participants became surer of themselves. One asked the organizers to circulate the citizens' vote not only nationally but also internationally in order to stop the German government from simply continuing its policy of justifying embryo-expending research by any means. The citizens seemed unaware that Germany's 1991 Embryo Protection Law was already one of the world's most restrictive of its kind.

The participants' initial confusion about their role and what was expected of them is understandable. They had met for only a few weekends to discuss background materials and to review statements from experts before the meeting. Moreover, even the experts were sometimes undecided. Although most took a clear position in one direction or the other, one admitted during his presentation that he had been thinking about the legal protection of embryos for five years and was still uncertain about the best course of action because both sides had such powerful arguments in their favor. It seems that the citizens' opinion, which they would deliver on the last day of the meeting, could only be provisional and partial as well, representing not *the* voice of the people, but only *one* voice among many.

One goal of the citizens' conference was to generate dialogue between citizens and experts, or between consumers and producers of knowledge. Rather than producing better opinion surveys, as the director of the museum put it to the audience, the aim of the dialogue was the acquisition of knowledge (*Wissensaneignung*) and the formation of opinions (*Meinungsbildung*). But although the citizen participants were ostensibly in a dialogue with science, this dialogue ran up against important obstacles. The relation between experts and citizens was made uneasy not only by questions over both groups' authority and legitimacy, but even by their physical positioning vis-à-vis each other. Their spatial arrangement was like that of a jury and witnesses. Participants consistently complained that the staged setup created a physical distance too great for informal exchange. During the meeting that I watched, the participants often asked questions prepared in advance, while the experts had prepared their responses in advance, so that sometimes there was as much as an hour during which documents were read out loud by one side or the other for the benefit of the audience. This recital added another layer of artificiality to an already awkward performance of expert-citizen dialogue.[43]

Sometimes the physical distance between citizens and experts was so great that it could be bridged only with the help of technology.[44] At one point an expert was literally "called" to the witness stand when he was questioned by telephone. The telephone was placed on a table in the center of the room with a name tag next to it. The moderator dialed the number, reached the expert at home, and made the introductions. The participants then took turns walking up to the telephone and asking their questions into the receiver; they bent forward in a pose of awkward attentiveness, and then put the receiver next to a microphone so that the other participants could collectively hear the expert's answer. This "expert hearing" was almost comical as the disembodied voice of truth was transmitted into the room through the inanimate telephone, duly personalized with the expert's name tag.

What surprised me most was that nearly all of the fifty to one hundred members of the audience were interested in the citizen conference in some social scientific capacity. Almost none of the people I spoke to during the event were primarily interested in the issue of genetic testing. They were not there as "fellow citizens" concerned with the transparency of expertise, and they did not share the "official citizens'" anxieties over genetic discrimination or the exclusion of people with disabilities. Instead, they were interested in the process of citizen participation and the management of expert-citizen relations. Some were writing theses about the conference, while others were evaluating it for some other purpose. In other words, most of the experts involved in the *Bürgerkonferenz* never emerged on stage or took part in the dialogue, but instead watched the performance from the shadows. While the participants were aware of the group of experts they questioned, this other group of experts never openly revealed themselves. When I later learned that an Institute for Systems Technology and Innovation Research had been commissioned to evaluate the entire performance, I had the odd feeling of having observed an experiment on unsuspecting bodies. Where I had expected to hear the voices of citizens evaluating science, I saw instead a staging of citizenship for the benefit of science. It was as if these men and women were put on exhibit and science was allowed to observe and judge their performance as knowledge-absorbing citizens. And if the conference succeeded as a tool for educating citizens, then the observing experts would devise ways of turning it into a template for future communication between state, science, and the public.

The institute's evaluation report confirmed my impressions. Even before the citizen conference took place, the organizers had presented the project at twelve different conferences.[45] From what I could tell of the citizen conference's aftermath, the organizers gave much more attention to presenting their methodology than to presenting the substantive results, thereby interpreting the cultural imperative to be transparent in specific ways. One could almost conclude that the citizens and their opinions on genetic testing were only a means to the end of improving the efficiency of science communication, abstractly conceived.

The citizens who were chosen to participate had no vested interest in genetic testing, and they needed to be informed about the technique from a position of ignorance. As a pie chart in the evaluation report makes clear, the materials that the Hygiene Museum sent out were the participants' single most important source of information.[46] One even complained that the organizers had sent him too much material, with the result that he could hardly follow what was truly essential (*"Das eigentlich Wesentliche ist kaum noch nachvollziehbar"*).[47]

In the section on how the participants formed their opinions (*Meinungsbildung*), the report notes that all but one agreed on the need for a broad knowledge base. According to one participant, "with too narrow an information basis one easily overlooks details important for opinion formation." According to another, "only through broad knowledge, and thorough weighing of pros and cons of testing can I form an objective opinion." Participants complained that there was not enough time set aside for a thorough discussion. One citizen remarked that "the real discussion did not take place until the opinion was being written."[48] In their questionnaire answers they regretted that there was too little time for engagement in small groups: it usually took them about half an hour before they "really got started working" (*richtig ins Arbeiten kamen*).[49] My own impression of the presentations I witnessed was that there was almost too much time. Although the moderator kept telling people to keep their statements and questions short so that there would be "more time for dialogue," there were routinely ten to thirty minutes at the end of each day when there was nothing left to say.

Before the conference, 82 percent of the participants said they knew little or nothing about genetic testing; after the conference, 88 percent judged their level of knowledge about it to be "high."[50] All but one left the conference certain of their position on genetic testing.[51] One participant "confessed," "I had already made up my mind, therefore the time was sufficient." Another said, "I had a firm opinion, which was shaken by some of the things I read. But the expert interrogation made my opinion firm again." A third said, "It pleased me that in spite of all the pro and con statements, and the probing from all directions, my own ethical sensibility prevailed"—meaning, in this case, that his or her opinion became the majority opinion of the group.[52] The citizen conference, in other words, was not only important for acquiring knowledge, but also seems to have helped several participants "get in touch" with their own intuitions and achieve a better understanding of themselves. The conference literally "gave shape" to public opinions about science and in that respect was perhaps an unexpected success.

The state's concern over the efficiency of public communication was driven home by another feature of the evaluation report. The time that the participants spent on different tasks was measured with Tayloresque precision: the conference evaluator noted that of the time (fourteen hours) allocated to the participants during their first preparatory weekend meeting, 32 percent was spent on procedural questions, 27 percent on communicating knowledge, 25 percent on engaging with the substantive content, and 16 percent on getting to know one another. On the participants' side, 79 percent thought the time spent on procedure was well invested, while 21 percent would have preferred to spend more time on the content.[53] Things appeared to proceed more efficiently at the second preparatory weekend. There, 78 percent of the time was spent on discussing substantive issues.[54]

During the experts' presentations the moderator kept asking whether the participants had any clarification questions (*Verständnisfragen*). Significantly, they had plenty of such questions for the expert scientists and philosophers. They wanted to know again how "informed consent" works, or how preimplantation genetic diagnosis (PGD) functions. Matters were different when an experienced midwife described the fears of expectant mothers about their children's potential disabilities, when people with disabilities told of everyday discrimination, or when mothers of handicapped children expressed the joy their babies brought them. The moderator again asked whether there were *Verständnisfragen*, but none arose. The participants had understood *these* experts very well.

For their part, the expert scientists and philosophers generally seemed to have less of a feel for the participants' concerns. They were there to impart knowledge, to educate the citizens about the facts of science and the principles of ethics. One scientist gave an eloquent and well-informed overview of the status of stem cell research, but her frank defense of that research did not put the participants at ease. Indeed, some were visibly startled when she casually talked of "expending embryos for research." The trust she had been building throughout her talk seemed to dissipate in a single moment through the use of that one loaded phrase. A legal expert revealed his belief in the fragility of proper education or enlightenment (*Aufklärung*). He said that "one and a half hours of sensationalist reportage on TV could easily destroy years of educational labor (*Aufklärungsarbeit*) of the federal ministry."

Many experts, both professionals and laypeople, tried to make the risks and benefits of genetic testing "transparent" to the participants. They mobilized graphs, charts, and statistics to argue for or against PGD and prenatal diagnosis. They marshaled impressive amounts of knowledge to prove that ignorance is (em)power(ing) (or not). I watched numerous overhead slides and PowerPoint presentations citing statistics and providing anecdotal evidence just to show the benefits of knowing or not knowing. Even when they claimed the right not to know and demonstrated its value, the experts used scientifically produced knowledge and thus affirmed Enlightenment values of knowing.

Bodily experience seemed at times to be a better conduit for education than cerebral expertise. One of the layperson experts, himself disabled, gave an emotional speech in which he described his particular genetic abnormalities and plaintively asserted that he would not have existed if genetic testing had been available forty years ago. The participants were visibly moved. At the end of his presentation, they asked whether they might have his speech in manuscript form "because it is important." As I discovered later, portions of this statement found their way almost verbatim into the citizen opinion that was delivered on the last day, thereby underscoring the felt correspondence between his perspective and theirs. Throughout the conference I heard repeated statements such as "The disabled make a special contribution to society, without which we would all be worse off," and "Being disabled will be normal only when all others are seen as disabled as well." I got the impression that most of the participants completely identified with the disabled, and that one of the main currents of the citizen conference was a redefinition of "being disabled." Under this "normalizing" definition, everyone is disabled in some respects, and all are equal in their deficiency. The experts with disabilities actively and pragmatically encouraged this shift in perspective. One disabled couple implored those present to confront their own prejudices: "Approach the disabled! Ask them questions!"

This appeal underscored the pedagogical force of emotion. One evening, on the way back to the hotel where I was staying, I happened to be walking behind the disabled couple who had just "testified" before the group. They were talking about their rhetorical strategies in the statements they had made and asked each other how they had come across. They had both avoided reading their statements, they said, because although that might have made them seem more fluent and coherent, emotion and raw passion might have been lost. They clearly knew that the participants' sympathies rested not with the disembodied experts, but rather with those whose statements revealed the embodied and personal effects of genetic disabilities. By staging immediacy and viscerality, these two experts were styling themselves as members of the community of laypersons, just as other experts were distancing themselves from the participants through their own stylized forms of disembodiment and authority.

The Citizens Speak, but Have Not Heard Clearly

After two preparatory weekends and two long days of expert hearings, the citizens retreated into seclusion for a day to formulate their final opinion. Having reasoned with the experts for two full days, it was now time to resurrect the wall between themselves and those experts, to apply their newly enlightened understanding, and to arrive at an autonomous judgment. Again, the Institute for System Technology and Innovation Research provides an insider perspective, having been allowed to enter the citizens' private quarters and analyze this formative moment. According to the institute's report, the presence of an evaluator never became an issue. On the very first day of the first preparatory weekend, the participants approved the presence of an outside observer and granted him access to their plenary debates, although not to the small group meetings.[55]

Perhaps in exchange for the trust extended to him, the evaluator seems intent on getting readers to see the *Bürgerkonferenz* as a success, and his report focuses on all the things that went "right." His observations are telling. The citizens demonstrated admirable stamina throughout the writing process. The moderator, who was there to structure the opinion and to allay the participants' fear of writing, had hoped to conclude the meeting by 10 p.m., but the deliberations stretched into the small hours of the next morning[56]; some citizens even said that the moderator himself slowed down the writing, thus getting in the way of democratic expression. The participants also remained alert till the very end. The report tells how one female citizen noticed a typo in the opinion according to which abortion in Germany would have been permitted until the forty-eighth week, or the eleventh month. Finally, the evaluator notes that all of the citizens engaged in the process, thus providing a working model for full-blown democratic participation.

On the following day, the final day of the conference, the participants presented their opinion to the organizers and a group of politicians, accompanied by a preamble on the image of the human (*Menschenbild*) as an imperfect being and on the need for society to protect its weakest members. Here, the citizens emphasized that human dignity is inviolable, thereby underscoring once more the constitutional leitmotif of the first article of the Basic Law. They stated that each human being is a singular personality with particular abilities and weaknesses, and that it is those abilities and weaknesses that distinguish people from one another. They stated that death and illness are a part of life, and that the sensations of suffering and of happiness may altogether define being human. No human being may be the pliable object of others' wishes and images, nor may one be reduced to statistical values like those cited by many experts. Human life must be accepted in all of its varieties. By articulating a vision of the kind of society they wanted to live in, and of the kinds of bodies they wanted to inhabit, the citizens in effect gave themselves a physical constitution.

The citizens made this *Körperbild* explicit at both an individual and a societal level. They warned against genetic testing, rejecting most forms and recommending tight regulation of others, because they believed it would reduce bodies to their genetic makeup, with deleterious consequences for both individuals and society at large. Not only would individual bodies suffer due to likely genetic discrimination, but the social body as a whole would be diminished and impoverished, and its viability put at risk, by the selective sorting of bodies according to their genes and the standardization this entails. Disabilities might then be seen as something that the individual *could* have avoided, and their manifestation might therefore point toward a personal(ized) guilt.[57]

The citizens refused this reduction of human experience and voted instead for bodies that could feel pain, happiness, and all the emotions in between. Instead of advocating for improved performance and an optimized population, the citizens spoke for the disabled and affirmed the solidarity principle. Perhaps because of the importance and visibility of the issue, the citizens opted to include an unsolicited opinion on stem cell research in their final verdict. The breakdown of their vote on that topic followed the proportions of the vote on preimplantation genetic diagnosis almost exactly, with a bare majority voting against the legalization of stem cell research.

The final opinion perpetuated the uneasy relationship between different claims to authority, and its text shows a body (politic) wrestling with its own internal divisions. On the one hand, the citizens accorded great authority to science and the state. Rapid progress in gene technologies, they wrote, requires education (*Aufklärung*) beginning in the early school years in order to produce "*mündige* citizens" who are capable of deciding freely on these matters. On the other hand, the citizens were deeply distrustful of the state and said that it must never institute mandatory genetic testing, even when dangers to public health are imminent, because citizens have a right not to know their genetic constitution.

In general, the dangers of new gene technologies were seen to outweigh their benefits. The citizens in several instances called for more *Aufklärung*, not about the uses of new medical technologies, but rather about more traditional medical practices, such as midwife-assisted birth and counseling in order to dispel parents' fears of giving birth to children with disabilities, which they saw as a return to more natural ways. Despite their internal divisions, the participants argued against those forms of transparency that would result in segregation and segmentation of the social body—in short, in its loss of solidarity. They feared genetic testing for its potential to make the social body transparent and thus to render it susceptible to cleansing of that subset whose genomes deviate from a posited norm.[58]

The participants saw the perceived "right to know" (which disease will strike whom) as tending to mutate into a "duty to know." This shift would separate people from one another by giving each person the illusion that illness is no longer a matter of unpredictable and unavoidable fate, but rather the end of a causal chain whose beginning was formed at conception or even before, and whose trajectory can be known—indeed, *should* be known—from a genetic test.[59] Instead, the participants reaffirmed the view that illness or death could strike anyone at any time. By reinscribing fate (and asserting that fate does need to be reinscribed in a world of science and medicine bent on control and management), the participants wanted to reassert a solidarity born of shared vulnerability.

Unlike the experts, who could speak with many voices, the citizens were thought of by the organizers as a collective unit, capable only of speaking with a single voice. The moderator's attempt to individuate the participants by drawing attention to their heterogeneity in outer (social) circumstances only underscored their supposed homogeneity in internal (cognitive) qualities. Yet the assumption of collective homogeneity was refuted when a male participant read portions of the opinion aloud. As one of the eight men who had voted in favor of genetic testing, this retired police commissioner misread the text several times, in a faltering voice, as he read through the majority position. When he came to the minority opinion, which he and seven other men had underwritten, he became fluent again, and in a thick Berlin dialect, he spoke with the deep conviction of a civil servant who is sure of his place in the social order.

In the end, the disabilities of individuals, rather than their physical integrity, became the ordering principle of a robust body politic. The majority report closed with the sentence, "The position of people with disabilities and their integration into our society is of great concern to us."

Expert Reactions

Empathy, in the citizen opinion, trumped the objective rationality of science, medicine, and philosophy—if only by a small margin. For some, this was not a complete surprise. I heard experts in attendance bemoan flaws in the opinion's reasoning and its disregard for the virtues of genetic testing. Several newspaper articles criticized the majority's lack of sophistication, its strong emotionality, and the crude style of its written opinion, but most added that one could hardly expect more from a panel of laypersons.[60] The evaluating institute also concluded that while the citizen conference was a suitable instrument for disseminating knowledge, for forming an opinion, and for solidifying it, there was no one-way street from increased knowledge to increased acceptance of the thing one has learned to know. In this case the expensively educated opinion was, if anything, much more critical of genetic testing than the prior "gut opinion" (*Bauchmeinung*).[61] In other words, some deemed the pedagogical effort deficient because it had not induced the laypersons to speak or think like the experts.

Other reasons for dissatisfaction were also found. Some of the experts emphasized the vote's small margin, while others pointed to the citizen panel's lack of representativeness. The official state patron of the event, the secretary of the federal Ministry of Education and Research, admitted that citizen votes typically have little impact on parliamentary politics, in part because of poor timing. To be sure, he reassured the conference participants that their timing had been perfect and that their opinion would meet a different fate. Yet the opinion was hardly mentioned during subsequent legislative proceedings. Judging from the small amount of press coverage of the conference and its lack of political resonance, the scientists' judgment appeared more likely to prevail than the citizens', and the barrier between expert reason and public understanding was likely to remain intact for now.[62]

Participating in this experiment in democracy, however, turned the participants themselves into experts of sorts. One was interviewed by a clinic director on his opinions regarding genetic testing. Another was asked by a friend whether a physician had prescribed the appropriate therapy. A third was asked to write an article on genetic diagnosis and citizen involvement for her high school newspaper.[63] Evidently democratic participation had turned the participants into amplifiers of expert knowledge, thereby contributing to the "process of education (*Aufklärung*) and democratization 'from below'" observed by the evaluator.[64]

In speculating about future citizen conferences, the evaluation report cites two obstacles. First, the cost of 189,000 euros was high, given the elusive benefits of such conferences. The report points out, however, that "expensive" is a socially defined term, and that the cost is not excessive, but "adequate" (*angemessen*), in a democracy if the citizens wish to participate in this form of collective reasoning. The second obstacle was the difficulty of timing such conferences so that they could in fact add citizens' voices to a particular debate.[65]

Conclusions

German craftsmanship is evident in the engineering of both the Transparent Human and the transparent body of citizens. Like democracy, transparency proved in each context not to be a fixed or fixable entity, but instead an ongoing accomplishment in need of continuous crafting and recrafting. Transparency comes with built-in associations, and the transparent architecture of both the government quarters in Berlin and the citizen conference in Dresden seems to promise something beyond mere visibility.[66] Transparency aims to offer information, availability, accessibility, health, frictionless functioning, and truth. Through these promises, transparency claims to produce objectivity, so that observers can end up believing in the truth of what they are seeing. Yet transparency, we have seen, is not the starting point of democratic politics, but rather the outcome of political processes. It is less a resource than an accomplishment. And even when the quality of transparency can be produced, if one is to understand what is before one's eyes, one needs to be trained to "see."

The citizen conference was an attempt to give a voice to the missing masses, the body politic, in contemporary debates on gene technologies. As an observer, I felt as though I was watching a performance of Schiller's ideas. The *Bildung* of citizens I witnessed, under the aegis of a museum dedicated to Enlightenment values, closely followed some of the principles set forth by Germany's great educator of the senses. By asking questions and acquiring information with the help of the experts, the participants developed an understanding of and formed an opinion about genetic testing. Finally they were deemed capable of making a judgment that could be hailed as the "voice of the citizens." But in seeking to consult the citizens, the organizers generated in microcosm the very body that they took to be the foundation of democracy, the "informed citizenry." The citizens themselves were products of careful selection and pedagogy. They were not fully citizens at the beginning of the conference, but only at its end; in effect, they *became* citizens in the process of fulfilling their civic responsibility. And in so doing they realized the organizers' vision of citizenship by making their own learning process transparent.

The evaluation report records this transformation. As this document makes clear, the citizens were asked how much preexisting knowledge they brought to the conference, how much knowledge they absorbed while they were there and through what media, and how that knowledge confirmed or changed their opinions. In effect, the participants were tested on how well they performed as knowing and participating citizens. They had been conceived (and in fact selected) as "unknowing" laypersons, but at the end they were released back into the world as "knowledgeable citizens." They were now *mündig*—capable of deciding for themselves. This transformation, moreover, occurred in a reservoir of transparency. The citizen conference itself had turned into a kind of glass laboratory, factory, and motor all in one, and it became possible to ask scientifically how well citizen conferences like this one would function as parts of the machinery of democracy.

But were these citizens mere passive participants in a transparent fiction of democracy? The answer is a cautious no. The educational undertaking I observed had attempted to create perfect harmony between the mind and the body, between reason and emotion, but it seemed that mind and body refused to become perfectly harmonized. This does not mean that there was no learning through the citizens' exposure to expertise. Rather, the citizens learned to value their imperfections and offered reasons for why they wanted to retain them. The attempt on the part of the organizers to select blank slates that could be inscribed with pertinent views resulted in a rebellion of the emotions against calculating rationality. In other words, the body politic, so circumspectly constructed and so carefully trained, turned out to be a feeling body. The majority relied on empathy and emotion rather than on objective knowledge and facts in making its recommendations. Or rather, it turned its own subjectivity into a foundation for objectivity, seen as a union of multiple subject positions.

The conference offers evidence that political transparency is always in the making. It is not the mere removal of a screen that separates the seer from the seen. In configuring the place, the participants, and the process, the Hygiene Museum continued its historical mission of forming citizens (or giving them *Bildung*) as an extended arm of science and the state. The stylized citizen conference may be seen as yet another instance of the Kantian problematic illustrated in Kahn's drawing of "the human as an industrial palace": how to have minds and bodies—reasoning experts and understanding laypersons—not sitting in separate chambers but instead coming to a judgment together. More than just a static form of materiality, the conference was a dynamic attempt to create mutual understanding. As it played out, however, the conference may have reinforced the lines between expert and public reason.

The official tenor of the conference's last day was celebratory. Yet the democracy being celebrated had changed its meaning ever so slightly through enactment in this novel form. Democracy, as the conference showed, does not exist apart from its localized instantiations. It is a self-referential concept whose success is contingent on its repeated re-inscription. The conference was, in effect, an attempt to put a vital organ—the citizens—back into a transparently political body. The participants saw that democracy works by seeing themselves as indispensable to that democracy. In effect, the citizen conference acted to co-produce the knowledgeable individual and public knowledge—making science public, disseminating a geneticized natural order, and at the same time creating the sorts of persons who can live in, with, and in consequence of that order. The ontologies of autonomous and understanding citizens were not there in advance for democracy to rest on, but rather needed to be reproduced through repeated performances of the attributes of democracy.

When I revisited the museum in the summer of 2004, I found a bed of wildflowers in front of the building. The museum itself was encased in scaffolding, and the Transparent Human was contained behind a glass wall in a cabinet off to the side of the main room. A model of an oversized chromosome had assumed the most prominent place in the exhibit. Taking center stage, with two "limbs" firmly on the ground and two more raised up to the skies, the X-shaped chromosome echoed the human figure in centrality, form, and size. The transparent statue created almost a century ago stood off to the side, looking gloomy as if from disregard. But although this human-made counterpart to the transcendent ideal of transparency is worn by time, it nevertheless persists as a model of enlightenment. The Transparent Human was once seen as an ideal of comprehensibility, of purity, and of progress. The citizen conference functioned in a similar manner: it made democracy transparent, with its parts practically blinking, its composition and inner workings exposed to science, the media, and the occasional curious anthropologist. Although it too had its moment on a pedestal, it remains to be seen whether, with all its opacities and imperfections, the Dresden citizen conference will become a viable technology of democracy—or at least one small component of it.

4 Conscientious Objections

This chapter and the last, taken together, articulate culturally specific relations between self and society, between citizen and state, between individuals and institutions, and between reason and affect. While the last chapter illustrated the complex understandings and practices of transparency that mediate these relations, this chapter shows how they are also mediated by equally particular understandings and practices of *Gewissen*, typically translated into English as "conscience." We have already seen that conscience played an important role in the parliamentary debates surrounding the Stem Cell Law. We saw how the individual consciences of ethics commission members became interwoven with the commissions' claims to speak legitimately and authoritatively on ethical matters, and we saw how Germany's elected parliamentary representatives framed the need for a national response to the moral dilemma of research on human embryonic stem cells as one of forming and articulating a national conscience, built on taking and defending an autonomously attained position. But how is a decision of conscience arrived at in practice? In this chapter I show that the formation, expression, and testing of conscience in Germany is a complicated process, understood to be deeply personal, but also generalizable across all of society. I show some of the background assumptions and work that hold together the ideas that a decision of conscience is a personal one, that it establishes a relation with one's moral self, and that it is an inherently "right" decision. I show how the inner voice that tortures those who refuse to listen to it becomes a tool in the hands of the state and how the inner thought-world of citizens is invoked, regulated, and thus constituted *as* a public thought-world in dialogue with the state.

In Germany, the freedom to follow one's inner conscience, as I will show, is not simply a matter of private thinking and consideration, but also a matter of public confrontation and deliberation. A decision of conscience is as much the result of a confrontation between private reasons and public reasoning as it is the product of private contemplation. The inner voice is deeply entangled with outer voices. In other words, it is not that there is an inner judge of conscience who presides over an adversarial process of self-questioning, but rather that conscience is constituted in the process of a public confrontation between individual and society. Conscience, then, is not the starting point of private reflection so much as it is the end point of public debate. Conscience is the individual internalization of public reasoning.

What kind of person inhabits the space of conscientious reasoning in Germany? Protecting the exercise of conscience while also eliciting and controlling it both requires and produces an ethical actor and an ethical self. In this chapter I show what is involved in producing persons capable of making moral judgments. I show, in other words, that the making of proper judgments is inseparable from the making of the judges.

The space where conscience is formed in Germany is, to begin with, a constitutional space. The Basic Law expresses a grave concern for the preservation of individual autonomy while at the same time expressing distrust of individuals who decide too much for themselves. The question for the state becomes how to craft individuals who see that they are autonomous, but who also make moral decisions in predictable and governable ways. How does one craft private reasons (and reasoners) that are in congruence with public reasons (and reasoning)? One solution is to have a public space that is subject to ethical legislation, and then to internalize this space in such a way that individuals take it to be their own inner law. In this way, conscience becomes an inner (because internalized) sense of right and wrong that can at the same time be mapped onto outer laws of rightness and wrongness. Freedom of conscience emerges as a relation between individuals and the state, as the latter decides who may decide for whom and who can be trusted to decide on their own. We will see that the version of enlightenment (*Aufklärung*) that Kant proposed more than two centuries ago contains just such a vision of the relation between private and public law. Let us therefore turn first to two deeply German histories: of constitutionalism and of Kantian reasoning.

Constitutions of Glass—Transparent, or Merely Fragile?

Walking along the river from the Bundestag toward Friedrichstrasse, one sees the rows of monumental glass structures that house the workplaces of parliamentarians and the administrative offices of government. One can only imagine the destruction that occurred here, in the heart of the city, to clear the space for this colonizing architecture. Most of the buildings are very new, as evidenced both by their pristine appearance and by the small sizes of the trees that have been planted all around them. Their trunks, the planners may have thought, will one day become as sturdy as German democracy itself.

On the way toward Friedrichstrasse one day, just before crossing Wilhelmstrasse, I passed the row of contiguous plates of security glass sunk into the ground on which are etched the first nineteen articles of the Basic Law[1]—the section spelling out the "basic rights" (*Grundrechte*) of individuals in relation to the state—in their 1949 version. The etching is light, and the letters could be easily read only from certain angles; they were almost invisible from others. On the fourth plate was Article 4, which states that the freedoms of faith, of conscience, and of religious and secular confession are inviolable. Farther down on the same plate, I read that no one may be forced to perform armed military service against one's conscience. The positioning of these rights in the early articles of the Basic Law, lodged between the equality of all before the law in Article 3 and freedom of expression in Article 5, underscores their importance.

The Basic Law was written only a few years after a war that had devastated the country materially and morally. Cities lay in ruins, millions were dead, the country itself had been carved up among the victorious nations, and entire populations were migrating from the East into a diminished national territory. Germany was faced with the incontrovertible fact that it had engineered and executed the Holocaust, in which millions of Jews, Gypsies, and other minorities lost their lives in depersonalized and mechanized ways, a horror wrought as much by commanding bureaucrats behind their desks[2] as by obedient executioners in concentration camps. Germany was in shock.

Postwar reconstruction meant rebuilding a citizenry along with a nation. The aim of the Basic Law was to equip the state with strong democratic institutions as well as strong individuals to operate them and to make them function accountably. The writers of the law crafted the document knowing that it would be a challenge. Margot von Renesse, whom I interviewed at length, recounted how as a young girl she had walked through the gardens of bombed-out Berlin dreaming that all of politics, which to her seemed only threatening, would just come to an end. Then, however, she began to follow the parliamentary debates on the radio for hours, and these debates fascinated her and drew her into politics after all. At the time of our interview she had just retired from a distinguished career, which had included the presidency of the Enquete Kommission on Law and Ethics in Modern Medicine.

Constituting Conscience

The early articles of most national constitutions lay down the legal, juridical, and administrative foundations of government, thus constituting the state. The first article of Germany's Basic Law, however, asserts the inviolability of human dignity (which the citizens at the conference in Dresden articulated in the preamble to their opinion) and sets the tone for the basic rights protected in the first nineteen articles. Significant among these is the right to freedom of conscience.

To understand the prominent place of conscience in the German constitution, it is necessary to go back a bit in history. Legal commentaries and other narratives of democracy read the history of German constitutions as one of progressive individual self-assertion, despite numerous and sometimes devastating setbacks, first against feudalism and absolutism and later against monarchy and Nazism.[3]

The ascendancy of the modern conception of conscience, at least in the German cultural context, may be traced back to Martin Luther. When Luther was asked to recant the ninety-five heretical theses that he had nailed to the door of a church in Wittenberg, he said that he could not do so under any circumstances because he had to follow his *Gewissen*. According to one biographer, Luther replied, "Unless I am convicted by Scripture and plain reason—I do not accept the authority of popes and councils, for they have contradicted each other—my conscience is captive to the Word of God. I cannot and will not recant anything, for to go against conscience is neither right nor safe. . . . Here I stand. I can do no other. God help me. Amen."[4]

Although Luther believed in the separation of church and state, the Reformation did not demand the right to freedom of religion. Only in 1648 did the Peace of Westphalia curtail the right of the *Landesherrn* to determine their subjects' faith, forcing them to tolerate subjects of divergent confessions in their territories. Under the rules established by this treaty, those who were not permitted to live according to the precepts of their faith were permitted to emigrate to a place of their choosing or to follow their faith privately at home. The right to exercise one's faith in private, however, came with the obligation not to display it in public. One was not permitted to spread one's faith in such a way that the dominant faith, and with it public safety and order, would be jeopardized. This rule for religious minorities makes the first mention of "*conscientia libera*," or freedom of conscience.

Not everyone saw this interpretation of conscience as an assertion of individual subjectivity in a positive light. For Thomas Hobbes, writing in England, it was an interpretation to be regretted, and he couched his objection in etymological terms. Hobbes noted that the term "conscience" had undergone a shift in meaning from "commonly known" to "innerly opined." In *Leviathan* he writes:

> When two, or more men, know of one and the same fact, they are said to be CONSCIOUS of it one to another; which is as much as to know it together. And because such are the fittest citizens of the facts of one another, or of a third; it was, and ever will be reputed a very Evill act, for any man to speak against his Conscience; or to corrupt, or force another so to do; Insomuch that the plea of Conscience, has been always hearkened unto very diligently in all times. Afterwards, men made use of the same word metaphorically, for the knowledge of their own secret facts, and secret thoughts; and therefore it is Rhetorically said, that the Conscience is a thousand witnesses. And last of all, men, vehemently in love with their own new opinions, (though never so absurd,) and obstinately bent to maintain them, gave those their opinion also that reverenced name of Conscience, as if they would have it seem unlawfull, to change or speak against them; and so pretend to know they are true, when they know at most, but that they think so.[5]

In other words, conscience in earlier times was a higher instance of truth because it was based on shared cognition and knowledge. Later, people corrupted "conscience" by applying the term to everything they held personally sacred and dear. No longer shared knowledge, conscience became private opinion.

Nearly two centuries later, Germans similarly regarded the unfettered individual conscience with suspicion. In *Madness and Civilization*, the historian Michel Foucault quotes and paraphrases Johann Kaspar Spurzheim, who in 1818 found in its liberation the cause for the madness that drove the English to apparently unreasonable suicide in disturbing numbers: "Madness, 'more frequent in England than anywhere else,' is merely the penalty of the liberty that reigns there, and of the wealth universally enjoyed. Freedom of conscience entails more dangers than authority and despotism. 'Religious sentiments . . . exist without restriction; every individual is entitled to preach to anyone who will listen to him,' and by listening to such different opinions, 'minds are disturbed in the search for truth.' "[6]

On the continent, the Peace of Westphalia of 1648 led the state of Prussia to treat all faiths equally. The Preussisches Landrecht of 1794 codified their equality. By withdrawing from the sphere of the religion of its subjects, it aimed to enable citizens of all religious affiliations to identify with the state. The liberation wars against Napoleon of 1813 to 1815 that led to the German Confederation solidified German consciousness, and people started to think of themselves as a nation. In 1848 the German National Assembly (Deutsche Nationalversammlung) convened 330 representatives in the Paulskirche in Frankfurt. They discussed a constitution for the entire empire as well as a catalogue of citizens' basic rights (*Grundrechte des deutschen Volkes*). All persons were to be free and equal before the law, and individuals' dwellings were to be inviolable. There were guarantees for postal secrets and for freedom of the press, confession, and conscience. The constitution of the German empire of 1849 still conceived of *Gewissen* as a religious term. The freedom of conscience was the freedom to exercise one's faith. Only later was conscience gradually secularized. The Paulskirchenverfassung, the most progressive German constitution of the nineteenth century, permitted for the first time the formation of religious communities and the public collective practice of religion, but the Nationalversammlung famously failed to assert this constitution against the ruling aristocracy, and it never actually went into effect. The constitutions of the individual German states met similar fates.

On July 19, 1870, France declared war on Germany and so began the Franco-Prussian War. Both north and south German states supported the Prussian armies, and thus united, they beat the French quickly and decisively. The 1871 constitution of the now unified German empire failed to guarantee any personal rights of citizens, except the right to vote for members of a powerless parliament.

The state theorist Hugo Preuß and the economic theorist Max Weber advised the drafters of the Weimar constitution of 1919 and supervised its writing. Both men saw the same reasons for the tragic development of the German state after 1848: public representation in politics had become meaningless at the same time as the emperor's executive authority had become all-powerful. Both agreed that Bismarck had excluded the masses from governance, had disempowered citizens, and had choked off all parliamentary life. Bismarck, in their view, had left behind a nation whose population lacked political education and was thoroughly estranged from the affairs of the state. The way forward, Weber believed, was to create a state apparatus that would be wholly rational. Industrialization was booming, and the spread of steam engines and the recent invention of electricity transported images of carefully calibrated machines that worked according to (natural) laws that could be grasped through reason and through fine-tuning of their regulatory mechanisms. Weber imagined the state as such a machine. The Weimar constitution that Preuß and Weber devised gave citizens many rights, but limited their role in shaping the state. Citizens were left with few political rights other than the right to vote, and it is likely that for most of them the state remained an alien power.

The constitution of 1919 was the most liberal the German nation had ever had, extending from collectives to individuals the right to publicly practice one's faith. The voice of the *Volk* had been strengthened to a hitherto unprecedented degree. In 1933, however, the Reichstag disempowered itself by giving legislative authority to the executive branch of government. When Hindenburg died, the top executive offices were united in the person of Hitler. The subsequent twelve years of Nazi rule all but destroyed the nascent German democracy. More important for our discussion here, the period also marked a radical setback for any notion that individual conscience could effectively resist political tyranny.

When the Basic Law was written in 1949, the disastrous effects of a totalitarian state on its citizens were clear to everyone. Unlike the United States and many other democratic nations, Germany did not have a constitutionally anchored tradition of resistance to authority.[7] No revolution by the German people against a government had ever been successful. The Basic Law was an attempt to strengthen the individual—thus inadvertently *constituting* the individual—by granting an unprecedented degree of codified freedom. The "basic rights" of individuals were now at the forefront of constitutional concern, and the "inviolability of human dignity" in particular became the leitmotif of German constitutional thought. The Basic Law also codified the human right to the freedom of confession and of conscience (Article 4, Section 1), protecting not only religious faith but also secular worldviews.[8] The effort to protect the individual's freedom of conscience may be read as an attempt to create strong individuals where there had been too few before.

Among the new liberties were the freedoms of opinion, of religion, and of conscience. The writers of the Basic Law considered these freedoms necessary because recent experience had shown the extent to which individuals could be crushed by an overpowering state. Because the state can fail, the individual needs to be allowed to set things right. The conscience of the individual is that which cannot err. Hence the voice of the individual, and especially the inner voice, must be free to assert itself. The postwar German citizen not only has rights to be asserted against the *Rechtsstaat*, but also becomes an active interlocutor of the state in shaping it. State and individual, one might say, were co-constituted.

What does the Basic Law protect when it protects the freedoms of faith and conscience? These freedoms, which presuppose both a certain kind of individual and a certain kind of state, result in a paradox: the individual who needs to be protected *from* the state at the same time needs protection *by* the state. According to one legal commentary, these freedoms protect "the nuclear elements of the personality, that is, those elements that individuals consider norm-setting (*massgeblich*) for themselves. That which individuals consider constitutive for themselves, and with which they normatively (*massgebend*) identify themselves for their existence as moral persons." Personality is here understood as an "internal complex of values, wishes, and expectations that determine the characteristic behavior of an individual."[9] The framing of those freedoms as individual ones marks an official break with earlier understandings of "con-science" as collective knowledge.[10] Yet I will argue below that individual conscience is collectively constituted, under official auspices. The tension Hobbes talks about also marks the development of conscience as a constitutionally protected attribute in postwar Germany.

Kant's Conscience

In his *Foundations of a Metaphysics of Morals*, Immanuel Kant famously marvels that "two things fill the mind with ever new and increasing wonder and awe, the more often and the more intensely our thought is drawn to them: the starry heavens above me, and the moral law within me."[11] Elaborating on his analogy, Kant writes that the sight of the infinite mass of worlds in the night sky practically annihilates his individual importance by showing him the transience of his existence as an animal creature that must return to the planet the matter from which it was made and to which life has been lent for but a moment. The second spectacle, on the other hand, raises the human's value to infinity, as an intelligence and as a person. "Infinite value" is Kant's way of saying "human dignity," which designates for him that inviolable quality of human beings that renders them unique. The awareness of the moral law within opens up to each person a life that is altogether independent from all animality, and even from the sensual world (*Sinnenwelt*). It gives human beings a purpose (*Bestimmung*) that is not confined by the conditions and limits of life, but rather extends to the infinite.

How, according to Kant, does the inner moral law work? Conscience, he writes, is the awareness of an "inner court" in every person. As a court decides matters of right and wrong, so conscience sits in judgment over human behavior. Like the idealized law, conscience is inescapable. From within, it watches what one does, threatens if necessary, and maintains around itself an aura of respect. Conscience continually asks whether a person is acting in the way his reason prescribes and his duty demands. No one can judge a person more harshly, and more unerringly, than that person's own conscience. One can escape judgment for a while, at the cost of a "bad conscience," but conscience follows like a shadow, and sooner or later one is forced to confront it. One can fool all judges except one's own inner judge.[12]

To paraphrase Kant further, a conscience cannot be acquired, and no one has a duty to acquire one; rather, every human being, as a moral being (*sittliches Wesen*), has such an inner judge from the very beginning. Conscience is the practical reason that reminds humans of their duty to follow or disregard a given law. When one says that "this person has no conscience," one merely means that "this person does not listen to the voice of his conscience." An erring conscience is unthinkable. Although one can be wrong in one's objective judgment of whether something is a duty or not, there can be no erring in the subjective judgment. To act according to one's conscience cannot in turn be a duty because then there would have to be a second conscience to be aware of the actions of the first. One can only have the duty to cultivate one's conscience, to sharpen one's sensibilities to the voice of the inner judge, and to apply all possible means in order to give this judge a hearing.[13] Acting conscientiously, in other words, means having an open ear for the voice within.

Yet, paradoxically, this voice within is ineffective unless it also functions as a voice without. Although conscience is every individual's inner judge, the instrument with which one confronts oneself and one's own moral standard, this conscience nevertheless has to be visible to do its work so that others can follow the moral decision. Since conscience is one's own inner reason, other reasoning beings must be able to share it for it to be persuasive. When one asks others to respect one's conscience, one's own reason-ed conscience must be opened up to inspection and checked for *Nachvollziehbarkeit*; conscience, in other words, must be made transparent.[14] Only a perceptibly conscientious being can be accepted into the moral community.

An underlying assumption of Kant's notion of *Gewissen* is that everyone has the same inner compass, which implies that all must consider the same acts to be good and right. Thus individualized conscience, as for Hobbes, refers to a universally shared moral space. Kant's categorical imperative states that one should "always act in such a way that the maxim of [one's] actions can at any time be made the foundation of general law." Moral action thus requires that one's reasons for acting be valid at all times and everywhere. Moral action is universalizable action. Deleuze, in his reading of Kant's *Critique of Practical Reason*, writes that "any action is moral if its maxim can be thought without contradiction as universal, and if its motive has no other object than this maxim.... The moral law is thus defined as the pure form of universality."[15]

This universality is achieved in part through training. Germans are trained in empathy from a young age. In both private and public conversations, I repeatedly heard the injunction to identify with others: "How would you feel if someone did that to you?" or "How would you cope if that happened to you?" In such daily micropractices citizens are asked to imagine the situation of the other and thus to apply the categorical imperative in their daily lives. One result of repeating this process over and over is that in some sense each citizen becomes capable, at least in theory, of deciding for society as a whole. Every individual, having been asked to consider his own actions from the other's point of view, can eventually stand for the larger collective, and by implication for the higher moral law as well. The moral law, modeled on the outer law, has been internalized. To be an individual in Germany always means to be, in sympathetic outreach, everyone else at the same time.[16]

When someone does something that disagrees with other people's conscience, or with some broader consensus, then that person does not have a "different conscience"; rather, in German terms, the person "lacks a conscience." Researchers too are expected in general to follow their collective conscience. When politicians talk about unconscionable research, they talk in effect about exceptions from a universally shared moral standard. Those exceptional researchers who claimed to have cloned a human being were called *"gewissenlose Forscher,"* and the implied parallel to research done under National Socialism was plain.

Native Theories of Conscience—Kant as Germany's Moral Gold Standard

Immanuel Kant is not an obscure reference in Germany. Many Germans know some facts about his life and recognize literary allusions to him. They can explain his categorical imperative, and they know his injunction never to treat human beings as means alone, but always also as ends.

When I asked my grandmother one day what she meant by *Pflicht* (duty), she told me that *Pflicht* is "something that one must do unconditionally" (*etwas, was man unbedingt tun muss*). When I asked her how one knew what *Pflicht* was, she at first thought that I was making fun of her and asked, ironically, invoking Kant: "What am I supposed to say now? 'The moral law within me'?" After I reassured her that my question was serious, she offered the following examples: If I had broken my foot, it would be her *Pflicht* to take me to the hospital. If I were crying, it would be her *Pflicht* to console me.

Her examples had two things in common. First, they projected onto me the needs that she herself was longing to have fulfilled. She had broken her foot just a few weeks ago, and it was healing very slowly. She had also been crying as she was practicing with me a short speech she wanted to make on the occasion of her eighty-second birthday, when many of our family members would be gathered around the table. And, second, her everyday examples of *Pflicht* relate to the idea of *Pflege*[17] in the sense of caring for someone. The moral imperatives of empathy are felt viscerally and immediately. The moral law compels in the most mundane of circumstances.

Appropriately enough, Kant himself wrote from a very circumscribed location and achieved universality from there. His life is emblematic of the order he found within himself, and he spread this order far beyond his own life. According to legend, his famous punctuality led his fellow Königsbergers to set their watches by his regular comings and goings. When one day his reading of Rousseau's *Emile* made him late for his daily walk, his neighbors thought their watches were off. Although Kant is said never to have left his hometown of Königsberg, his works and ideas traveled far in time and space. In particular, his conceptualization of the relation between citizen and state became foundational for German philosophy and politics.

Quoting Kant is a recognized marker of education, or *Bildung*. At a widely publicized congress on cloning that I attended in Berlin in May 2003, the Munich philosopher Friedo Ricken summarized Kant in this way:

> The human being and every reasonable [*vernünftig*] human being is an end in itself. The human being does not merely have a value [*Wert*], but also dignity [*Würde*]. This absolute inner value cannot be considered against anything else. The human being does not follow immutable moral laws, but he gives himself rational laws that he has a duty to follow. The dignity of human beings does not rest on what they want, nor on what they do, but on the duty they may or may not fulfill. Dignity may be recognized [*anerkannt*] or not recognized [*aberkannt*], but it cannot be attributed [*zuerkannt*] to a person. In other words, dignity is not transitive. Dignity is independent of particular qualities of people. It is there from the beginning of development. It is part of belonging to the biological species of the human.

The enduring importance of Kantian ethics is especially evident in the philosophy of Jürgen Habermas, contemporary Germany's most prominent and influential public philosopher. Habermas is explicit about his goal of extending Kantian duty ethics by incorporating his own insights into the communicative structures of intersubjective argumentation. According to Habermas, moral arguments that originate in an "ideal speech situation," and which further follow particular rules of discourse, will result in rational principles and moral norms that all can consider acceptable and that therefore possess universal validity. Habermas's so-called "discourse ethics" may thus be seen as a kind of categorical imperative of moral argumentation, sans the transcendent elements that Kant had posited.[18]

In the May 2001 parliamentary debate on gene technologies, several people quoted Kant: his three critical questions ("What can we know? What shall we do? What can we hope for?") were cited as foundation stones of the Enlightenment generally, and his prohibition against instrumentalizing other human beings was invoked as a frame for the German bioethics debate. Kant, along with Habermas and other philosophers, also found his way into the parliamentary debate on stem cell research on January 30, 2002. Speakers cited the categorical imperative and referred to the concept of human dignity more than a hundred times in those five hours. Even a decision of the Federal Constitutional

Court affirming the absolute human dignity inherent in the unborn was brought into alignment with Kantian views. Kant's is the reason people seem to follow in their public decisions. He is at once internalized into the German consciousness and externalized into forms of political participation. Kant, we may say, *is* the German conscience.

Public and Private Reason

Enlightenment, according to Kant, is humanity's exit from a self-imposed immaturity.[19] To accomplish enlightenment, humans must have the freedom to make use of their reason in public. While the public use of reason must be free at all times and even be encouraged, the private use of reason may be curtailed without sacrificing enlightenment.

Kant's use of the terms "public" and "private" inverts present-day usages. In his view a person uses public reason when addressing the world at large, whereas a person uses private reason when acting within a civil post or office. Private reason, in other words, implies an externally imposed mandate and is therefore restricted, while public reason is free of mandates and may be exercised freely. Kant gives examples of this distinction: An officer must obey when he receives an order from a superior, but he may speak out to a larger audience on the flaws of the military. A citizen must fulfill his civic duties, such as paying taxes, but he may nevertheless complain publicly about their unfairness. A spiritual leader must obey the church that employs him and represent its dogmas faithfully when he addresses his congregation, but he may dissent when he addresses the world at large. It is clear that for Kant unrestricted *freedom of conscience* is possible only in the realm of public reason, when one is reasoning as a member of a larger humanity; in the realm of private use of conscience, when one is subject to any kind of official mandate, as a cog in a machine, one must simply obey.

Enlightenment will arrive, according to Kant, when men need no longer be told to obey, but can instead be told that they can reason all they want, so long as they obey. What does Kant mean by this? Kant regarded Friedrich II of Prussia as an enlightened ruler because he recognized the distinction between the public and private uses of reason. By leaving his subjects free to reason in public, the ruler ensured that their capacity to reason would develop in such a way that it would grow congruent with the reason of the state, provided the state's laws were themselves rational. Foucault reads this as a "contract of rational despotism with free reason": "The public and free use of autonomous reason will be the best guarantee of obedience, on condition, however, that the political principle that must be obeyed itself be in conformity with universal reason."[20] Kant hoped for completely rational government and offered in exchange the reassurance that the ruler need not be afraid of allowing his subjects to use their own reason, since the reason of the enlightened ruler would eventually converge with that of the enlightened subject. Kant's view of enlightenment was consistent with contemporary understandings of the relationship between rulers and subjects. It was Friedrich II who had subordinated himself to the higher reason (i.e., the law) of the state and pronounced himself that state's first servant. Today the role of servant to the state is parceled out among a large cadre of officials—the *Beamte*, or German civil servants.

Beamte—*Delegated Conscience Then and Now*

Two million of Germany's eighty million citizens have delegated their conscience to the state and entered into a web of mutual rights and obligations with their *Dienstherr* (liege).[21] They have given up some of their liberties, and even some of their constitutionally guaranteed rights, in exchange for sometimes lifelong care and consideration from the state. These Germans are in a *Beamten*-relationship, either in public service or in the privatized service sector.

The rise of the *Beamte* caste, as Max Weber and others have shown, is a product of rational occidental modernity, with early manifestations reaching as far back as the late Middle Ages.[22] Local principalities required servants who did not care for personal enrichment and would dutifully and competently serve their masters. These administrators, for whom the word *Beamter* appeared more and more frequently, were seen as keepers of peace and benefactors of general welfare. Mediating in a sense between the prince and the people, they saw themselves not as governors, but as shepherds of the people. Then, primarily under the influence of the French Revolution, the political structures of Germany changed, and law gradually displaced tradition as the primary form of legitimate domination. The prince no longer embodied the state, and the state no longer served the prince. Instead, the state in itself was its own highest purpose; its reason for existence was to guarantee the welfare and safety of its citizens. The regent was merely the head of the state. Friedrich II, the enlightened monarch of Prussia whom Kant implicitly addressed in his essay "What is Enlightenment?," famously referred to himself as the "first servant of the state" (*erster Diener des Staates*). The *Beamte* likewise became lifelong servants of the state and of the law. Over the nineteenth century their professional status became firmer.

While they were personally tied to the monarch, their chief allegiance to the law largely removed them from his direct influence.

By the turn of the twentieth century, the *Beamte* had come to be considered a conservative political force. Bismarck, in uniting the German empire in 1871, had done what he could to homogenize and disempower the administrations of the individual states and to craft a type of *Beamte* who would follow orders from above without questioning them. This blind obedience to orders from higher authorities, modeled on the structure of the armies of the individual states, later did its part to obstruct the democratic processes of the Weimar Republic. Personally identified with the apparatus that had supported the German monarchy through its fall in 1918, the *Beamte* were largely opposed to parliament and the republic. By disregarding the authority of parliament, they contributed to the fall of the republic in 1933, after barely fifteen years of democratic government.

During the Nazi years the *Beamte* caste helped to stabilize the totalitarian and authoritarian regime. The Law for the Reinstitution of the *Beamten* Status (*Gesetz zur Wiederherstellung des Berufsbeamtentums*) of April 1933 not only expelled all non-Aryans and other potentially disloyal persons from public service but also obligated the remaining *Beamte* to follow orders even if they contradicted existing laws. The *Beamte*, who had both sworn to and learned to obey the state so long as the orders they received were valid in a formal juridical sense, supported many of the measures that constituted the Holocaust.[23] Then, as before and after, they were crucial instruments of the state.

After the defeat of Nazi Germany in 1945, the Allies tried to eradicate all traces of the former regime by democratizing, demilitarizing, and denazifying Germany. The prosecution and rehabilitation of *Beamte* after the war highlighted differences between how Germany and the victorious Allies thought about the concept of *Beamtentum*, obedience, and the relation between the state and its civil servants. One of the Allies' first actions was to abolish the *Beamtentum* altogether; they relieved 53,000 civil servants of their duties. Their aim was to create a new state on a new legal foundation by annulling all laws passed under National Socialism and by prosecuting the engineers and executors of the Holocaust. The previous German state, constituted by its laws, its norms, and the functionaries who upheld them, simply ceased to exist. In order to prosecute those who claimed to have acted in accordance with, and in obedience to, the law, the Allies introduced the legal category of "crimes against humanity" in 1945. These crimes included murder, mass extermination, slavery, deportation, and any kind of persecution for political, racial, or religious reasons. Torture, which Germans today consider one of the worst crimes against humanity and human dignity, was notably absent from that list.

In the Nuremberg trials the Allies tried and convicted many of the primary architects of the Holocaust. They then attempted to denazify those professions whose guilt was considered less significant. The new category of "crimes against humanity" allowed the Allies to prosecute the vast majority of *Beamte* who claimed they were "only following orders." In the Allies' understanding, the overthrown German state and its legal system had been nullified, and it could now be redefined not as a *Rechtsstaat*, but as an *Unrechtsstaat*, a state of unlawfulness. Civil servants were seen as people who had flexibility in interpreting laws and resisting them if necessary. The Allies were unsympathetic to the idea of an "unconditional duty to obey" with which many *Beamte* explained their execution of orders.

Prominent German jurists, however, saw matters in a different light. The conservative law professor Rolf Stödter, for example, argued in 1948 that laws and norms are external to the true essence of the state. For him it is the *Volk* that always holds sovereign power. In Stödter's view the German *Volk*, which he saw as an ethical-political entity, still had sovereignty and merely had been unable to exercise it for the duration of the Nazi occupation. Since the German *Staatsvolk* still existed, he saw the present state (and all future states on German territories and inhabited by a German *Volk*) as continuous with what had gone before. Nothing would change this so long as the *Volk* continued.[24] Consequently, he argued, the state apparatus and some of its structures remained legitimate; the goal therefore should be not to create a new state, but rather to reform and reorganize the existing one.

A year later, Carl Heyland, a specialist in civil service law, developed the so-called *Staatsidee*, according to which civil servants served the idea of a state, regardless of whether that state was a democracy or a dictatorship. *Beamte*, he argued, served the state as a totality and worked for the general welfare. Thus the *Beamte* served an object that was independent of any current or particular political system. Heyland also argued that the *Beamte* could not have resisted the state, either as individuals or as a professional class: "The individual citizen's or the people's right to resist the unlawful exercise of state power is an unknown institution in the positive constitutional law of the German state."[25]

While in other Western democracies the right to resist the state has been indelibly inscribed in both constitutional law and civic consciousness, in Germany this right disappeared from state law around 1650 and from German consciousness no later than the failure of the German revolution in 1848. As Heyland wrote in 1949, the "right to resist . . . which for the English, the American, and the French citizen has been a self-evident legal tool, has been, for the German citizen, and even for the German jurist, a rather unknown and strange legal institution for the past one hundred years."[26] How did the German state become so hard to resist? Heyland explains that in the transition from an absolutist to a constitutional state, there arose the question of whether the constitution should favor the protection of citizens' rights by allowing for resistance to the state or whether it should favor the *Rechtssicherheit der Staatsgewalt*—that is, order within the state (*Ordnung im Staate*) at the expense of that right to resist. Unlike virtually every other nation, Germany opted for the latter.[27] If citizens did not have the right to resist the state, then *Beamte*, a fortiori, had even less right to do so.

Opinions like those of Stödter and Heyland excused the wrongdoing of those who had morally neutered themselves by claiming that they were "only following orders." The idea that *Beamte* in particular had been constitutionally barred from making any subjective assessment of National Socialism, and instead had to subordinate their own judgment to a collective and hence more objective authority, was so influential that a few years later even the Federal Court of Civil Affairs (Bundesgerichtshof in Zivilsachen) confirmed it: "The *Beamtentum* fulfills functions that are always present in a state and that are largely independent of [the state's] changing forms and the political forces expressed in these changes."[28] The law thus lent credence to the defense of *Beamte* who claimed that they had sworn their oath of office not to the particular person of Adolf Hitler (which would have made them "guilty by association"), but rather to the head of the state they were serving, who at that time just happened to have been Hitler. In effect they opted for Kant's prescription: they might have reasoned all they wanted, but in their official role they had to obey.

Denazification officially ended in West Germany in 1951. All *Beamte* appointments had been annulled by law when the war ended. As we have seen, numerous conservative *Beamte* and jurists vehemently protested this law, claiming that it incorrectly posited a personal and institutional allegiance to the person of Hitler rather than an impersonal identification with the state. The Federal Constitutional Court (Bundesverfassungsgericht, or BVG) upheld the law in December 1953, but a few months later, in May 1954, the Federal Court of Justice (Bundesgerichtshof, or BGH) sided with the *Beamte* and claimed that these officials had served the state rather than the National Socialists. By that time, however, the point had become moot, since the vast majority of *Beamte* (by some accounts 52,000 of the dismissed 53,000) had returned to public office—a move hastened by the beginning of the Cold War and a felt need to tie an administratively stable West Germany more firmly to the Western world.[29]

East Germany, by contrast, did not reinstitute the *Beamtentum* after the war, and after reunification the number of *Beamte* remained significantly lower there than in the former West Germany. While denazification in West Germany was halted and is generally considered to have failed, only to be revived by the so-called 1968 generation, the process in East Germany was much more thorough. At the same time one should not idealize East Germany: the right to protest was never a part of the general culture there, and as party functionaries, state employees in effect had to delegate their conscience to the East German state.

What does *Beamtentum* mean today? In the beginning of their careers *Beamte* are sworn into office with an oath to serve the German *Volk* and to uphold the constitution (*Diensteid*). Once sworn in, *Beamte* are obligated to stand up for the liberal-democratic basic order (*freiheitlich-demokratische Grundordnung*) at all times. They are completely committed to their office by law. They cannot quit their jobs since they are not hired on the basis of a contract, but appointed.

To ensure their neutrality, *Beamte* are not permitted to be active in political organizations, which amounts to a limitation on even their basic right to freedom of speech. *Beamte* must further maintain themselves in a state that ensures their ability to do their work. They may not, for example, place themselves at risk in ways that could inhibit their ability to perform their duties. *Beamte* also have a *Pflicht* to inform themselves about their field at all times and to read up on the rules and regulations governing their field of work, although this duty surely characterizes the notional ideal type more than the actual, everyday practice of *Beamtentum*. Thus continually, at least in theory, self-educated in their office, *Beamte* are further duty-bound to advise their *Dienstherrn*.

Other behavioral prescriptions for *Beamte* are intended to preserve the prestige of their office at all times. *Beamte* must take care to lead exemplary lives even outside the office. By governing the extra-official conduct of public servants, the law in effect extends the meaning of "official" to all of life. The state governs everywhere, and *Beamte*, as extensions of the state and its laws, are also governed in nearly every aspect of their lives. Just like the state, *Beamte* are effectively on duty everywhere and at any time, with all of their minds, bodies, emotions, and souls. The right of *Beamte* to use their official titles in public and, if applicable, to wear a uniform only underscores their complete identification with the state.

Although the law governing *Beamte* says that they must always act in accordance with the law, they have a fundamental duty to obey their superiors, which in practice often overrides basic rights, such as the constitutionally guaranteed freedom of conscience.[30] If *Beamte* think they have received an order that contradicts existing law, then they have a right and a duty to remonstrate. If, however, the superior confirms the order, then it must be carried out, unless it is punishable by law or in blatant violation of the constitution. This requirement of absolute obedience effectively presumes that the orders themselves will be lawful in every case. The possibility of unlawful orders is not considered.

If the court concludes that a *Beamter* has committed a crime in following an order, then the court typically invokes Paragraph 17 of the Penal Code, which defines the extenuating and even exculpating *Verbotsirrtum*: "The perpetrator acts without guilt, if at the time of acting he was not aware of wrongdoing, and if there was no way he could have avoided the error." This paragraph is typically applied in cases in which the legal situation is so complex that even the court has a difficult time evaluating it and in which the perpetrator is presumed to have acted according to "best knowledge and conscience" (*nach bestem Wissen und Gewissen*). The use of this paragraph to relieve *Beamte* of possible culpability is significant because it acknowledges that they committed a crime not according to their own knowledge and conscience, but in accordance with the knowledge and conscience of the superior. A higher power has silenced the *Beamter*'s own inner voice.

During my research I saw numerous instances of *Beamtentum* in action. To give just one example, after my pass to the Bundestag had expired, I nevertheless continued to visit the meetings of the Enquete Kommission. I thought to myself that this was not entirely illegitimate because I had been an intern with the commission for several months and had not bothered to renew my pass properly in the past either. I was not aware of breaking any explicit rules. The commission's first secretary had just been made a *Beamte*, and in accordance with her new duties, she had been studying the procedures and regulations of the parliamentary administration of which she was now a part. One day she noticed my expired pass and told me that according to the Bundestag rules my behavior was unlawful. She added that she had now made me *bösgläubig* ("in bad faith"—a legal neologism modeled on the word *gutgläubig*, which means "in good faith"). By pointing out to me the punishability (*Strafbarkeit*) of my behavior, she gave my (in my eyes) petty transgression an ethical dimension. Whereas before her intervention I had been acting in good faith—that is, in ignorance of the rules—I was now transgressing the rules knowingly and willfully, and in that sense acting in bad faith. By alerting me to the distinction between right behavior and wrong behavior, she had acted as my displaced inner moral judge and activated my own conscience. If I now continued to attend the commission meetings, I would be acting against my inner knowledge of right and wrong. The rules of the state had become, through the good offices of this new civil servant, *my* rules.

Although I probably could have renewed my pass at that point, it seemed impractical to do so because the commission had nearly concluded its work. To restore my "good faith," I should therefore have stopped attending the commission meetings. When I asked the first secretary if she had made me *bösgläubig* because she minded my presence at the meetings, she looked confused and apologized. She realized then that her conscientiousness had effectively barred me from the commission's remaining meetings, since as a *Beamte* she would be compelled to enforce the laws that she knew I was breaking. Neither she nor I could continue to claim *Verbotsirrtum* (unawareness of wrongdoing), which was already an unlikely excuse for *Beamte*. She explained to me that, in telling me, she had merely followed her own conscience, which had dictated to her that one must act according to the best of one's knowledge. For any *Beamte*, the best kind of knowledge is that written in the rulebooks. By studying those rulebooks in every free moment, and by internalizing the laws they contained, she had formed her own conscience in just the way all *Beamte* are in theory expected to do. She had made herself into an embodiment of the state and into its instrument.

Tortured Conscience

How does Kant shape the culture of moral reasoning in Germany today? Ironically, his distinction between public and private reason is perhaps best understood by looking at a case of torture—an emblem of what is commonly held to be unreasonable and outside the bounds of human morality. This legal case highlights the stark difference that exists, and that Germans so strictly uphold, between the public and the private—between the violence of the state against its citizens and, by implication, the violence of individuals against one another. This case illustrates where people may exercise their conscience and where they may not. The case also shows that it is in the interplay of state reasoning and individual reasoning that conscience is formed and manifests itself.

In February 2003 a story broke that sparked a national wrestling with conscience. Four and a half months earlier, in October 2002, a police officer had supposedly tortured a suspect to elicit information about the location of an abducted child. Eleven-year-old Jakob Metzler had disappeared while on his way home after school. Shortly thereafter a letter demanding a ransom of one million euros was deposited in his parents' mailbox. The police followed Magnus Gäfgen, then a twenty-seven-year-old law student, who picked up the ransom and drove home. After waiting several days in vain for the boy's release, the police apprehended the presumed abductor and took him into custody. Although they had found the ransom money in his apartment, Gäfgen at first refused to reveal the location of the boy, but during the interrogation he quickly changed his mind. He had killed the boy, he confessed, and he directed the police to the body, which he had hidden by a lake. Months after the interrogation had taken place, around the time he took his orals for the bar exam in custody, Gäfgen claimed he had been tortured. The persons charged with this offense were Wolfgang Daschner, the vice president of the Frankfurt police, under whose supervision the "torture" had been carried out, and Ortwin Ennigkeit, the police commissioner who conducted the interrogation.

What was the torture that Gäfgen experienced at the hands of the police? The news site *Spiegel Online* described what he told his lawyer as "a story from other times or other places." First an interrogator had threatened to punch out his teeth. Then the interrogator threatened that two large black men were already waiting for him in his cell, ready to rape him. Finally, Gäfgen said, his interrogators had threatened to cause him "pain like he had never experienced before" and told him that a pain specialist was already on the way by helicopter. It was unclear whether the police had ever physically touched Gäfgen. Of his three initial claims, only the third survived: the threat of unprecedented and untraceable pain inflicted by a physician.

The German police, who enjoy the status of *Beamte*, have some specified liberties in coercing a suspect: They may conduct "tiring interrogations" with up to twenty-four hours of sleep deprivation. In such long interrogations, the *Beamte* do not need to give the suspect coffee or cigarettes, even if he asks for them. They may play "good cop, bad cop" routines, and they can work with "shock methods" such as confronting the suspect with the dead body of his victim (impossible, of course, in this case). They may also confront the suspect with the consequences of his silence in a coercive way, such as by threatening him with removal from the country if he is a foreigner or with the prospect of prison in all its gruesome details. They may *not*, however, beat or hurt the suspect physically or give him truth serums, nor may they threaten to resort to those measures.

Daschner, the police vice president, said that early appeals to Gäfgen's conscience had been fruitless. A police psychologist had recommended an extreme measure to reach the abductor's conscience—namely, confronting Gäfgen with the sister of the abducted boy in order to get him to talk. But Daschner justified threatening the suspect more directly by claiming that "the republic would not understand if we waited any longer." Daschner, in effect, acted on his own, against or without the consensus of other police officers, and he ordered an officer to issue verbal threats. He remained a *Beamter*, however, noting down every action in his official logbook and publicly admitting his actions after the fact.

Daschner's colleagues were surprised at Gäfgen's quick confession, but had mixed reactions after they heard how it was obtained. One expressed consternation, since those means had no place in his imaginative repertoire (*Vorstellungswelt*). He said he was glad not to have been in Daschner's position because it was an almost impossible decision to make. Another said he had learned in his training that one must not cause pain during interrogations. According to Daschner, however, the accusation of "torture" was absurd. He said that he had ordered "immediate coercion" (*unmittelbaren Zwang*) since the life of a child was at stake. Legally, there was wide agreement that Daschner would have to be accountable to a judge. Morally, however, many conceded that the situation was less clear.

The two reporters who independently brought the torture case to public attention were awarded a prestigious media prize, the Wächterpreis, for the year 2004. The prize is sponsored by the foundation "Freedom of the Press." The foundation's website[31] details how the two journalists were able to gather the information that led their uncovering of the "torture case." The prize has been awarded since 1969 to "courageous reporters" who, in "recognition of their duties as citizens . . . take up the fight for a clean administration [*eine saubere Verwaltung*]," who research "infractions of bureaucracy and other powerful groups . . . [and who] . . . uncover societal wrongs without regard for names or existing circumstances." It goes to journalists who fulfill the "public mandate of the media," those who "selflessly uncover societal wrongs and criticize them"—in other words, to *Wächter* (guardians). One of the journalists, Jürgen Schreiber, had received the prize once before when he wrote a revealing portrait about the student days of Germany's foreign minister, Joschka Fischer, and started a broad debate on the generation of 1968. The death of the child, Gäfgen's victim, was nearly forgotten as newspaper coverage of the torture case quickly eclipsed the coverage of the actual murder. Daschner's actions represented the state that has a monopoly on violence, and people were much more concerned with the deviance of a state official than with the murderous act of a private individual. We will return to this point below.

Many of those who followed the story were outraged. The most common reaction was that there is no such thing as a little bit of torture (*"ein bisschen Folter gibt es nicht"*[32]) and that even the smallest step in the wrong direction would mean an unstoppable return to barbarism. Letters to newspapers liberally used analogies to the period of National Socialism, when life and dignity were at the disposal of a state out of control. There was near consensus that the duty of the police, especially in their role as *Beamte*, was to uphold the law at all costs, even if it meant the death of a child. One letter said that a perpetrator, no matter what his crime might have been, has a right to feel safe in the hands of the police.[33]

Daschner's mistake was attempting to coerce a person into doing that which a person can only do freely. It was up to Gäfgen's conscience to tell him what to do with respect to confessing his crime. And even though his conscience failed him, no one, least of all the state, was allowed to *force* him to do what others might have thought the right thing to do. Dignity is conceived as an inviolable attribute of all persons. Dignity, according to the understanding in the Basic Law, is what makes one human; for Kant, it distinguished people from animals. Torture violates human dignity because it intrudes upon individual autonomy. As the expression of personal autonomy, one's conscience is inviolable. The moment when the state's threatened force overwhelmed Gäfgen's inner will thus became torture in German eyes—even to the point of constituting a formal violation of the UN Convention against Torture.[34]

The UN Convention against Torture, which bans inflicting great pain on the body or the soul, and the European Convention on Human Rights are often described in the German press as great accomplishments and as steps forward from earlier and more barbaric times. Barbaric pasts are vividly present in Germany today, particularly through graphic reminders of the Holocaust. Education in the horrors of that period is part of every German schoolchild's upbringing. All German schoolchildren are exposed to pictures of the piles of dead bodies that the Allies found when they opened the concentration camps after World War II.

In *Wages of Guilt*, Ian Buruma writes about how Germans encounter their past; older Germans, in particular, sometimes come to see it as if for the first time. When I visited Buchenwald, a concentration camp near Weimar, in the summer of 2005, I was struck by the virtual absence of other adult visitors. Instead, guides led classes of students aged fourteen to eighteen through the camp and conveyed to them the enormity of what had taken place there. Inadvertently listening to their explanations, I noticed the repeated association of torture with the descent into barbarism. For example, the guides spoke feelingly of the experiment chamber where Jews underwent "treatments" at the hands of physicians and of the highly mechanized ways their bodies were disposed of afterward. State pedagogy constantly reminds even its most *unmündige* citizens, schoolchildren, of the persistent danger of abuses of state power.

In its efforts to police its conscience, Germany observes the UN Convention against Torture so strictly that even the threat of physical pain stands for that pain itself. It is almost certain that neither Daschner nor any other police officer touched Gäfgen, the "torture victim," at any point. By making the physical aspect of torture insignificant, the victim's subjective experience of having been tortured became all-important. The readiness of the German media to accept the victim's claim testifies, on the one hand, to their readiness to identify with a victim of the state and, on the other hand, to their belief in an inevitable slippery slope from a state-sanctioned threat to a complete breakdown of order.

In Gäfgen's case it did not matter whether torture was "real" or whether it was "only" threatened, in part because of Germans' understanding of the slippery slope—what they often refer to as *Dammbruch*. *Dammbruch*, or a break in the dam, is thought of as inevitable if certain preconditions are fulfilled. In this case, the *Dammbruch* argument held that the smallest step toward police torture would inevitably lead to the conditions of the 1930s and 1940s, when the state tortured and killed countless innocents. In Germany the ends cannot justify the means because it is thought that certain means, once accepted, will soon be used to justify other, more nefarious ends. Utilitarian justifications for or against torture—such as those advanced in the United States[35]—have no place in contemporary German legal and political thought.

The courts declared Daschner guilty, but did not punish him. This, according to newspaper commentators, represented a solomonic judgment. It warded off a moral *Dammbruch* because the judgment made clear that an infraction had occurred and that the principle of no torture, even by implication, would be upheld in the future. At the same time the judgment showed sympathy for Daschner by refusing to punish him for an understandable exercise of personal judgment.

Conscience and Resistance

Bernhard Schlink, the Berlin law professor and author of the 1995 bestseller *The Reader*, wrote an insightful article for *Der Spiegel*[36] in which he obliquely addressed the problem of individual conscience and legal authority. In June 2004 the Bundestag enacted a Law for Aerial Security (*Luftsicherheitsgesetz*). This law was a response to the terrorist attacks of September 11, 2001, on the United States. The most controversial part of the law, Article 13, Section 3, empowered the air force to shoot down passenger planes if terrorists had taken control of them and were threatening to use them to wreak havoc. The logic behind the law was plainly utilitarian: a smaller number of lives may be sacrificed in order to save a larger number of lives, even though such an action is a blow against the dignity of those who thereby lose their lives.[37] This logic was a novelty in postwar Germany. The Basic Law absolutely prohibits violating human dignity, much less disregarding it altogether, and it also prohibits any cost-benefit calculations that involve human lives. Schlink quoted a decision of the Constitutional Court to this effect: "The protection of the individual life may not be given up when one aims to save other lives. Every human life . . . is equally valuable and may therefore not be valued differently in any way, much less subjected to quantitative considerations."

Parliament affirmed the inviolability of human dignity and the absolute right to life in its debates on the protection of embryos and the importation of human embryonic stem cells as recently as 2002. Yet the new Law for Aerial Security regarded the priorities differently. In the latter context human dignity and innocent lives could be sacrificed by the minister of defense if he decided that the danger was acute. Although parliamentarians had qualms of conscience during the debate on that law, they eventually calmed those doubts by saying that they were not authorizing shooting down airplanes, but merely authorizing the minister of defense to make that decision if the air force saw a credible cause. In other words, parliament had merely delegated responsibility for deciding whose lives and dignity could be sacrificed for a greater good.

Is this always wrong? Schlink asked. What if a greater danger really *can* be averted at a lesser cost? What if shooting down an abducted airplane, or torturing the setter of a bomb, is the last hope of saving hundreds or thousands of lives? Schlink recounted how parliamentarians argued repeatedly that the responsibility for deciding to shoot down a plane must not be placed on the fighter jet pilot. Rather, so the debates went, parliament must assume that responsibility and must delegate it to the minister of defense. Schlink found this logic unconvincing. The responsible pilot would be tortured by his conscience regardless of whether he fired on orders or not.[38] It was illusory, and even demeaning, to think that laws and orders can save a person from the pangs of conscience.

At the same time, it was treating the law too deferentially to think that legal rules always and inevitably reflect what is morally right in every situation. The conflict between law and morality, in Schlink's view, can never be eradicated entirely. When such a conflict occurs, then a person who decides to break the law for a higher cause—that is, who is acting on grounds of conscience and whose reasons are not identical with the reason of the state—may or may not find understanding and sympathy, as Daschner did in the judgment of the Frankfurt court. That person may be pardoned and perhaps even admired, but the act remains an infraction of the law, and the law remains intact. The widely feared *Dammbruch* and the subsequent return to barbarism are thereby avoided.

The Federal Constitutional Court, however, disagreed with these lines of reasoning. In February 2006 it declared Article 14, Section 3 of the Law for Aerial Security unconstitutional.[39] The court ruled that although parliament may use its legislative powers to combat the effects of catastrophic disasters that are virtually certain to occur, it may not legislate that specifically military force be used for that purpose. Moreover, such use of force is unconstitutional to the extent that it would violate the paramount rights to life and human dignity of any innocents who might die in the action.

Schlink's proposal broke with the tradition according to which resistance by public servants to the order of the state, as codified in its laws, was not only illegal but also unthinkable. His position accepted a possible lack of congruence between the inner moral law and the law of the state, and it required individuals to train their own conscience rather than delegate it to a higher authority. Schlink, in other words, wished to develop the individual's resources for resisting the state and to generate a culture of potential civil disobedience. The personal tragedies that might result from such conflicts would also be part of life.

Daschner, in Schlink's telling, was precisely such a tragic figure. Like Antigone caught between the law of mortals that prohibited her from burying her brother and the law of the immortals that commanded her to do so, Daschner was caught between a legal imperative that prohibited any and all uses of coercion and a moral imperative to locate a child who might still be alive, but who might die at any moment from hunger or cold. In threatening force, Daschner made a decision that German law would have made differently. He took upon his own conscience a decision that the law claimed it could make for him under any and all circumstances. Daschner refused to let the law override his conscience the way a German *Beamter* ought to. He opted not for the conscientiousness of common knowledge, but for the dissenting voice of conscience deep within him.[40] He refused to equate public law and private morality, but instead realized that the two could be in profound conflict. According to Schlink, Daschner acted in these respects as precisely the "strong individual" that the Basic Law hoped to call forth.

Schlink, finally, asked us to think differently about the *Dammbruch* argument, according to which one small step in the wrong direction means inevitably going the whole way toward irreparable moral breakdown. Once opposition is justified through countervailing reason(s), that view holds, then nothing can stop that reasoning. It will unfold inexorably. Thus, if a *Beamter*, an embodiment of the state, breaks the law a tiny bit in a single case, then pretty soon the whole edifice of the state and its laws will come crashing down. In the reasoning of the state, law, moral law, and the social structure of *Beamtentum* all reinforce one another. Schlink and others were seeking to undermine and disrupt the rigidity of this conceptual system from within. For them, Daschner's transgression and the ensuing debate were small steps in that direction.

Conscientious Objectors

As we have seen, the *Beamte* uphold the state by delegating (part of) their conscience to the state and effectively become extensions of the state they serve. It is not for nothing that the *Beamte* are collectively called the *Beamtenapparat* and that administrative departments are officially referred to as *Staatsorgane*—organs of the state. The *Beamte* are the hands and feet of the state, and they are in that sense prohibited from acting independently of the head of the state. There existed until recently, however, another group of people vital to upholding the state who served not by becoming one with the reason of the state, but on the contrary, by exercising their conscience to resist the state. In the case of these so-called objectors to military service on grounds of conscience (*Kriegsdienstverweigerer aus Gewissensgründen*), however, "conscience" refers to an institutionalized conscience that mediates between the state and individuals.

One of my Enquete Kommission co-workers had grown up in a district in Franconia (in northern Bavaria), close to the border of the East German state of Thuringia. He told me that when he was growing up, people in his town firmly believed that the Russians could invade their homes within fifteen minutes of a declaration of war, an event thought possible at any given moment. He said that his townspeople were shocked when he refused to perform his compulsory military service on grounds of conscience. It was still relatively difficult at that time, in the mid-1980s, to be certified as a conscientious objector. The authorities had tested him with the following scenario: "Imagine that a stranger jumps out from behind a bush and threatens your girlfriend. You happen to have a weapon in your hand. What do you do?" The purpose of this test was to get the would-be objector to admit that he was not a complete pacifist after all, that there were imaginable circumstances under which he might use a weapon, and that he was therefore fit to join the military.

When I myself was finishing high school in the early 1990s in a city in Franconia, not far from my co-worker's hometown, many of my fellow students were applying for exemption from military service as conscientious objectors, opting instead for alternative civilian service, which was one-third longer than military service. That additional third, according to one court ruling, was considered an additional "test of conscience." Some of my classmates refused to bear arms because it contradicted a deeply seated ethical sensibility, some saw military service as a waste of time and as a poor career move, and some were concerned that the unpredictability of military planning might take them away from their homes and friends for long periods. Some, more instrumentally, were interested in entering the caring professions and saw the alternative civilian service as a more attractive career option.

Initially conceived as a nonmilitary alternative for those who refused to take up arms, civilian service gradually became a staple of the German welfare state. The number of conscientious objectors grew rapidly. In 1971 a mere 14,000 objected to military service. The number had quadrupled by 1985 and had risen to 80,000 just before the Wall fell. The number then crossed the 100,000 mark (in part due to the increased number of eligible young men after German reunification), and the stream of conscientious objectors became so steady that many hospitals, hospices, retirement homes, and other caregiving institutions relied on an annual contingent of such evaders of armed service. They would not have been able to perform their services adequately without this "army" of practically unpaid labor to pick up the slack. For a time it seemed as if the state's peaceful survival virtually depended on there being large numbers of young men who exercised their conscience and refused to defend that very state. The number of conscientious objectors serving in social institutions remained high for several years, but then dropped to about 40,000 by 2010.

Following reunification, which both increased the number of men eligible for military service and reduced the threats perceived to emanate from the socialist East, the length of compulsory service had been gradually reduced from eighteen months to six months. Effective July 1, 2011, the German government suspended mandatory military service altogether, and alongside it the alternative civilian service option for conscientious objectors. To make up for the missing social services, the federal government introduced a replacement program that encourages men and women of all ages to serve a year doing unpaid volunteer work in their communities. This program aimed to attract 35,000 people annually, and it was declared a success after just a few months. Because military service, as well as the possibility of objecting to armed service for reasons of conscience, remains part of the German Basic Law and can be reintroduced at any time, the issue of conscientious objection remains relevant.

In East Germany too one could object to military service on grounds of conscience. Those who refused to serve were not interrogated or imprisoned. They were based in specially chosen locations, given uniforms with a spade on the shoulder, and were dubbed "*Bausoldaten*" or "*Spatensoldaten.*" While such objectors did not experience problems while they were in service, they might subsequently experience certain "disadvantages" with respect to their professional careers. Access to educational benefits, for example, became one of the state's leverage points for exacting personal commitments to state ideals and institutions. Those who refused to publicly demonstrate commitment to the state by joining the FDJ (Freie Deutsche Jugend—a socialist youth organization) or entering military service might be refused admission to a university. Those who did not serve their state were not, in turn, served by their state.

The German constitution of 1949 specifies in Article 4, Section 3 that no one may be forced to do armed service against one's conscience. In 1956 the constitution was modified to include provisions for a service alternative for those who objected to armed service (Article 12a, Section 2). This *Ersatz* service had to be completely disconnected from the armed forces, and a 1960 law specified that this labor had to serve the general social good (*Allgemeinwohl*). On April 10, 1961, just four months before the Berlin Wall went up, the first 340 conscientious objectors began their twelve months of service.

In December 1960 the Federal Constitutional Court declared both compulsory service and alternative civilian service constitutional. The court emphasized that the duty to reunify Germany spelled out in the West German constitution did not contradict the institution of compulsory military service. The court also held that conscientious objectors would be recognized as such only if their rejection of military service was total. An argument for exemption based on particular circumstances (*situationsbedingt begründet*) would not be recognized. This meant, for example, that one could not be recognized as a conscientious objector if one objected merely to fighting the East German army but not all others, or if one refused to take part only in wars using nuclear weapons. If one was willing to fight at all, Germany's highest court held, then one should be able to fight anyone, anywhere, and in any way. Conscientious objection had to be universal.

In an interesting minority opinion, articulated in a related case in 1985, two judges of the Constitutional Court held that disregard for the particular situation of conscientious objections was unconstitutional. The two judges, one of whom later declined Chancellor Schröder's 2001 invitation to serve on the Nationaler Ethikrat, argued that a decision of conscience was always grounded in a particular situation, rather than being made in the abstract, true for all times, and uninfluenced by the circumstances of action. A conscientious objection, they said, need not necessarily be based on the absolute judgment that all war is always morally wrong (*sittlich unerlaubt*). One could just as consistently base a conscientious objection on a judgment related only to modern warfare (e.g., because it is always total war) or related only to a war using nuclear weapons. Either view could lead to an unconditional judgment of conscience, prohibiting one, albeit contingently, from serving in the military.

The Constitutional Court in 1978 upheld the principle that made alternative civilian service longer than compulsory military service. The court argued that civilian service did not represent a simple alternative to military service that could be chosen easily and freely. Rather, civilian service was intended only for those who were exercising their basic right to freedom of conscience as articulated in Article 4, Section 3 of the constitution. In other words, decisions of conscience were not to be arrived at easily, but only after long reflection and careful testing by the objector himself and by a commission representing the state.

The Civilian Service Law, a 1983 law regulating the process of conscientious objection, specified that objectors need to submit a curriculum vitae, a police record, and a document stating their reasons (*Begründung*) for objecting. The Office of Civilian Service would then check to make sure that the petitioner had based his decision on grounds that the constitution recognizes as technically valid reasons of conscience, worthy of protection. The credibility of one's conscience, on the other hand, was considered sufficiently demonstrated by one's willingness to make the extra time commitment that the alternative civilian service represents. The law, which now included environmental protection as a field where objectors might be deployed, went into effect in January 1984.

In 1985 the Federal Constitutional Court affirmed the constitutionality of the Civilian Service Law. The majority argued that the constitutional commitment to having an effective military defense apparatus sets limits on the basic right of freedom of conscience. Here again there was an interesting dissent. Two judges argued that it is impermissible to conclude from the law as written that the effectiveness of Germany's defense of its borders has a higher constitutional status than the basic right of each individual to freedom of conscience. Arguments based on the constitutional requirement to defend Germany cannot be used to set limits on individual conscience, the dissenters wrote, especially when it is precisely the defensive function of the state to maintain armed forces against which the basic right is directed. Immanent limits on basic rights cannot be derived from the state interest to which these rights are opposed; put differently, the implicit claims of the state may not override the explicit rights of its citizens to counter those very claims. In this case, this meant, according to the dissenters, that the sovereignty of the state ought to be subordinated to the sovereignty of the individual.

But how did one object in practice? Many young Germans were apprehensive about serving in the military, and many weighed their options for alternative forms of service. It is not surprising that something of a "service industry" sprang up to help these young men avoid armed service. Many NGOs offered free legal, emotional, and bureaucratic aid, and exemplary form letters detailing individual reasons of conscience could be downloaded from the Internet. Statements meant to convey precisely that most personal and private exercise of moral judgment—an objection for reasons of conscience—were available in standardized form for mass copying and modification. The demonstration of personal afflictedness (*Betroffenheit*) that one needed to make became a mass commodity, circulating in forms that anyone's conscience could be taught to follow. Let us look at two such form letters that detail the kinds of life experiences that can justify a pacifist position.

The first letter is by a young Christian who declares that his individual conscience is the final moral authority and that his faith dictates that human life may not be destroyed under any circumstances. He claims that he can make the decision to shoot at others, or even to participate in activities that the military euphemistically calls "securing peace" or "creating peace," only in accordance with his own conscience. The author implies that he would refuse the order to shoot if his petition were to be turned down. While these sound like purely subjective assertions, the putative author also states that his view of war is the outcome of years of education and "conscience formation" (*Gewissensbildung*). His mother had taught him to solve conflicts peacefully, and although he had played with toy soldiers as a child, the meaning of death had become clear to him only when his beloved grandmother died.

Later visits to concentration camps (of the kind I witnessed in Buchenwald) had also formed his conscience and fortified him in his view that never again must a war be initiated from German soil (*Boden*). Films like *Platoon* and television shows like *M*A*S*H* demonstrated to the author the painful pangs of conscience felt by soldiers forced to fight in Vietnam as well as the suffering of victims of war. Accounts of personal peaceful engagement in supporting an antiapartheid group in South Africa, along with his attendance at a Christian school and membership in a pacifist organization, increase the writer's credibility.

The author also learned that the use of force is always wrong and that conflicts should be solved through communication. He grew up with two younger sisters, and his physical superiority obligated him not to resort to violence. Interestingly, the author drew the same lessons from his physical inferiority in school. Because he was weaker than his classmates, it was "reasonable" (*sinnvoll*) to use words, not aggression, to resolve conflicts. In other words, whenever the author experienced a power differential, he felt that resorting to force was the unethical choice—whether he was the stronger or the weaker party.

The second letter also grounds the author's conscience in his Christian and nonviolence education. He writes that he cannot join the military because it is a training ground for killers, whereas his faith dictates that he should love his neighbor and regard life as the highest of all values. Being God's creature means that all decisions over life and death must be left up to the Creator.

This author also refers to his relatives' suffering through war. His great-grandmother was widowed at a young age, and her son returned with severe wounds. One grandfather lost several fingers when his hometown was bombarded. His great-uncle never recovered from psychological damage suffered during his four years as a Russian prisoner of war. A visit to a military graveyard, the reading of Bertolt Brecht's *Mother Courage*, and the innumerable photos and films of wartime destruction that he was confronted with in high school confirmed his attitude toward war and violence.

The author concludes that he would never be able, for any reason at all, to harm or kill another human being. He writes that the possibility of having to destroy human life would burden him with a moral guilt from which no one would ever be able to free him.

These personal confessional statements provide pedagogical models for conscientious objectors. Both letters link the development of individual conscience to memories of a German state out of control, but importantly, those memories are instilled in the authors through objectively describable processes of experience and education. The letters suggest that an untutored conscience is not a reliable conscience. At the same time, however, conscientious objectors must be taught how to object. Not only is their conscience trained, but so is its proper exercise. Interestingly, the genealogical connections to grandparents or great-grandparents that are here deployed to link the objector's biography to the trauma of World War II are also used by the same generation in other contexts to say, "I was not yet born during the war, hence I am not personally responsible for it." Later in this chapter I show how World War II becomes an untouchable moral resource from which one can construct not only a *personal* but also a *national* ethical position both for and against war.

Other institutions offered still more constructive advice on how to jump through the hoops set up by the state. A guidebook for conscientious objectors distributed by the Protestant Church of the state of Saxony, for example, recognizes that a person's conscience cannot be examined directly, but has to be displayed in certain ways to bring about the desired recognition as a conscientious objector. The booklet is meant to help potential objectors answer the question (emphases in original): "Why must *I* refuse *military service* with a *weapon* to satisfy my *conscience*?" It then goes through the highlighted terms one by one. Since the decision to object centers on one's own personality, one should write in the first person, the *Ichform*. Next one must explain what military service would mean for the applicant. The booklet encourages objectors to empathize with soldiers in a situation of war, and then to describe how this imagined reality of war stands in stark contrast to their own values and norms. Then the objector must clarify his position with respect to violence and the use of weapons. A series of questions get at whether direct contact with a weapon makes a difference, whether weapons of mass destruction are particularly reprehensible, and whether one might not also use a weapon to protect that which one wants to preserve.

Finally, the applicant must display an understanding of the concept of conscience: its genesis in his particular case, instances of its prior activation, and the feelings that accompanied the awakened "voice of conscience." In effect, one had to lay bare one's inner life and present oneself as a bearer of legitimate conscientious objections. Even if people *felt* that military service was wrong, they still needed to find the *reasons* that would convince others. Only on the basis of such *Vernunft* can one exercise conscience. By demanding public justification from those who defend it, the state (and affiliated institutions, like churches, that presume to understand the reasons of the state) invoked almost a codified statement of a person's inner sense of morality. Becoming a conscientious objector was therefore at the same time an exercise in forming a reasoning conscience.

Conscientious Abortions

In this section I will show that a similar logic governs women's decisions to undertake abortions under German law. This argument is particularly important in the context of bioethics because (as in the United States) the moral status of the embryo has been debated most explicitly and at greatest length in connection with abortion law and practice. Paragraph 218 of the Penal Code regulates abortions. The law punishes anyone who performs an abortion after the fertilized egg has entered the uterus with prison terms of up to three years. The pregnant woman can be punished only if she performs the abortive act on herself. An amendment (Paragraph 218a) provides exceptions: the abortion is not punishable if the pregnant woman demands the abortion, if she can prove that she has been counseled at least three days before the abortion, if the abortion is performed by a physician, and if conception occurred no more than twelve weeks before the operation.

The following paragraph, 219, regulates the counseling that the pregnant woman is required to receive before she can get an abortion. The counseling serves to protect unborn life, and it must be directed toward encouraging the woman to continue the pregnancy by offering her positive perspectives for a life with the child. The counseling is meant to help her make a responsible and conscientious (*gewissenhafte*) decision. The woman must be made aware that unborn life in every stage of pregnancy has a right to existence of its own and that a pregnancy can be terminated only if it results in heavy and exceptional burdens for the mother. In other words, women, although their intentions are good, are not trusted to make responsible and conscientious decisions on their own, but can do so only after a state institution has counseled them and made sure that they are acting for the right reasons. The invitation to be counseled is an invitation one cannot refuse.

Although the law directs the counseling agency to convince the woman to continue her pregnancy, counseling agencies try to reassure women that counselors will not advise them in a particular direction, but instead will be nondirective in their help and seek to do what is in their best interest. A brochure by the prominent agency Pro Familia states that counseling is an offer to help by talking through reasons and problems, but that the decision rests with the woman alone.[41] The certification that a woman receives as proof of counseling must not contain any information about the content or the direction of the counsel given. If the counselor thinks that the talk should be, or needs to be, continued, then he or she can withhold the certification unless this would mean that the deadline for an abortion would pass by. Only when the woman has come to an autonomous decision can she get an abortion. Yet the woman cannot act on subjective grounds alone; her autonomy is supervised by the state, making it in effect a duty.

In practice, I have been told, a somewhat self-selecting process occurs in counseling agencies. Often women who staff such institutions favor the woman's autonomy over the fetus's rights. Some agencies ask their clients to refer those gynecologists to them who they think "will not give women a bad conscience." In other words, unlike *Beamte* teaching respect for the law, they cannot make women *bösgläubig*. Churches at one time were a trusted and frequently used source of counseling. Against much protest from German Catholic churches, the Vatican forbade them from handing out the certificates that allow abortion, in effect mandating that the churches counsel against abortions if they offer counsel at all.[42]

Margot von Renesse, who had been centrally involved in shaping the current abortion law and who had served as a family judge for many years, explained the logic of the law to me. According to her, everyone has the obligation to take other viewpoints into account and to avoid anything that would violate another person's conscience, even if one disagrees with the objections raised by the other's conscience. Since everyone views the world from a different position, one must not universalize one's own partial perspective, but must recognize where the other's vulnerabilities and reservations stem from. A law is workable not if a 51 percent majority supports it (the classic American "winner take all" view), but when it is the product of a genuine search for a consensus on values that all share in common.

Renesse told me that the Federal Constitutional Court upheld the current abortion law (Paragraphs 218 and 219), which outlaws abortions but makes them unpunishable under certain circumstances, with a line of reasoning that she had pioneered. This "deadline-based solution" (*Fristenlösung*) involves a different conception of "protection." Renesse argues that abortion, from the perspective of the woman, is not an act of killing another life, but rather, ontologically, an act of omission, or of refusal. The pregnant woman does not intend to take the child's life per se; she simply refuses motherhood. And crimes of omission must be treated very differently from crimes of commission. Since the woman is not obligated by law to become a mother, she cannot be guilty of committing a crime if she refuses to do so.

It is as if, for Renesse, the embryo in the womb is a developing citizen and the woman, in effect, is a conscientious objector. She may refuse, after declaring her reasons, to put her life, the integrity of her body, and her autonomous will in the unconditional service of another—but, as in the case of national defense, her reasoning must be acceptable to the state. When the law asks the woman to make a *gewissenhafte* decision, it places her in the same category as those who are formally required by law to fulfill a certain duty. The woman, in one sense, has a duty to bear children for the state; in another, she has a duty to be an autonomous citizen.

A consequence of Renesse's reasoning is that the state is now caught in a double bind: it cannot punish abortions (or it can do so only in rare cases), but it must protect human life. According to Renesse, the duty to protect is now converted into a duty to console, to help, and to encourage the pregnant woman. In order to fulfill its duty, however, the state must know when a woman is pregnant in the first place. As a four-time mother, Renesse is well aware that in the first month no one knows about the pregnancy unless the woman reveals her condition. Hence the state, in order to fulfill its duty to promote live births, has to invite the woman to tell the state about her pregnancy. In exchange for this revelation, the state must offer the woman what she wants: the abortion. In other words, the *Fristenlösung* is necessary for the state to be able to protect anyone at all, either the prospective mother or the unborn child.

Renesse admits that her logic requires thinking around several corners, but she finds her solution justified by its apparent success. During our interview she pulled a document from a pile of papers and quoted supportive numbers to me. In 1992 about 5,000 women between the ages of twenty and thirty-nine in East and West Germany were asked how, in theory, they would respond to an unplanned pregnancy. In the West 64 percent said they would keep the child, and in the East 40 percent said they would. In 2002, after the introduction of the *Fristenlösung*, a similar sampling showed that the numbers had gone up by 14 percent in both East and West.

Renesse sees these changes in attitude as evidence that once women have the piece of paper that permits them to have an abortion, they are for the first time free to think for themselves about whether this is the course they truly want to follow. In other words, by easing the struggle of women against a state that wants them to carry their unwanted children to term, the state puts women in a position of greater autonomy, which in turn helps them envision a future as responsible mothers. By empathizing with the position of pregnant women, by assuming and thereby validating their perspective, the state accomplishes its own goal of lowering abortion rates, at least in theory.

The woman must make a decision of conscience not by herself, and not only according to her own conscience, but *with* someone else, in cooperation with that other's different logic. Only when her decision is made with others' advice does it become truly *gewissenhaft*. In other words, as in the Hobbesian sense, transparency (between persons) produces a legitimate conscience. And as long as the process is transparent, the outcome is *gewissenhaft* by necessity. If the woman decides to have an abortion after counseling, then she has exercised her autonomy by resisting the demands of the state; if instead she decides to continue the pregnancy, she has followed the collective conscience embodied in the law of the state and so upheld its moral order. One might say that in the counseling situation the individual conscience enters into a dialogue with the conscience of the collective. The resulting decision will reflect the woman's decision to accept or reject the reason of the state, but it will be *gewissenhaft* in either case.

Constraints on Conscience

Individuals, we see, may refuse to participate in warfare, in military training, or even in motherhood if they claim certain reasons of conscience in ways that the state recognizes as valid. But what if parliament as a whole is to act as the nation's conscience? Days after the attacks of September 11, 2001, Chancellor Gerhard Schröder, like other national leaders, promised "unconditional solidarity" (*uneingeschränkte Solidarität*) with the United States, and on September 19, 2001, almost all members of the Bundestag gave support to Schröder's promise in a parliamentary resolution (*Bundestagsbeschluss*). Schröder's moment of truth came in early November 2001, when the United States asked Germany to contribute 3,900 troops to support missions in Afghanistan for the duration of one year. For the first time since World War II, German soldiers would be taking part in missions outside Europe. The promise of unconditional solidarity now became a grave domestic political issue, dividing the nation.

Opponents of participation in the war in Afghanistan justified their position with reference to German history. Germany had been allowed to rebuild its military during the Cold War with the understanding that its army would be purely for defensive purposes and that it would maintain close ties to NATO. East Germany's military, likewise, was conceived for defense only and had been part of the Warsaw Pact. The horrors of two world wars had produced a pan-German consensus that military aggression must never again emanate from German soil. Proponents of participating in the war also emphasized German history, but they argued that the situation in 2001 was different: Germany was reunified, the Cold War was over, and the Warsaw Pact had been dissolved. German participation in international peacekeeping efforts was a logical consequence of the nation's "growing up" (*Erwachsen-werden*) to the point at which it had to assume responsibility in foreign affairs and not leave all military missions to its allies. No longer permitted to act like an adolescent that could avoid responsibility, buy its way out of physical harm, or hide behind its history, Germany had to face the fact that it was a major player on the world political stage and do its part to ensure international security. There was even talk of another German *Sonderweg*—the concept that, in refusing to fight, Germany was yet again veering off the path taken by other civilized nations and thus slipping into uncharted and perhaps dangerous territory.

Numerous members of parliament asked that the vote on this issue be made one of conscience—that is, not subject to party pressure (*Fraktionszwang*)—but Schröder refused to make the decision on Afghanistan a matter of individual conscience. For him this was not a vote of conscience at all, but a necessity required by both domestic and foreign policy. Faced with waning support from all parties, Schröder took up the challenge and represented a vote against sending soldiers to Afghanistan as a vote of no confidence.[43] He made this very controversial move, he said, because a chancellor cannot govern effectively if he cannot be certain of the confidence and support of his coalition. Schröder, in effect, used his personal conscience to override the conscience of parliament, or rather, to get parliament to go along with his conscience. Having thus made his conscience truly national, he garnered moral authority to speak for the nation as a whole. By forcing the Social Democrats to declare their support publicly if they wanted him to stay in power, Schröder turned the party into the reference point of conscience. Participating in a war as a nation was no longer a matter of individual conscience, but a matter of maintaining a collective political advantage granted by the electorate.

On another level, however, a parliamentarian's surrender of personal autonomy is a means of securing the collective autonomy of the nation. Personal autonomy and political sovereignty mirror one another. Just as one's conscience ensures that one maintains integrity as a person, parliament, as the agent of collective moral judgment (and through its power to authorize military deployment), ensures the integrity of the nation. Both are responsible for preserving and enforcing boundaries between an inside and an outside. The clear demarcation of those boundaries, individual and national, enables and preserves the sovereignty within. An individual is sovereign to the extent that she can make decisions based on conscience, while a state is sovereign to the extent that it can conscientiously protect its borders and keep those within from harm. The more clearly the boundaries are drawn, the greater is the sovereignty within. Foreign policy becomes a reflection of individual self-protection on a societal scale, and vice versa. The surety of the moral law within is the surety of the legitimacy of one's national community writ small. Moral law and national law become extensions of one another: national law *is* moral law. We will see in the next chapter what happened to the understanding of moral law when the two Germanys, equally claiming sovereignty but diverging in national laws, unified to form a single nation, with one shared ethical sensibility in matters of conscience.

What counts as a legitimate question of conscience in the political sphere? To answer this question I talked to a number of people who were former or current holders of political office. One of them, Katrin Grüber, was trained as a teacher of biology and chemistry. She had spent ten years in the state parliament (*Landtag*) of the German state of North Rhine–Westphalia. After leaving politics, she had founded the Institut Mensch, Ethik, Wissenschaft (IMEW), an organization funded by several advocacy groups for people with disabilities and devoted to developing alternatives to the official bioethics discourse, in 2001.

Recalling her time in state parliament, she began by telling me that according to the constitution, each person is obligated only to follow his or her conscience (*"nur seinem Gewissen verpflichtet"*). When I asked her what counted as a question of conscience, she replied that it had to do with ethics. I pointed out that in public discussions the term was applied almost exclusively to medical issues, and she answered that this was because there were certain questions that one could not explain from within medicine. In other words, medicine alone was not equipped to answer the questions it was raising; it needed larger social reflection and the application of a "collective conscience" to its dilemmas.

As we sat in her office on Warschauer Strasse in the former East Berlin workers' district Friedrichshain, she admitted that she had never consciously considered what defined a "question of conscience" and tried to think through possible definitions. In the course of our conversation, she narrowed the definition down to something into which values enter (*"wo Werte miteinfliessen"*), something personal (*"etwas persönliches"*), and something related to a person's fate (*"wenn es das Schicksal des Menschen betrifft"*). Although one's conscience is supposedly the only authority to which each person is finally accountable, it seemed very difficult for her (and for others I asked as well) to pin down what precisely was meant by this. At last she expressed doubts about her own authority on the topic of conscience by disclaiming her own ethical expertise (*"ich bin ja keine Ethikexpertin"*). She thereby redefined ethics as a realm of experts (perhaps inadvertently discounting her own role in such an arena) and also made it the basis for deciding where mere politics ends and conscience begins.

If each person's conscience is the ultimate guide to moral behavior, then how can an ethics commission define morality and thereby take over the role of the legislator's conscience? Furthermore, how can members of parliament represent their constituencies and serve the voters who elected them if they are obliged only to follow their own conscience? What kind of sovereignty does the parliament exercise when the voters ask for explanations? What kinds of conflicts arise between the individual's freedom of conscience and the sovereign conscience of parliament, as etched in glass near the Reichstag?

When I questioned Margot von Renesse, the former head of the Enquete Kommission, on these topics, she told me that, according to the German constitution, the conscience of the elected representatives in parliament is the standard for every citizen's conscience. Yet the parliamentarians and their consciences, she adds, are constantly in need of correction, and they must be willing to be corrected. In her view one of the problems in the Enquete Kommission was that half of its members were not elected representatives and therefore did not have a mandate, yet they found themselves on a pulpit from which they were preaching morality to the public. And that was wrong, she said; that was not even the job of the ethicists. Not even the Catholic or Protestant churches can claim to stand in the light and see precisely what the correct path is. Rather, it is each person who needs to come to a judgment for herself, then apply it to the situation at hand.

One's conscience can be wrong, however. That is why argument—that is, public reasoning—is so important. The purpose of argument is to test all aspects of an issue and to approach what is right, if that is objectively possible. Renesse used an analogy to illustrate: "There is one animal that can turn its head so that it can almost see 360 degrees, and that is the owl. But we are not owls, and we need to know that. We are limited in our ability to perceive, and even if we don't share the other person's view, we need to respect it."

There are people, Renesse said, who "do not give an account to themselves of the origins of their own moral positions and their own philosophical and ideological foundations. They are dangerous because they believe that they are following strict scientific logic without seeing that they are just as ideological." She gave the example of "substance ontologists" (*Substanzontologen*) in the commission who believe that the natural is also that which ought to be (*was natürlich ist, sei auch das Gesollte*).

Although Renesse acknowledged that conscience could feel like a kind of better inner self, composed of old parental strictures that have become autonomous in one's interior, she refused to explain conscience in such psychologizing terms: "*Den ganzen Freud lassen wir mal sein.*" Instead she began with etymology: "Conscience hangs together with *science*; that is, knowing." (*Gewissen . . . hängt zusammen mit Wissen.*) From this she inferred that whoever refuses to take in new and potentially uncomfortable knowledge does not live up to his duty to train his conscience. One's conscience is not something one can make up in private, or for all time. That would be childish. To claim a position of conscience implies accepting the consequences that follow from the fact that a majority may not share one's decision. This means that one needs to be willing to accept a personal risk, which in Germany is very small to begin with. She added that while conscience is something very individual and personal, it is also related to the fact that we live in groups to whom we owe things. The question of *Fraktionszwang* is one example: "What do we (she was now speaking as a member of the Social Democratic Party) owe to our chancellor?"

Renesse recounted the attempts just before the stem cell vote in January 2002 to pressure the Social Democratic Party into voting with the chancellor. Renesse saw this pressure as dangerous because it could easily lead to a defiant reaction in the opposite direction. Yet, she added, everyone also knew that they owed certain things to the group they were part of. Not least, it is a legitimate political goal to maintain and hold on to the power that the electorate has given one, since the party needs the instruments and the leverage to turn its ideas into reality. A vote of conscience is potentially damaging to this goal and is therefore never a simple matter of individual choice. Decisions of conscience are complex, and Renesse saw it as a problem that "most people decide from the gut." She added that sometimes people also decide on trust, and recounted a story in which she herself had signed a petition that she secretly hoped would be rejected. She then overheard others admit to signing the petition as well because "if Renesse's name is on it, then you can sign it too."

When I asked her whether there was still any form of *Fraktionszwang* in parliament, even when there was a decision of conscience, she said that *Fraktionszwang* is omnipresent. She laughed as she remembered that she used to have a small apparatus that she had named "*Fraktionszwang.*" This electrical device worked like a radio with which the party could tune into every representative's office at any moment. And whatever one was doing, it was impossible to turn off the device and the voice it transmitted. Every now and then, she recalled, one heard the voice calling: "*Genossinnen und Genossen*! In a moment we are having a vote on XYZ. We still do not have a majority. Please come so that we can proceed with the vote." This voice, she said, kept the party leadership constantly connected with all its members. It was for her the embodiment of *Fraktionszwang.*

Renesse then told me about another provision that is written into the guidelines of every party: if the majority has decided to vote a certain way on a certain issue, and if someone is of the opinion that he or she cannot vote that way for reasons of conscience, then that person must declare the intended deviation from the party line to the party leader a few days before the vote. Conscience, Renesse emphasized, comes into play not in decisions over right and wrong (*richtig und falsch*), but in decisions over good and evil (*gut und böse*). Conscience, she said, does not mean "I know better." The requirement for a declaration is reasonable, she added, because it gives the party leader a chance to check on whether the vote can proceed without the dissenter or whether it would fall short of a majority. In addition, it gives the party leader the opportunity to act upon (*einwirken auf*) the person with objections. If people think of *this* as *Fraktionszwang*, she said with mild disdain for those reluctant to debate in public, then they think it is a *Zwang* to make oneself available to reasoning (*dass man sich Argumenten aussetzt*). She herself thought of this as only normal. In other words, only a publicly tested conscience is a good conscience, and the more rigorously that conscience is tested, the better it is. Kant writes that "virtue needs an enemy." The greater the temptations (e.g., to do stem cell research), the more virtuous are those who resist them.

In the stem cell decision of January 30, 2002, there was no *Fraktionszwang*; that is, no requirement to vote along party lines. Nor was there any *Fraktionszwang* when it came to writing and passing the Stem Cell Law itself. Still, when I observed the stem cell debate in parliament, I saw members of the different parties walking from seat to seat in an effort to gather votes and help the undecided make a decision. Angela Merkel of the Christian Democrats was particularly active in this respect. Voting according to one's conscience, then, really cannot be separated from external constraints. One does not decide matters of conscience purely on the basis of some internal moral sensibility or an inner sense of right and wrong. Rather, laying one's reasons open to questioning is the precondition of true and legitimate exercise of moral judgment. One can decide what is ethical *for* oneself in Germany, but not *by* oneself.[44]

When I mentioned Kant's well-known saying about the fixity of the moral law within him ("Two things fill the mind with ever new and increasing wonder and awe . . ."), Renesse offered an interesting interpretation. According to her, the marvelous thing is that we experience this inner moral law as fixed when in fact we know how malleable it is and how much it is subject to historical and personal experiences and contingencies. She said that every individual life produces its own individual conscience, and that often one is unclear on the layers that make it up. She remembered this particularly from jurisprudential discussions among lawyers and judges, who all assumed that their own actions were logically rigorous and consistent. In reality, she said, one does not convince anyone with one's arguments because all of one's arguments are based on one's own subjectivity. In her experience, the more her colleagues insisted on the logical rigor of their own reasoning, the more they were children of their emotions. The more they claimed that "this has nothing to do with emotions," the more they were the playthings of those very emotions. "Reasons and emotions are interlaced with one another [*durchdringen sich*]," she said. "Everyone has a goal he wants to accomplish, and one's reason [*Verstand*] is flexible enough to deliver the logical arguments to get there."[45]

Renesse then quoted Kant's categorical imperative to me: "Always act in such a way that the maxim of your actions can at any time be made the foundation of general law." In her opinion, many of her Bundestag colleagues seemed to understand this imperative differently. They acted as though it meant the inverse: "You can at any time make the maxim of your actions into general law. Since you are good, and since I am good, this can be the guideline that we can also prescribe for everyone else." For Renesse, on the other hand, Kant must be understood to marvel at the fact that humans have developed a conscience at all, since they could as easily have remained in a lawless state.

To better understand how commissions decide questions of conscience and how they connect private and public reasoning, I went to the University of Münster to talk to the philosopher Ludwig Siep, the head of the Central Ethics Commission for Stem Cell Research (Zentrale Ethikkommission für Stammzellforschung, or ZES) mandated by the 2002 Stem Cell Law. Siep attempted to clarify the relation between morality and the law for me by citing the classic case of a physician who must decide, alone or with an ethics committee's help, whether to turn off the machine that is keeping a patient alive. If the law is not clear about the matter, or if one finds the law inhumane, or if there is an emergency of conscience (*Gewissensnotstand*—also the term used by Daschner's defenders), *then* one has to make a decision of conscience. In clinical research, however, one is not forced to use one's conscience. Research is regulated by laws or by professional codes of ethics, which one is supposed to apply, but which often leave some leeway. Instead of using one's conscience, one would begin with ethical considerations: one could use a Kantian argument, one could argue on the basis of human dignity, or one could orient oneself in relation to what one takes to be some social consensus, something that is articulated somewhere in the first articles of the Basic Law. The ZES is different, Siep said. In that commission one is tied even more closely to the law because the law has precisely specified the commission's functions. This means that one simply cannot do things differently from how the law intends them to be done.

Siep told me that many consider him to be in a difficult position. He had publicly stated that he would not regulate many things as strictly as the German laws were doing and that he considered the British solution, which permits research on embryos until fourteen days after fertilization, acceptable. Siep acknowledged that he cannot say those things in the commission itself. In other words, he distinguished between an unconstrained public self (in the Kantian sense) and a private commissioned self. While he can still speak his mind in the daily papers, he must obey the law when he acts in an official capacity.

Siep said that as a philosopher, he may express opinions that diverge from the law, and he may even ask questions that go beyond the constitution. But when he holds an office, or fulfills a function, to which the legislator has appointed him, then he is tied to the law's intention. If he spoke his mind there, he said, "that would be cheating . . . Then I wouldn't be allowed to serve on the commission." As a commissioner, Siep said, he has to decide on the ethical acceptability of research, but within the limits set by the law. Since the law makes reference to the basic rights protected by the constitution, he cannot discount those directives. And because of that, he said, he does not have to make any decisions of conscience. Siep told me, in effect, that because he holds an official position, and because the law he implements is very clear in its intentions and specifications, the possibility of a matter of conscience does not in practice arise for him.

A moment later, however, he qualified his earlier statement, saying that there *were* occasional questions that he had to decide on conscience, but that he could not tell me which those were because he would have to talk about stem cell importation applications in a way that would violate confidentiality (*Vertrauensschutz*). He hinted at an example by saying that one applicant wanted to implant human stem cells into animals. The question was whether every human embryonic stem cell inside an animal might be classified as a chimera, an entity prohibited in Germany. Surely, he said, a single human stem cell in an animal is not yet a chimera, but what if one implants human stem cells into animal embryos? The further back in development one goes, the less one knows about what will come out of this fusion, and then difficult problems arise. Biological potentiality, it seems, can open up areas of ethical uncertainty that even the clarity of the law cannot wholly shut down.

Whatever constraints the commission, and its governing law, may place on its members, it cannot fully determine the choices or actions of legislators or researchers. Richard Schröder, a theologian and member of the NER, used the analogy of a map to indicate how an ethics commission's reasoning may guide individual behavior without binding it:

> In German one can make a play on words to say that such ethics commissions serve to make the appropriate decisions "conscientiously." That means, with a broader perception of the possibilities, dangers, alternatives, and problems than normal people could muster. That is the point. We can be thorough in the place of [*stellvertretend für*] others. But we cannot decide in the place of others. That's it. And that is also the case with other ethics

councils. One can ideally offer a map of the problems. One can also say that such and such a path ends there. But then please take the map into your own hands and decide which path you want to take. And that is as it should be, isn't it? Everyone knows that one should not ignore maps even when one wants to find one's own path. But those maps don't force you. Unless, of course, one considers the force exerted by seeing that such and such a path ends at a rock face. Then you can either climb up the rock, or you can take another path. And in that sense a landscape also marks boundaries: this is how an ethics council can warn of certain things. But the lawgiver is free. And each person in his individual decision is bound to right and law. The *other* questions concern the individual's moral identity, and there one decides in the remaining field of possibilities.

Conclusions

Decisions of conscience are assumed to be deeply personal in nature. The consultation of one's conscience is that near-sacred moment when one is responsible only to oneself. And yet in Germany, when conscience is applied to decisions that concern the welfare of the collective, it must be made visible and demonstrated to the public. One must be able to defend the exercise of conscience through reason. In a sense, then, the purpose of invoking conscience is to strengthen one's own individuality in place of demonstrating unquestioning conformity with the collective. An assertion of conscience contains within itself the potential for resistance. It is the power of the small and weak individual against a stronger force, such as the state—and in Germany, as I have tried to show, the force of the individual's conscience develops only in dialogue with the voice of the state.

In the United States, when George W. Bush appointed the President's Council on Bioethics, he wanted it to act as the "conscience of the nation." Since Bush's views on pressing issues like stem cell research were already well known at the time, this mandate meant, to many, that the commission should amplify the president's views of good and evil. Heavily politicized from the beginning, the council was in effect charged with finding grounds for supporting the political imperatives that the president's operatives wanted to satisfy. Fortunately for Bush, the council's president, Leon Kass, shared his belief that preconceived notions need not be displayed and justified to others. In his well-known article "The Wisdom of Repugnance," Kass argues that an inner sense of revulsion alone suffices to support a negative moral judgment on such issues as human cloning or stem cell research. In his view repugnance—what he calls the "yuck-factor"—can be "the emotional expression of deep wisdom, beyond reason's power fully to articulate it."[46]

Conscience in Germany, by contrast, means at once a "personal voice" that tells each individual what is right and wrong and a sense of obligation to others. Neither is negotiable, and neither is subject to the "gut reactions" of individuals or political parties. In Germany a decision of conscience *must* be "rationalized" in a particular way. It is the outcome of a dialogue between individual reason and state reason. Conscience is the outcome of training and testing, and individuals expose themselves, in forms of argumentation that collectives develop, to countervailing reasons. The idea of *Nachvollziehbarkeit* culminates, ideally, in the wholeness of perspective, or what Renesse referred to as the owl-eyed view. After a collective conscience has formed in this way, individuals can position themselves in relation to it, but on the strength of reasoned argument rather than subjective feeling. Where only personal opinions existed before, there is now a legitimated conscience (in the Hobbesian sense of a common judgment) that can ground, or at the very least orient, ethical and unethical behavior.

In conclusion, let us return to Immanuel Kant, whose concepts of human dignity, of duty, and of conscience have shaped German thinking about ethical matters for two centuries and are keenly felt to this day. Kant, we recall, felt equal wonder and awe at contemplating the starry heavens above him and the moral law within him. We might in turn wonder what awed him so. Was it the certainty of his knowledge of what is in fact right and wrong—knowledge as firm as the laws that guide the stars in motion? Or was he marveling at the fact that, even in the absence of universal moral authorities, a secular conscience can still ground our actions *as if* they flow from natural law?

5

A Failed Experiment

On May 8, 2005, the day that the media marked as the sixtieth anniversary of the liberation of Germany from fascism by the Allied powers, I stood near the newly restored Brandenburg Gate, the preeminent symbol of both German division and reunification. On its eastern side, in the former East Berlin, a large semicircular panorama painting showed the scene of destruction at the end of the war as seen from that very vantage point. In sharp contrast to the colorful new buildings put up by banks and embassies around Pariser Platz and the blue sky above, this larger-than-life-size black-and-white image showed rubble lying several feet high, cars turned over on the street, badly damaged buildings, and the Reichstag with its half-destroyed cupola. On the ground in front of it was a marked spot from which the top of the beautifully refurbished stone gate perfectly fitted its ghostly painted image. The viewer could thus take in the image of utter destruction capped by one of perfect restoration. The past and the present, as discontinuous as they may have seemed, and still seem in the collective experience, here became contiguous as the canvas of historical memory mapped onto the open skies of the present.

A group called 180 Grad Berlin had sponsored the installation. The group is working to create similar paintings for other cities destroyed by the war. While the name gestures toward the semicircular perspective of the canvas itself, the contrast between the painted rubble and the blue sky also suggests the 180-degree turn that Berlin and all of Germany have taken from 1945 to 2005. But where in this juxtaposition was there space for East Germany, the state I was born into, but whose existence disappeared in this seamless rendition of the German phoenix rising from its ashes? The canvas represented the two defining experiences of the German *Volk* in the twentieth century: Nazism and reunification. It left out the time when the *Volk* was divided, making separate histories. The "failed experiment" of East Germany, as I have often heard it called, did not even need to be recorded in this success story of the new Germany.

As I was taking in the spectacle, I overheard a middle-aged man explaining to others how to make sure that today's youth would never again be converted to the right-wing ideology whose devastating effects were there before our eyes. "Young people need warmth," he said, pausing, ". . . and *Bildung*." This *was* Germany. Proper education was the key to preventing societal disasters from recurring.

The back of the canvas offered *Bildung*: a series of images and texts described the history of the Brandenburg Gate from its first installation in 1734 through its capture by Napoleon in 1806 and its restoration in 1814. In 1814 the goddess of peace on top of the gate was recoded as the goddess of victory, and the eagle with oak wreath above her was supplemented with an iron cross. After World War II, both Berlin governments collaborated in reconstructing the heavily damaged gate. The East Berlin city council removed both eagle and cross, which represented Prussian militarism, in 1957. When the Berlin Wall was built in 1961, the gate became impassable. It was opened again when the Wall fell in 1989, and the eagle and cross were then restored to it.

Reading the historical texts counterchronologically, I arrived at an image, taken in 1957, of a lone uniformed man standing near the gate on an otherwise deserted Pariser Platz. The caption read that before the building of the Wall, traffic through the gate was still possible, though officers of East Germany's *Volkspolizei* would check passersby. As I stood reading, an older man on a bicycle, his worker's cap shading lively eyes, stopped near me and said, "It's not true what's written there!" He was not addressing me specifically, but as I turned toward him, words came streaming from his mouth. "It is always the same," he said. "When West Germans depict the inner-German border, they always present it as guarded by the police." In fact, he explained, it was mainly customs officials who had patrolled the borders. Yet almost all histories omit this fact. West German historians, he suggested, wanted to suppress the inconvenient fact that East Germany had not simply been a repressive *Polizeistaat*, but rather a separate, legitimate, and sovereign state, as evidenced by the fact that goods traversing its borders were routinely taxed. This man, it turned out, had worked as a customs official from the late 1950s until the *Wende*, and he had witnessed firsthand the policy changes between the two Germanys and their reflections in customs law. He remembered swearing the oath of loyalty to the state required of East German customs officials, soldiers, police officers, and others whose duties were closely linked to protecting the state's sovereignty.

As we stood there, it occurred to me that the Brandenburg Gate had functioned as a customs control point from its eighteenth-century origin, when items that passed between Berlin and surrounding Brandenburg were taxed there. The gate was now in the center of Berlin only because the city had grown larger over the centuries. West German visitors to the East, my interlocutor went on, were annoyed that they had to pay taxes on the goods they thought they had bought so cheaply. So-called *Intershops* in East Germany sold products made in capitalist countries for West German money. This arrangement permitted the East German state to collect hard currency while some privileged East Germans gained access to coveted foreign goods. Since *Intershops* sold their wares at a lower price than West German stores, West Germans would sometimes shop there and then try to avoid taxes by smuggling their purchases past East German customs officials.

The former customs officer complained that nearly everyone wrongly gave the border police credit for revealing the hiding places of people trying to smuggle goods (and, after the Wall was built in 1961, sometimes persons) across the border, when in fact customs officials like himself had uncovered these attempts. Even though the 1957 photograph seemed superficially benign, ostensibly serving to create common memories of postwar reconstruction, to this man its caption stood for erasure and forgetting. He wanted to tell a different story. He was an educator in search of an audience—a dispenser, in his own way, of *Bildung*.

In this chapter I will explore the complexities of postwar East and West German relations. I will begin by tracing the ways in which each Germany developed a national narrative that discriminated against, and even excluded, the other side. Reunification became a moment when the need for a single coherent German history led not only to the silencing of East German voices but to the erasure of East German institutions, practices, and experiences. In its efforts to assert its moral authority, West Germany imposed its own structures and ethics on the preexisting structures of East Germany. We will also see how notions of transparency and conscience were reconfigured to help produce a single German moral sensibility.

Abwicklung und Aufarbeitung

Soon after reunification parliament established two Enquete Kommissions, one focusing on the inner workings of East Germany, the other charged with trying to understand the time after the *Wende* (literally "turn," and used to denote the complete turnaround from socialism to capitalism). These commissions found that East Germany had been a dictatorship from its very inception, lacking any political legitimacy, but that the vast majority of East Germans had not been corrupted by their regime. The conservative CDU member Rainer Eppelmann, who presided over both commissions, concluded that life in East Germany had amounted to "life in prison" and that the *Aufarbeitung* (i.e., processing, or coming to terms with) of East Germany was the fundamental task of reunification.[1] What, then, had East Germany been, especially as a source of ethical sensibility, and of biopolitics, for those who lived under its umbrella for forty years? And where did it disappear to? To understand how and why East Germany disappeared, we first need to look more closely at its appearance in the early postwar years. How did the division between the two Germanys come about in the first place? How did the two states come to terms with the preceding thirteen years of Nazi rule, during which millions were killed by a state that had turned on a part of its own population? How did each Germany claim legitimacy as the rightful Germany? How did a shared national history become, from a certain point onward, what historian Jeffrey Herf calls "divided memory"?[2]

Herf sees writing history as "a matter of reconstructing the openness of past moments before choices congealed into seemingly inevitable structures."[3] Focusing mostly on policymakers and political leaders, Herf analyzes their interpretations of the Nazi years and the divergent memories and histories they created after the war. In the immediate aftermath of the war, both East and West Germany initially claimed legitimacy through rhetorics of denazification, but their narratives quickly diverged. In the 1950s the West German response to the Nazi crimes was ambivalent. Germany's first postwar chancellor, Konrad Adenauer, saw the need to bring justice to the mostly Jewish victims and to pay reparations to the new state of Israel. At the same time, as Herf shows, Adenauer was concerned that public trials would uncover the full extent of ordinary Germans' involvement in the Nazi crimes and thereby hinder collective efforts at building a strong democracy. The demands of the Cold War further called for reintegrating known ex-Nazis into vital areas of government and for proceeding slowly with the prosecution of suspected ones.

In East Germany, on the other hand, the official "antifascist" ideology no longer exclusively targeted Nazism but trained its sights on cosmopolitanism and capitalism as well. While West German heads of state saw themselves as heads of one of the Third Reich's successors, East German leaders saw themselves as victims of the Nazis, who had also persecuted Communists.[4] Their project was to build a true alternative to all the Germanys that had gone before. To avoid social unrest, and to bring about a better future, the East German state saw itself as justified in restricting the freedoms of its citizens, and official state doctrine upheld the often invasive means employed by the East German secret police (*Staatssicherheitsdienst*, or Stasi). East Germany was also closely tied to its occupying power, the Soviet Union, which confirmed East Germany's full sovereignty in 1955.

When the two Germanys wrote their respective constitutions in 1949, each claimed to speak for all of Germany. Each part acted as if it were the whole nation. West Germany assumed that the vanquished German state had not ceased to exist as a legal entity in 1945, and the Basic Law defined as Germans all those who possessed German citizenship or who, following their expulsion from what were now Polish and Soviet lands, had found refuge in the territory that was Germany on December 31, 1937, before the *Anschluss* of Austria. The East German constitution of October 7, 1949, also claimed to speak for the entire German *Volk*. Article 1 stated that Germany was an indivisible republic and that there was only one German citizenship. The implication was that East Germany represented the whole republic and that reunification was unnecessary. We will see throughout this chapter how the need to tell a single story has shaped whose voices are included, and whose are excluded, in the emergent narrative of Germany.

While Herf writes about the division of the two Germanys and their development into two distinct political communities, the anthropologist John Borneman shows that throughout the Cold War, East and West Germany defined themselves in opposition to one another not only in the realms of political memory and foreign policy, but also in terms of social and family-related policies.[5] In *Settling Accounts*, Borneman analyzes the difficulties entailed in reunification.[6] He argues that in peaceful transitions from socialism to liberal democracy, accountability must be central to the self-definition of the transitioning states. Accountability, moreover, must include a form of "retributive justice," by which Borneman means a justice that "seeks to compensate victims for moral injuries."[7] Retributive justice will hold wrongdoers responsible for criminal acts and punish them. Retribution, embodied in the rule of law, is the moral center of democratic polities; it ultimately legitimates the *Rechtsstaat*. In order for the *Rechtsstaat* to be accomplished, however, crimes must be clearly defined and criteria for accountability must be constructed. Borneman describes the challenges that were involved in bringing about (a sense of) justice in East-Central Europe, where not only East Germany but other relatively recent socialist regimes were intertwined with more traditional European notions of the rule of law.

My own story builds on and extends Herf's and Borneman's accounts. While Herf shows how moral narratives diverged during the Cold War, I show how, in the process of reunification, one national moral order overwrote another. While Borneman shows how crimes were defined and criteria of accountability established to compensate victims for moral violations, I show how the re-formation of East Germany's internal moral intuitions to conform to Western notions of transparency and conscience resulted in what many perceived as a moral double standard.

How did West Germany deal with East Germany, and what would be the place of East Germany in the new Germany? And how is this story linked to an ethnography of transparency, bioethics, and citizenship? The process of reunification formally joined East and West Germany in a Unification Treaty, but it was much more: I will show that it was also a process of purifying East Germany and of turning an unlawful state, an *Unrechtsstaat*, into a state where the law ruled, a *Rechtsstaat*. History was written at that moment, and it had to be written in such a way as to maintain the new state's moral authority. Reunification was a proving ground where Germany could assert its moral maturity by demonstrating, in effect, that it was capable of dealing with an unlawful state, an *Unrechtsstaat*, on its own. The implicit contrast is, of course, with the Nazi era, when all of Germany was corrupted and the rule of law was completely perverted. After the war, help in restoring moral order had to come from outside. Now, with the reintegration of East Germany, the new Germany could show that it had matured and learned from its earlier failures. It could apply its own moral standards and resources in crafting the terms of reunification. In reeducating East Germans and integrating them into a self-consciously democratic order, Germany could forge from within itself the moral authority to pass judgment. The new Germany did not need outsiders anymore to tell it what was right and what was wrong; it knew for itself what to do. Germany could prove its moral capacity by responding with special rigor to the East German dictatorship. And the rigor of the response would in turn vindicate the value of all the moral soul-searching that West Germany had gone through, which did not allow for the acceptance of other value structures arrived at by other means. Writing a single history meant developing a single moral sensibility.

I will argue that in order to prove the maturity of its own system, West Germany needed first to show that East Germany was in fact an unlawful state. And the darker East Germany could be made to look, the brighter West Germany would appear. As we have seen in the two preceding chapters, transparency and conscience are cultural resources with which the German state fashions a moral order. Reunification was a moment when the deployment of these resources became visible. In that moment, West German understandings of transparency and conscience were superimposed on preexisting East German understandings—with repercussions, as we will observe in this chapter, for bioethics and citizenship.

Erich Röper, a Bremen scholar of parliamentary law, analyzes the processes by which East and West Germany permanently differentiated and isolated themselves from each other.[8] As early as 1950, more than a decade before East Germany built the Wall, West Germany closed its borders to East Germans. Those who sought to move to West Germany permanently needed permission, which was granted only on a demonstrable claim of political persecution. When the Social Democrats proposed to admit even those who were not persecuted, the conservative Adenauer government refused, with words that would later be used to turn away foreign immigrants: "The boat is full."

West Germany excluded East Germans in many other ways. Church representatives from the East were not invited to German religious events. In 1968 the Grand Coalition of Christian Democrats and Social Democrats got rid of the *Gedenkstunde* (memorial hour) on June 17, with which West Germany had commemorated, in an initial show of solidarity, the day of the East German popular uprising in 1953. In the mid-1980s the Federal Administrative Court denied the German citizenship of 800,000 East Germans (that judgment was later overturned by the Federal Constitutional Court). East German television programs were not given preference over those of other EU countries in apportioning broadcast times and frequencies, and when West Germany built nuclear reactors near the East German border, East Germans were not permitted to participate in court proceedings that challenged the reactors (even though citizens of other neighboring countries could do so).

In postwar West Germany, moreover, many civil servants who had been removed from their posts in the process of denazification returned to work (though with a lower rank) as early as 1951. East German civil servants who had crossed the border into West Germany, by contrast, could be rehired by their new government only if they had fled their previous country and fulfilled conditions like those normally demanded of asylum seekers. For West Germans, Germany unequivocally meant the Federal Republic.

This pattern persisted after reunification. Those former East German officials who were not immediately "*abgewickelt*" (dispensed with) were thoroughly tested for connections to the Stasi. Although reemployment of East German civil servants was supposed to be contingent on their behavior both before and after reunification, most were not rehired, in part because the *Abwicklung*, unlike in 1951, was meant to reduce the number of overall positions.

Judges provide a particularly prominent example of the contrast between the two eras. Although civil servants and judges were not permitted to remain in their positions after 1945, the lack of "qualified" personnel meant that almost the entire judiciary were soon back in position. In some West German courts the percentage of (ex-) Nazi judges was higher than it had been under National Socialism. Since those judges sentenced others for crimes committed under National Socialism, the punishments tended to be very mild. The politics of state continuity and discontinuity have also led to the paradoxical situation that former East German military officers are not permitted to use their titles, whereas former soldiers of the *Wehrmacht* are allowed to wear both their titles and their military medals (though skull and swastika must be removed).

The discrimination against East Germany extended to the constitution. To maintain its self-understanding as the one and only rightful Germany, the West even reinterpreted its 1949 Basic Law: where Article 146 had once left it up to a reunified Germany to give itself a new constitution that would replace the Basic Law, the Unification Treaty simply referred to "the joining of the DDR to the territory of the Basic Law" (*Beitritt der DDR zum Geltungsbereich des Grundgesetzes*) and the consequent completion of reunification. By the time of reunification, the political system under which a reunified Germany would live had already been decided, more or less unilaterally. The laws of West Germany were inviolable; those of East Germany, the *Unrechtsstaat*, could be disposed of. Those who had been governed by an illegitimate system did not need to consent to their incorporation into a *Rechtsstaat*. And, of course, the fiction was that the Basic Law already "spoke for" them.

Germany, I suggest, is still engaged in producing an ethical present purged of an unethical past. To become a viable state, reunified Germany has to overcome Herf's states of divided memory: it must craft for itself a single "correct" history and a single moral sensibility. Without necessarily forgetting either, present-day Germany seeks to establish its moral credibility by distancing itself not only from its Nazi heritage, but also from the taint of East Germany. It needs to discredit and delegitimate both regimes in order to cast itself as the sole rightful moral voice. We will see later in this chapter just how West Germany conceived the task of that *Aufarbeitung* as the overwriting of East Germany's moral order with its own ideas of transparency and conscience.

One Volk, *One History? — Writing History Together*

The first of the Monday demonstrations that culminated in the peaceful November revolution of 1989 took place in Leipzig on September 4. East German citizens' rights groups had for the first time been able to prove that officials had systematically falsified results of the May 7, 1989, communal elections. On September 9, thirty activists from eleven East German districts founded the Neues Forum in order to create a new political platform for East Germany. They adopted the slogan "We Are the People," which challenged the Socialist Unity Party's claim that it alone was able to represent the people's interests—indeed, that it was acting on the people's behalf at all. When conservatives and liberals from West Germany distributed West German flags among the protesters, however, the slogan "Wir *sind das Volk*" (*We* Are the People) famously turned into "*Wir sind* ein *Volk*" (We Are *One* People). The narrative of German unity was thus overlaid on the call for democratic representation in the East.

Soon after the Wall was breached on November 9, 1989, the two states of the German nation and the four Allied powers began to negotiate Germany's reunification. On October 3, 1990, less than a year later, the six parties ceremoniously signed the Unification Treaty and the formerly divided states were officially rejoined. Reunification was seen as a power struggle not only between Cold War opponents, but also between an adolescent Germany striving to be recognized as a grown-up and its long-established political "parent." It was, however, as much a marriage ritual as a coming-of-age—a ceremonial occasion on which two unequal partners were lawfully joined together, through the mediation of legitimating authorities, in the presence of assenting witnesses. In age-old fashion, the feminized East passed entirely into the male West's control, and all property belonging to East Germany was vested in West Germany; the female member thereby joined the male's more established household. One newspaper article even described the pristine Baltic beaches as the *Tafelsilber* (table silver) that East Germany had brought, as dowry, into this marriage of nations.[9]

East Germans saw the more powerful West as having suppressed the voice of the people, beginning what would be seen as a long story of condescension. In a move that angered many East Germans, Helmut Kohl declared the date of bureaucratic unification a national holiday, while the date of their popular uprising was consigned to history. October 3, while unobserved by the general public, has become something of a media event, with television shows commemorating German unity. November 9, by contrast, though officially unobserved, remains freighted with memory. It is the anniversary of the declaration of the Weimar Republic (1918), of Adolf Hitler's first failed attempt to seize power in Munich (1923), and of the destruction of synagogues and Jewish stores and apartments that became known as the Kristallnacht (1938). After 1989, it became a day when right-wing neo-Nazis and left-wing protesters took to the streets in the center of Berlin, under the watchful eyes of the police. When I wandered through central Berlin on the night of November 9, 2001, the extent of the police presence surprised me. Walking home through the former Jewish quarter, I saw crowds around several Jewish institutions, while the police were sweeping up broken glass from the pavement. Next day, the newspapers downplayed the significance of the previous day's events.

Although the concrete Wall was torn down in November 1989, many people maintain that a "wall in the heads" still persists. While it was easy to talk about *ein Volk* when a wall stood in the way of unity, such talk became much more difficult when the two Germanys came face to face. A talk by the Dresden social psychologist Hendrik Berth that I heard in May 2005 presented empirical data illustrating the divisions. One of the more interesting numbers Berth mentioned was the percentage of West Germans who had visited the former East Germany. In 1999, a decade after such visits became routinely possible, only 30 percent of West Germans had crossed the former border. By 2005 the number had risen to 50 percent. Ironically, people have told me, perhaps half-jokingly, that "foreigners" (i.e., residents without a German passport) appear to cross the former border most often and most easily. Germany's large Turkish population in effect acts as the ambassador between two halves of a nation that have grown apart and often fail to communicate. Foreigners do the active work of cultural exchange. Germany's troublesome question of cultural identity seems resolvable only for those whom Germans regard as culturally different. For those people, Germany really is *ein Volk*.

Other East Germans whom I questioned about the small number of border crossers pointed out that visits between East and West have no bearing on a sense of unity; it was as if one were trying to measure the unification of north and south Germany (formally accomplished in 1871) by counting the number of Bavarians traveling to the north German *Land* of Schleswig-Holstein.

For East Germans, the significant numbers and stories lie elsewhere. Newspapers are filled with stories of attempts to re-create East Germany as a place for an imagined influx of West German visitors and capital that then fails to materialize. As a result, cities in the East are shrinking as young people try to escape unemployment by going west. For example, the population of industrial Schwedt, the once booming town close to the Polish border where I spent part of my childhood, declined from 55,000 to 35,000 in the fifteen years after the Wall fell. When I visited in summer 2005 for the first time after more than twenty years, many apartment buildings had been torn down and their foundations covered with grass. Many of the identical ten-story "matchbox buildings," as my father called them, that had lined Thomas-Mann Strasse, where I lived for six years, were in various stages of deterioration and disappearance. Those that were still standing had an empty look about them, with gaping dark holes for windows. Others had been demolished so recently that cranes and Caterpillars were still sitting atop piles of rubble, as if they were proud to have conquered these sad remnants of socialist architecture. Still others had vanished entirely, and all that was left of them were gaps in the urban landscape that would hardly be noticeable to someone who had not spent part of his life there. My old school was scheduled for demolition. Apparently it had left few fond memories, and many of the building's windows looked as if they had been willfully broken. When I took a picture of the now empty site where my old apartment complex had stood, a local couple told me not to photograph the ugly parts of their town. Instead, the elderly man emphatically impressed on me, "One must look at what's beautiful" (*Das Schöne muss man sich ansehen*), and he pointed at the supermarket where my family had gone grocery shopping when I was young. Once a social center, the supermarket now carried Western goods. It too was partly boarded up and bore traces of vandalism.

The fall of the Wall changed virtually everything for East Germans. They received a new constitution, a new political system, a new currency, new freedoms to travel—and new insecurities about employment that undermined the very foundation of their existence. By contrast, hardly anything changed for West Germans. The main markers by which they experienced reunification were a solidarity tax of 3 percent of their gross income, new categories of tax-deductible expenses, new zip codes, and eventually a new capital. With so few perceptible changes, it seems almost understandable that by 2005, 50 percent of West Germans had not yet set foot on the territory of the former East Germany, as though the Wall were still standing. There was nothing in those regions that could contribute to West Germans' understanding of their own identity.

One of the most interesting sociological findings Berth mentioned came from a longitudinal cohort study of East Germans born in 1973, roughly my own age group. After 1989 they were asked annually how long they thought it would take for "inner unification" (*innere Einheit*) to be accomplished. Every year the estimated time increased. Right after the fall of the Wall, the respondents had estimated six years; in the 2005 questionnaire they said, on average, fifteen years.[10] Germany, it seems, is growing together, apart.

One of the reasons for "East Germany's" persistence may be precisely that surveys like this one continually re-create and reinscribe it. Examples of similar studies abound. Polls in newspapers and magazines continually try to articulate and measure differences between East and West Germany. The apparent need to see how far East Germans have come by comparing them with the more advanced West not only normalizes a particular trajectory of development, but also re-creates and regenerates East Germany's relative deficiency.[11]

Those days in May 2005 also marked the opening of the controversial Memorial to the Murdered Jews of Europe, designed by the New York architect Peter Eisenman for a vacant lot near the Brandenburg Gate. The memorial consists of some twenty-seven hundred stone columns sunk into the ground to varying depths and arranged geometrically to form straight and narrow paths for visitors to walk on.

On May 8 I walked around the fence that protected the memorial from the curious until its official opening the following day. My overwhelming impression was of a concrete jungle. The blocks that constitute this place of memory grow higher toward the center, while the walkways through them are lowered so that, in the middle, the blocks tower over one's head. Along the fence I noticed several signs for the disabled. A plaque at the memorial's corner told me that there were thirteen special pathways for people in wheelchairs on which the slope did not exceed 8 percent. The same plaque also laid down the rules for visiting the monument: There was to be no shouting, no music, and no movement that exceeded walking speed. People were told not to jump from block to block, nor to sunbathe on top of the blocks, nor to have barbecues among them. Visiting the memorial itself served to discipline memory, in the way that artifacts routinely discipline their users.[12] In an effort to ensure that the opening ceremonies of May 8 and 9 would proceed without interruptions, the police had told residents whose windows overlooked the area that they would not tolerate persons on balconies, or even permit them to open windows. Citizens were also told that they might have to put up with unannounced apartment searches.

A few days later the memorial opened, and outraged journalists reported that children were indeed jumping from stone to stone. On the day I visited the memorial, the atmosphere resembled a playground for adolescents. Schoolchildren daringly jumped on even the tallest stones and disobeyed their guardians' requests to climb down again. One reporter claimed to have discovered a small swastika drawn on one of the stones, but his description of its location proved hard to trace. And who would be able to check the more than twenty-seven hundred stones regularly for graffiti?

As if in answer to this thought, the stones were covered with a graffiti-resistant coating, which had involved another interesting scandal. The German chemical company Degussa had manufactured the coating. At some point in the construction process, it came to light that a Degussa subsidiary had supplied the poisonous gas Zyklon B, which had been used to murder the very victims whom the memorial was seeking to commemorate. The question arose whether a company linked to the Holocaust in this way could legitimately participate in the work of constructing the memorial. Construction was briefly halted, but then resumed. Wolfgang Thierse, then president of parliament, said that a German company like Degussa could not be excluded from a German project. In a newspaper interview, Israel's former ambassador Avi Primor added, "If one wants to exclude Degussa, then one would also have to exclude the German government as a successor of the Nazi government from building the memorial." Thierse said that using a different supplier would have significantly increased the memorial's costs, and he announced that a plaque would commemorate Degussa's role in the Holocaust. Eisenman was happy with the decision and with the discussion around it. The purpose of the memorial in the first place, he said, was to initiate a discussion among today's Germans. And besides, he added, "not forgetting" should not be equated with "not forgiving": "One must always forgive. That is the essence of this memorial."[13] When I went to the memorial after a rainfall, drops of water had collected on the stones, but they did not stick to it. In fact, when I pushed the water around with my finger, it rushed around on top of the stone like beads of mercury, not wetting the surface underneath.

Making East Germany Transparent — and Seeing an Unrechtsstaat

In order for Germany to assume its place in the world as a *Rechtsstaat*, it first had to demonstrate that East Germany had been an *Unrechtsstaat*, whose dissolution and incorporation were consistent with the rule of law. It did so by displaying the injustices that the East German state had committed against its citizens and by commemorating them in particular ways. John Borneman has written about the problems and ambiguities that emerged in establishing criteria for right and wrong when one political system displaced another.[14] Borneman's focus is on the rituals of cleansing and retribution, achieved through law, that are preconditions for avoiding violence. I, by contrast, am interested in how communal standards of lawfulness come into being in the first place. I want to show just how an entire state disappeared and where the society that was living in it went, with its spoken and unspoken rules, its ideals, and its abuses.

How did East Germany come to be framed as an unlawful state—an *Unrechtsstaat*? In this section I first show how the East German secret police, the Stasi, became one of the defining images of East Germany. I argue that the Stasi came to fill this role in part because it inverted the logic of transparency by which the West German state claimed to function: instead of making itself transparent to its citizens, the East German government had made its citizens transparent to the state. Revealing the inner workings of this machinery of inversion—the East German secret police—became an obsession for East Germans and West Germans alike, with the result that the East German government, completely identified with the Stasi, emerged as an utterly tainted entity. We will see, however, that the process of denying East Germany's lawfulness also negated, and eventually dissolved, forms of solidarity that had emerged in the East, potentially threatening the reunified government's monolithic sense of law and its pervasive gaze.

The East German secret police came to serve as the least ambiguous symbol of a state turning against its citizens. The reconstruction and *Aufarbeitung* of its files is a prime example of how the East German state was made transparent—and East German history rewritten. What could illustrate the corrupted nature of East Germany better than to reconstruct in detail that state's surveillance of its citizens? And to highlight those citizens' acts of resistance, such as escapes, or the political imprisonment that followed failed attempts? We will see too how the Stasi, and East Germany, became museum objects, thereby legitimating the authority of the "curators" of the new state.[15]

Immediately before the Wall fell, when the end was felt to be near, Stasi officials began destroying the most volatile and incriminating files they had accumulated over the decades. They succeeded in destroying a great number of them; when shredders broke down, they continued to tear sheets apart by hand and stuff them into bags. Soon after the East German regime crumbled in November 1989, citizens' rights groups took over Stasi offices in several cities to make sure that everything stayed where it was until the files could be transparently processed. On January 15, 1990, they occupied the main offices in Berlin. There alone they found 17,000 bags filled with torn-up papers and files, unraveled films and audiotapes, cut-up photos and negatives, and the notorious olfactory samples (*Geruchsproben*). These samples, which typically consisted of items of clothing worn against the skin, had been collected in clandestine apartment searches and stored in glass jars. If a suspect needed to be located or identified, the samples were given to dogs, which would then sniff out the suspect. After the citizen groups had sorted out and destroyed all the completely unusable materials, they were left with about 6,500 bags of reconstructable papers from the main Stasi offices in Berlin alone. Altogether they recovered roughly 16,000 such bags from all over East Germany, which the Stasi had planned to destroy, between October 1989 and January 1990.

When the Stasi files became accessible, an unprecedented soul-searching began in East Germany. On August 24, 1990, the East German parliament (Volkskammer) crafted and passed a "Law on Securing and Using the Personal Data of the Former Department of Stasi," which provided for the data to be processed in special, decentralized archives. When these provisions did not appear in the draft of the Unification Treaty, East German civil rights groups, some of whose members were also members of the Volkskammer, demanded clear regulations on what would happen to those files. Some of the "archive squatters" even went on a hunger strike until the drafters of the treaty added an article mandating a law based on the bill passed by the Volkskammer. On September 18 a provision was added to the Unification Treaty stating that the Bundestag would quickly pass a law to deal with the Stasi files. In early 1991 East German citizen groups drafted a Stasi Files Law (*Stasi-Unterlagen Gesetz*), which was modified over the course of the year. The Bundestag passed the law in November 1991, and it went into effect in December.

On October 3, 1990, the official reunification day, a federal authority was appointed to process the Stasi files. The day before, on the last day of East Germany's existence, the Volkskammer had nominated Joachim Gauck to head the authority, and on October 3, President Richard von Weizsäcker and Chancellor Helmut Kohl confirmed Gauck, a former East German Protestant minister and civil rights activist, in his new office.[16] In some sense, establishing the framework of transparency was East Germany's last official act, just as it was reunified Germany's first.

With the establishment of the Federal Stasi File Authority—commonly known as the Gauck authority—the East German revolution became institutionalized in the form of a federal authority that would search for the truth of the East German state as emphatically and as thoroughly—indeed, as obsessively—as the Stasi had once searched the lives and minds of East German citizens. The central aim of the Stasi Files Law was to fully disclose the files that the secret police had collected over the forty years of East Germany's existence and to make the information from them available to the citizens from whom it had been collected.[17] Historically, this was probably the first time that an entire population gained access to the information that a secret service had collected on it. No other country had a wealthy mate who could finance the kind of transparency that was felt to be necessary for securing the future legitimacy of any government located in East Germany. For the first time, a dictatorship had successfully dealt with its dictators, and over the years delegates from Bulgaria, Argentina, and Iraq visited the Gauck authority to exchange information and learn from the German experience.

When the Gauck authority began its work, it had 52 employees. Over the course of 1992 it hired an additional 2,150 employees, who began reconstructing the pieces of paper the Stasi had left behind. In the hiring process, 4,000 applicants were checked for prior Stasi affiliations, and 69 such affiliations were "uncovered." Not only could private individuals ask to see their files, but federal and state offices could also request security checks of potential employees. The results would form the basis for constructing new administrative structures, such as police departments, in the new German *Länder*. In order to participate in constructing a new state, one had to prove that one had not been affiliated with the old regime. Even before the law went into effect, almost 350,000 requests for Stasi checks had already reached the authority. The total number of requests (from private individuals and public offices) reached 1 million by April 1993 and 2 million by August of that year. By January 1996 that number had risen to 3 million, and by August 1998 to 4 million. By 2005, 7,000 more requests were still being filed every month.

The Gauck authority's findings startled everyone, bringing many unexpected facts to light. Even years after the files had been opened, official investigators talked about them as still holding many surprises and the potential for many political scandals. Perhaps most shocking was the extent of the general public's involvement in data collection. In 1989 an estimated 91,000 persons had worked full-time for the department that understood itself to be the "Shield and Sword of the Party" (*Schild und Schwert der Partei*). At least another 100,000 were estimated to have worked for the Stasi informally, meaning that about one in fifty East Germans (or one in three hundred in reunified Germany) had been working for the secret police. This meant that the absolute number of informants employed by the smallish East German state of 17 million outnumbered the Gestapo in larger Nazi Germany. The informal collaborators with the secret police were so numerous that the full extent of the Stasi seemed almost coextensive with the society itself. As we will see later, however, the mere number of informants tells neither a complete nor a reliable story about the operations of the Stasi.

Obsessive Transparency

Throughout its existence the Gauck authority remained overwhelmed by the sheer amounts of data it was responsible for. In early 1995 the office started a unique pilot project to speed up the reconstruction of torn-up files: twenty-four employees of the Federal Authority for the Recognition of Foreign Refugees (Bundesamt für die Anerkennung ausländischer Flüchtlinge) near Nuremberg began to survey, sort, and piece together some of the 1,100 bags of torn-up documents from the Stasi section that had been dealing with "state apparatus, church, art, culture, and opposition." On its website the authority describes the process of material reconstruction and the skills those twenty-four workers needed for piecing together, like a gigantic set of jigsaw puzzles, letter-sized pages from tens of thousands of scraps. One needs patience, the website says, a feeling for detail, and a sense for meaningful connections (*Sachzusammenhänge*). The procedure is thorough: First, a specialist gives the restorers background information on persons and events in the former East Germany. The workers then take the torn-up documents from bags in layers, in order not to destroy possible physical connections among them, and place them on a work surface. They first sort according to kinds of paper and the degree of destruction (i.e., how many times a piece was torn) and then according to the writing material (pencil, ballpoint pen, ink) and whether the text is handwritten or typed. Finally they look for connections in meaning. Like archaeologists, they try to piece together the layers of meaning that belong together. By 2001 this group had succeeded in reconstructing 480,000 pages from only 185 bags of material. The authority claims as a token of triumph a half-sized sheet of letter paper that workers reconstructed, like the British Museum's precious Portland Vase, from ninety-eight individual pieces.

Once the documents are ordered in archival form and registered, a courier drives them back to Berlin, where the documents are ordered and registered once more. Only then are they made available to citizens. When I received my own Stasi file in the mail in 2001, the photocopied pages were carefully numbered, and each page bore the stamp of the BStU (*BundesStasiUnterlagenBehörde*, or Federal Stasi File Authority). It had not been among the files that were torn up.

In 2000 the Bundestag asked the executive to replace manual reconstruction with computer-assisted methods. In response, it started a Europe-wide competition for the most efficient proposal. Spokespersons for the company that won the contract claimed that it could reconstruct the 600 million shreds of paper within five years, at a cost of only an additional 1 million euros per year. The company proposed to laminate the pieces in plastic sheets, scan them assembly-line style, and then piece together the data in a hundred parallel processing computers. The method is reminiscent of how the American company Celera, under Craig Venter's direction, speeded up the decoding of the human genome by introducing new forms of DNA sequencing. The German government called on private enterprise to undertake a similarly energetic decoding of East Germany's political genome, hoping to decipher its workings down to its most minute elements.

Despite all the faith in the information the reconstructed files would eventually yield, there was also an inevitable sense of futility about the undertaking. In her book *Stasiland*, the Australian author Anna Funder quotes a memo that the director of the Federal Stasi File Authority in Zirndorf, near Nuremberg, had shown her. In it we read a schoolbook-type calculation that shows at once the obsession with control of the data and the absurdity of trying to reconstruct it all:

> Time Required for Reconstruction:
> 1 worker reconstructs on average 10 pages per day

> 40 workers reconstruct on average 400 pages per day
> 40 workers reconstruct on average in a year of 250 working days 100,000 pages
> There are, on average, 2,500 pages in one sack
> 100,000 pages amounts to 40 sacks per year
> In all, at the Stasi File Authority there are 15,000 sacks
> This means that to reconstruct everything it would take forty workers 375 years.[18]

Funder writes that she was left speechless when the director told her that as of that moment (in 1996) there were only thirty-one workers in the office, not the forty used in his calculation.

Transparency on Display—The Stasi in Museums

State energies were not only directed inward, toward reconstructing and understanding the workings of the Stasi, but also outward, toward making it a public spectacle. Beginning in 1994 the various Stasi offices around East Germany initiated a "day of the open door," during which between a hundred and a thousand citizens came to each office to view the archives and the rooms in which the Stasi had worked. When the central headquarters in Berlin were opened for a day, 20,000 visitors came. Later the Berlin office opened a Stasi documentation center, which welcomed its hundred thousandth visitor in August 2002. Throughout the 1990s numerous exhibitions, and even a film,[19] demonstrated the methods the Stasi had used to spy on East and West German citizens. These exhibitions were soon followed by others that focused on the civil disobedience of East Germans and the ways they found to resist the Stasi.

Museums are part of the reunified state's armamentarium of transparency and of pedagogy. Thus it is not surprising that the narrative of East Germany as an *Unrechtsstaat* found its way into several Berlin museums that exhibit the workings of the Stasi. In these museums the descriptions of East Germany constantly remind the viewer of the illegitimacy of the East German state and of the unlawful means by which it enforced loyalty. On Wilhelmstrasse in the former East, for example, there is the Stasi Museum, which documents the surveillance methods of the state. This museum's focus is not on the triumph of freedom-seekers over a repressive regime, but rather on the methods by which the East German state kept dissenters in line. Near the entrance a fifteen-minute East German promotional video shows some of the training methods with which the Stasi operated. The museum attempts to show both the everydayness of surveillance and the invisibility of it.[20]

In March 2002 the Berlin Museum for Communication on nearby Leipziger Strasse opened "An Open Secret," its first exhibit to document the comprehensive Stasi surveillance of the postal and telephone services of thousands of East Germans. The interior of the museum was designed to convey both the sense of watching over others and the sense of being watched. The exhibit was housed inside a walk-in container. On the outside the visitors could see everyday East German scenes; inside, they could see the workings of the surveillance state. When I visited the museum in summer 2002, I was drawn to the display of the by then almost antiquated instruments of surveillance. On two old-fashioned-looking telephones one could listen to the sound of a telephone line being tapped. In East Berlin alone, a sign read, there were twenty-five telephone surveillance studios, from which up to 20,000 telephones could be monitored simultaneously and the conversations completely or partially recorded. The focus was on conversations across the border into West Germany. Next to this display of state intrusion, a small note reassured the museumgoer: "Of course your telephone call will not be recorded." (*Ihr Telefonat wird selbstverständlich nicht gespeichert.*) Museum visitors evidently needed to be explicitly reassured that the surveillance strategies that had been omnipresent in East Germany had now been abandoned. It was as though that aspect of the dictatorial past needed to be actively disavowed by the *Rechtsstaat* that had succeeded it.

Other signs inform viewers that an automatic letter-opening machine could steam open up to 600 letters per hour. In all of East Germany the secret police read an average of 90,000 letters every day, taken from mailboxes on the street and in private homes, and opened 4,000 packages. The numbers peaked significantly around the holidays. Items that were thought to subvert the state's authority were archived or destroyed, though this last drastic course was not common. It happened only a few times that letters or packages that my own father sent from East Germany, after the rest of my family had escaped to the West, failed to reach their destination. More often my mother received letters that had been opened and then resealed with tape. Apparently the state wished only to demonstrate its continuous background presence.

After reunification the German state went increasingly public in demonstrating its commitment to the privacy of mail and telephone communications. In a 2004 ad campaign in connection with planned revisions of laws governing surveillance (the central question was whose telephones could be tapped and whose could not), I saw several posters highlighting citizens' rights that were to be protected from state supervision. One poster showed a young woman chatting on the telephone, with a caption below: "Flirting, Complaining, Gossiping: And no one is listening in." In keeping with German sensibilities regarding law and textuality, the legal paragraph guaranteeing the secrecy of telecommunications was printed below the caption. It was marked "Decided on by the German parliament." Still farther down, another caption read, "Decisions for Freedom. The German Parliament." These "Decisions for Freedom," intended in some sense to (re)build moral order, drew an implicit contrast to East Germany, where private communication among citizens was not off-limits to the state.

Surveillance in practice contradicted the official rhetoric, however, and politicians from all parties were outraged at this obvious misrepresentation of facts on the ground. In an interesting irony, the image campaign coincided with the revelation by experts in the security industry that German authorities had officially tapped 20,000 telephones in the previous year, more than any other democratic nation; estimates of unofficially tapped telephones ran much higher.[21] The media had referred to Germany as *Abhörweltmeister*, or telephone tapping world champion. Some politicians said that the ad campaign bordered on criminal misrepresentation and falsification and warned that it suggested a degree of telephone privacy that simply did not exist in Germany.[22] Apparently governments on both sides of the former Wall were intent on listening in on private conversations.

But back at the Berlin Museum for Communication, one learns that surveillance in itself is not always illegitimate—not when serving the West German *Rechtsstaat*. *Rasterfahndung* (roster searches, or profiling) had led in the 1970s to the arrests of some members of the leftist terrorist group RAF, whom the later minister of the interior, Otto Schily, had defended in court. In 2004 it became public knowledge that Schily now planned to use *Rasterfahndung* to search for terrorists who might be connected to the attacks of September 11, 2001. *Rasterfahndung* is a controversial process. Certain suspect personal characteristics are fed into a computer, which then reads out the names of all those who fit the profile. The technique does not search for certain guilty individuals, but rather presumes that a certain category of person is potentially guilty (of a potential crime) and that its members can be exonerated only after investigation. *Rasterfahndung* in a sense violates the legal presumption of innocence until proven guilty. Next to the Internet display on the 1970s arrests, at a different computer, another website was available for learning about the *Rasterfahndung* used after September 11. When combing through 8.3 million data sets revealed that there were no sleepers whom the search could uncover, the documents that had been produced were destroyed. A computer CD with the data sets on it was then symbolically cut into pieces at a press conference. Having made a part of the population transparent in the search for dangerous individuals, the government (symbolically) placed a veil of anonymity back over it. Clearly the new Germany had its own forms of far-reaching surveillance, but their packaging and performance invested them with an aura of moral legitimacy.

One of the most famous museums commemorating East Germany is the Checkpoint Charlie Museum, also known as the Mauermuseum (or Wall Museum). Opened by the historian Rainer Hildebrandt in June 1963, less than two years after the building of the Wall, the museum (which on its website calls itself an "island of freedom"[23]) was located right in front of the Wall. In fact, through a small back window, *Fluchthelfer* (escape helpers) were able to observe what transpired at Checkpoint Charlie and to hatch new escape plans for those willing to take the risk. The museum dedicates itself to documenting in detail, and in four languages (German and the Allies' English, French, and Russian), the popular uprising of June 17, 1953, the building of the Wall, and the thousands of failed and successful attempts to escape from East Germany.[24] Escaped East Germans who support the museum's mission have told their stories and donated numerous original escape vehicles. Today the museum features homemade hot air balloons, escape cars (including a small reengineered car that fitted a person where the heater and battery used to be), an underwater diving suit and the miniature submarine that pulled one man through the Baltic Sea, imitation Soviet uniforms, and even one of the notorious automatic firing mechanisms installed on the border, which activists took at considerable risk to their own lives.

Learning to See Themselves as Victims

Joachim Gauck ran the Federal Stasi File Authority with a concern for balance. His main focus was on protecting the victims and prosecuting the perpetrators, but he wanted to practice forgiveness as well. He precluded the idea of a general amnesty, however, because such a blanket act would destroy the population's faith in the *Rechtsstaat*.[25] Instead, Gauck wanted to make sure that those in leading political and economic positions were untainted by Stasi affiliation.[26]

The Stasi File Law does not mention the terms "victim" and "perpetrator," yet these terms have become canonized in subsequent discussions of the Stasi. How did these terms enter the public discourse? Matthias Wagner, himself a Stasi member and, after 1990, briefly one of the files' keepers, provides an answer.[27] Wagner recounts that it was Gauck himself who introduced the terms on the evening the Stasi File Law went into effect. On a live television show aimed at clarifying what the law meant, Gauck was taking phone calls and answering questions. One young caller described his activities and confirmed that he had been an "unofficial informant." Gauck told him, on the spot, "*Sie sind Täter*" (You are a perpetrator). That term stuck. Those who had been spied on, by contrast, were *Opfer* (victims), a word that is often used to refer to the victims of the Holocaust. It is a strange echo of that greater tragedy that the Stasi was found to have accumulated an estimated six million files on the victims of its unlawful surveillance.

According to Gauck, everyone who had a Stasi file and who had not in some capacity worked for the Stasi was a victim. That classification was an enormous relief to many East Germans in official positions, Wagner writes. Since members of many East German official institutions had Stasi files, they could now all claim victim status. Those who had in their files the Stasi's approving sentence that they were "upstanding socialists in word and deed" could almost feel redeemed, as idealistic individuals who had worked for a better society.

When full-time Stasi employees had destroyed files in the weeks just before and after the fall of the Wall, they had predictably destroyed the most damaging files, which often included their own and those describing their activities in West Germany. What remained were largely the files on the unofficial informants and on those four million East Germans they had spied on. The selective destruction of files thus worked to incriminate the small-time, unofficial informants while it got many of the official collaborators off the hook. Perhaps their files are in the thousands of bags of shredded material awaiting reconstruction. In the meantime, the existing evidence of hundreds of thousands of unofficial informants presents a shocking image of an entire society trained to police itself, and the wealth of collected information provides the illusion that piecing the documents together again will lead to a coming to terms with East Germany's dictatorial past.

Wagner's description highlights the difficulties and tragic ironies inherent in using the Stasi files as an accurate divider between perpetrators and victims of illegitimate surveillance. He points out the flaw inherent in taking the Stasi files to be a complete picture of how East Germany functioned: "The files have their own internal politics . . . Having a file does not necessarily make one a 'victim,' and informing does not necessarily make one a 'perpetrator.'" What was sold as the "uncovering of the structures of a totalitarian state," Wagner writes, served instead to prepare the ground for dismantling East Germany.[28]

Gauck himself was acutely aware of the difficulty of separating the victims from the perpetrators, and he described the obstacles to making clear distinctions between them. For one thing, the Stasi had often pressured its victims to spy on other suspected enemies of the state, thereby turning victims into perpetrators themselves.[29] Furthermore, the Stasi's surveillance of its own informants was so thorough that the files of victims and perpetrators often became indistinguishable from one another.[30] In spite of these problems, Gauck concluded that the *Aufarbeitung* of the Stasi past was an absolute necessity for developing the *Rechtsbewusstsein* (awareness of "right") of East Germans.[31] For some, as we will now see, this in effect meant that East Germans needed to learn to see themselves, as well as the law, the way West Germans were already seeing them.

How German Was It?

Differences in cultural norms of transparency and surveillance crystallized when one of the most prominent Stasi victims turned out to be a West German rather than an East German: former chancellor Helmut Kohl. Kohl at the time was at the height of his career, but just about to fall. He had made history by bringing the two halves of Germany together again, but he was politically vulnerable because his promises of revitalizing East Germany within just a few years had turned out to be illusory.

It was well known that the Stasi files contained information about Western leaders and organizations. West Germany, after all, had quickly brought many such files across the border. But there had been a tacit consensus within the Federal Stasi File Authority not to give out illegally gathered information for political purposes. This changed with Kohl's party financing scandal. The Christian Democrat Party (CDU) was found not to have declared large sums in campaign contributions, and no one knew where the money had come from. When it came to light that Kohl had illegally accepted funds from an unknown source to finance his campaign, a search for the donor began. Kohl refused to say anything, claiming that he had given his "word of honor." Some people suspected that the Stasi might have been blackmailing Kohl because they knew the answer. Members of the Green Party, which was allied with civil rights protesters from East Germany, wanted to use the Stasi files to find information that would shed light on the scandal. In November 2000 Kohl sued the authority in order to prevent it from turning his Stasi files over to researchers and the media. Almost one year later, in July 2001, the court decided that Kohl's files would not be made public.

After the Greens had broken the tacit consensus, it seemed that a law was needed to reestablish the previous practice. In August 2001 Marianne Birthler, a former East German civil rights advocate and the authority's new head, asked the Bundestag to clarify the Stasi File Law in order to ensure that other files would continue to be returned to the people.

A debate between Birthler and Robert Leicht, a well-known and influential journalist and theologian who had been the chief editor of the weekly magazine *Die Zeit* and for many years one of its most prolific contributors, about opening Kohl's Stasi file highlights the multivalence and ambiguity of ideas of transparency and surveillance at the moment of reunification and shows what was at stake in the conflict. Where Gauck had tried mainly to inform the Stasi's victims of the extent of its spying, Birthler promoted a more comprehensive use of the files for the purpose of critically engaging with Germany's power structures, both eastern and western. She had consistently argued that the public had a right to know the details of Kohl's files, and she did not want the Kohl case to set a precedent for keeping files on public persons locked up.

In a 2001 article entitled "Every Medicine Has Side Effects: Why Helmut Kohl's Stasi Phone Tapping Protocols Must Not Be a State Secret" ("Jede Medizin hat Nebenwirkungen: Warum die Stasiabhörprotokolle in Sachen Helmut Kohl kein Staatsgeheimnis sein dürfen"), Birthler addressed the question of whether one should use information that the Stasi had gathered in violation of the Basic Law—the German constitution—if one's goal is to clarify the workings of the Stasi. Her argument consisted of two main strands: first, that Kohl, as a public figure, should be made as transparent as possible to the citizens he governed; and second, that Germans, as a people entitled to accountability from their state, deserved to know all that could be known about the Stasi's operations. If Germany was to make sure that a dictatorship would never again take root in German soil, it was necessary to know as much as possible about how it had happened in the case of East Germany.

Birthler portrayed the Stasi File Law as a success. She regarded the opening of the structures that had dominated East Germans, including censorship of press and politics, secrecy, and pretensions of all kinds, as an act of atonement. To her, it was only understandable that, following the peaceful revolution of 1989, revolutionaries would turn their attention to the Stasi. Just ten days before official reunification, on August 24, 1990, the first democratically elected Volkskammer had passed a bill specifying how the Stasi files should be dealt with. Like Birthler, they wanted to hold the East German state accountable, rather than dismiss it as entirely unlawful.

Opening the Stasi files, making them transparent, was always in partial contradiction to West Germany's legal notions of privacy and data protection. The Stasi File Law even has a provision that limits the applicability of the Data Protection Law. That is why Birthler saw the Bundestag's enactment of the law as a concession to the peaceful revolution in East Germany, generating new norms for the reunified state. The law, she wrote, "ties the tradition of the East German citizens' movement to the principles of West German data protection as well as to archive laws and to the principles of the freedoms of information and of science." In other words, in Birthler's view, the hybridization of East and West German laws and traditions produced a new, composite entity. Tensions arose, according to Birthler, because the Stasi File Law could not be explained solely with reference to the West German legal understanding before 1989.

Birthler further argued that if Kohl's file were closed to the public, then all kinds of other factors important to political behavior (such as motivations, alternatives, tacit understandings, or secretly arrived-at conclusions) would be off-limits to legitimate inquiry. Even information on the East German opposition would be out of reach, since "opposition" in that dictatorship tended to be "nonpublic" in nature. One would continue to look at the facade—at what the state had chosen to present and make public—but not at the struggles behind such decisions. In trying to make the workings of the Stasi visible, Birthler wanted a different kind of transparency. She was not content with looking at the state as it presents itself; she wanted everything of concern to citizens to be visible. In some sense, she wanted to show the Stasi in action by following its operators through society. Uncovering the workings of a secret service, she concluded, necessarily involves uncovering secrets—*all* kinds of secrets.

It was in her very attempt to redefine openness and secrecy that Birthler came up against resistance from West German quarters. Her positions on transparency and surveillance are thrown into high relief when we contrast them with those of Robert Leicht. In a 2001 response to Birthler, entitled "Citizen Kohl Also Has Rights: A Chancellor as Victim" ("Auch der Bürger Kohl hat Rechte: Ein Kanzler als Opfer"),

Leicht invoked the *Rechtsstaat* in all its majesty. He claimed that Birthler had used deeply flawed logic to legitimate her personal curiosity about Kohl's file while failing to make any persuasive arguments. For Leicht, the written text of the law is the touchstone of the *Rechtsstaat*, not some more basic ethical principle lying outside and possibly above the law.

Leicht himself used legal categories as dispositive of the argument over the meaning of transparency, and he proposed steps to bring order into the chaos created by Birthler's demand for disclosure. First, he clarified the inner logic of the Stasi File Law. The law, he wrote, divided victims from perpetrators, not East Germany from West Germany. Victims, regardless of which side of the Wall they were on, had the right to read their files and find out what was done to them. Perpetrators, again regardless of which side they lived on, could have their files exposed. Understanding the Stasi did not mean understanding its victims (that, after all, was what East Germany had been trying to do), but rather its perpetrators. Second, Leicht concluded that the law made it extremely clear that Kohl's file must not be handed to journalists and researchers. That is why, he claimed, Birthler cannot use the law's text to support her argument. If Birthler had looked at the law itself, he wrote, she would have seen that the intentions she claims underlie the law did not become legal reality. And if something is not found in the text of the law, then it has been discarded in the legislative process. In a form of legal positivism, Leicht assumed that the law as it was written expressed all there was to know about its meaning.

Moreover, Leicht emphatically argued that one could not retrospectively lose one's basic rights—which are taken as written on behalf of *all* Germany. It was clear before 1989 that the Basic Law protected Helmut Kohl, as a citizen, from having his telephone tapped. Basic rights such as this could not be taken away from citizen Kohl, especially not through the instrumental use of an organization that had specialized in systematically violating those basic rights.

By framing the conflict in purely legal terms, Leicht mobilized the law as a cultural resource to produce the morally right kind of transparency—in this case, a limited transparency that protects Kohl as a private person and victim. He referred more than once to the letter of the law and the logic of the law. He cordoned off the law from all polluting influences, including the messy history through which it came into being. In his conclusion, Leicht accused Birthler of being interested less in revealing the workings of the Stasi than in revealing the workings of Helmut Kohl, as if the two could be separated. Leicht's closing line, "tertium non datur," which refers to the absence of alternatives between letting Kohl's private life be private and using legal means to partially uncover it, might as well mean that there is no alternative between the *Rechtsstaat* of West Germany and the *Unrechtsstaat* of East Germany. There could be no third entity in which the two are merged, for to admit this possibility would undermine West Germany's authoritative moral position, based on a painstaking construction of a single postwar history.

In important ways, however, the Stasi File Law *was* an East German law. It was the Volkskammer's attempt to come to terms with a state that had spied on its citizens. It can therefore be read as an East German attempt to craft a *Rechtsstaat* using the materials available to it. The conflict between Marianne Birthler and Robert Leicht thus stands for a more general conflict between two different understandings of the law's role in *reunified* Germany as well as between two ideas of what principles to use in making the state accountable to its people. In their exchange we see, then, in almost stylized form, a confrontation between two understandings of the meaning of reunification itself.

The question both can be read to be answering is the one implicitly posed by Birthler in her article on "side effects": What is the sick body that needs the therapy offered by the Stasi File Law? Both Birthler and Leicht are concerned with making sure that German politics will never again be perverted. And both advocate transparency—up to a point. But Leicht, in some sense, has already figured out how to accomplish this. He knows what needs to be shown and what needs to be seen. For him, and for many West Germans, West Germany has already cured itself of its Nazi-era pathologies. It has passed a Basic Law that puts human dignity first, grants protection to German citizens and gives them rights, and prescribes a liberal democratic order with careful separation of powers. In Leicht's view West Germany has achieved wholeness on its own. The politics, and the surveillance, that happened on the other side of the Wall, therefore, cannot possibly have any bearing on West Germans, and the dictatorship there must not be permitted to defile the democratic order that has been so painstakingly accomplished. The pathologies of government that the Stasi files bear witness to are limited to East Germany and must not be allowed to spill over into reunified Germany. The sick political body in need of therapy is that of East Germany alone; the biomedical cure that Leicht advocates consists in surgically isolating the Stasi tumor. When the potential damage from the secret files threatened to spill over into West Germany, however, different standards had to come into play. Transparency had come up against its limits in West Germany.

Birthler's approach is different. She emphasizes that the Stasi did not care about borders. The Wall was not effective as a divider between healthy and sick politics. The Stasi, for one thing, operated in the West almost as easily as in the East. It tapped the telephones of many West German parliamentarians, and it even brought down several high-ranking politicians, most infamously Chancellor Willi Brandt, whose trusted assistant was discovered to be a Stasi spy. The Stasi successfully hired thousands of West German informants, and it influenced the professional and private lives of many more. Hubertus Knabe, a historian who spent several years working in the Stasi document office, estimates that between twenty thousand and thirty thousand West German citizens spied for the Stasi. In his opinion, the agency's operation was so pervasive that he considers West Germany to have been thoroughly subverted.[32] For Birthler, therefore, *everyone* in Germany is implicated in the kind of surveillance institutionalized by the Stasi, and the task of understanding how that massive violation of rights worked requires being open about where the files lead. If we want to understand just how the Stasi was able to function so well, Birthler effectively argues, then we need to understand the larger context in which it worked. To understand so pervasive a disease, one must not stop at monocausal and local explanations. Rather than finding a localized biomedical cure, we need to look at the larger picture and find a social, or epidemiological, explanation.

For Leicht, reunification is a localized process with a foregone conclusion, at least with respect to the Stasi; for Birthler, it is an open-ended process in which new rules must be crafted along the way. In her view West Germany does not have the right to foreclose those areas of transparency that it does not want to be seen. For her, the revolution of 1989 was a revolution not only for East Germany, but for *both* Germanys. Rather than taking as given whose law (and whose concept of the law) shall operate in unified Germany, Birthler sees the revolutionary moment as an opportunity for rewriting the rules about what society we (want to) live in.

Mauerschützen—*Suspending the* Rechtsstaat/*Erasing East German Conscience*

The idea of conscience underwent similar rhetorical and semantic shifts following reunification. On November 9, 2004, exactly fifteen years after the fall of the Berlin Wall and a date freighted with German national memories, the last trial of the men who had made that Wall insurmountable concluded. The judge found the final four defendants guilty, but did not punish them. Now aged between sixty-three and seventy-one, they had developed and maintained the notorious automatic SM-70 firing apparatus that the East German state had installed on its borders in the early 1970s. If a trip wire was touched during an escape attempt, a mine would spray the area with nearly a hundred small, deadly metal cubes. Those machines deterred most but not all would-be escapees. The metal splinters killed four men between 1974 and 1984 and severely wounded a fifth. The judge did not punish the machines' makers, he wrote, because "they showed remorse." They had meant only to deter people, not to kill them. This final trial recapitulated many themes from earlier trials. Prosecutors and judges spoke of transcendental values that any East German citizen ought to have understood: "*Naturrecht*" and "*Völkerrecht*" both prohibit killing human beings. Even East German military law, the trial disclosed, had wanted a "soldier with a conscience."

The process of dealing juridically with the East German state apparatus in many ways paralleled the process of denazification after 1945. As the values of a new regime replaced those of the old, many people needed to be tested for their integrity and ability to uphold the new order. The speed of the second round of trials, however, was unprecedented. While the judicial inquiry into Nazi crimes began only about fifteen years after the war, accounting for East German crimes was nearly concluded fifteen years after the fall of the Wall. In the postwar trials of Nazis, one of many acceptable excuses for past wrongdoing was *Verblendung*, literally being "blinded" by ideology. In no case did West German courts recognize socialist *Verblendung* as a valid excuse. While West Germans could collectively "recognize" that they lived in an *Unrechtsstaat* under Hitler and wake up from the nightmare of the preceding years, East Germans were punished for their "blind" adherence to the ideology of their state. In numerous instances the courts judged East Germany by a higher moral standard than they had used to judge Nazi Germany. It is almost as if disciplining East Germany swiftly and resolutely pacified the conscience of the West German legal establishment, which had done too little to weed out Nazi criminals after World War II and had by many accounts failed to master Germany's Nazi past.

How many in fact died at the Wall? Uncertainty surrounds the question, mainly because East Germany treated the numbers as a state secret.[33] The numbers keep rising, and they vary depending on who is doing the counting. The prosecutor for Berlin says 86; the president of the Berlin police says 92; the Zentrale Erfassungsstelle in Salzgitter counts 114; and the Zentrale Ermittlungsstelle für Regierungs- und Vereinigungskriminalität says 122. Alexandra Hildebrandt, the director of the Mauermuseum Checkpoint Charlie, who holds an annual press conference to announce the latest numbers, claims there were over 200.

The state where each killing took place handled the case. Starting in 1991, the Berlin Criminal Court handled a total of 270 cases. Of the *"Republikflüchtlinge"* killed, 237 were shot, while 33 were killed by mines; 109 of them were actual *"Mauertote,"* who died trying to cross the Wall around West Berlin. The court sentenced a total of 128 persons for these killings, some to long prison sentences. Of the 80 border guards who were found guilty and sentenced, 77 were placed on probation. The legal question was whether the defendants, most importantly the border guards, could have recognized that they were violating international law when they were shooting at escapees. Most escapees had in fact been killed not by the guns of border guards (*Mauerschützen*), but by mines and automatic firing devices. The defendants on trial claimed that neither the East German Politbureau nor its individual members had had the power to change the border laws. The Soviet leadership, they said, had unilaterally decided how the East German border would be guarded.

When three members of East Germany's Defense Council (Nationaler Verteidigungsrat) and one border guard appealed to the Federal Constitutional Court, the court made a significant decision. Shooting at escapees had become unpunishable in East Germany in 1982, when the Border Law (*Grenzgesetz*) explicitly permitted it. On November 12, 1996, the Constitutional Court declared that border guards could not excuse their lethal shots by claiming that they were acting according to East German law.[34] The order to fire, the court held, was criminal *Unrecht* and a violation of international human rights. Even East German law and orders from higher up could not justify placing state interests above an individual's right to life. *Nulla poena sine lege* (no punishment without law), a basic principle of the *Rechtsstaat*, can be invalidated, the court concluded, if there are grave violations of human rights. Punishing those who had killed would-be escapees on the border thus became a way for Germany to assert its moral authority against the failed East German state—this time by invoking a higher law.

In 1946, out of concern that Nazi criminals would go unpunished, the Social Democrat law professor Gustav Radbruch had developed the so-called Radbruch formula, which says that laws may be limited in their applicability if they contradict a more fundamental sense of justice. The court saw this transcendental principle as also anchored in the European Convention on Human Rights and decided that it applied in the case of the accused border guards. After two years of deliberation, the highly qualified Constitutional Court judges thus decided what a twenty-year-old East German soldier should apparently have known intuitively: that he should have disobeyed the orders of his state in the name of a higher law. In so deciding, the court diverged from the legal positivism that otherwise pervades German legal thought. Christian Social Union (CSU) head Theo Waigel hailed the Constitutional Court decision as being important for "strengthening the *Rechtsbewusstsein* (awareness of what is lawful) in both West and East." A speaker for the Alliance '90/Green Party said that the decision contradicted the complaint that only the small criminals were punished while the decision makers were let go: "Desk criminals should never again remain unpunished." (*Schreibtischtäter sollten nie mehr straffrei bleiben.*)[35]

But the Radbruch formula has a checkered history in Germany. The Bundestag in 1952 agreed to the European Convention on Human Rights, but the vast majority (the Communist Party abstained) expressed reservations because the convention contained the Radbruch formula. The Bundestag took exception to the formula because it did not want to recognize the sentences of the Nuremberg trials against military personnel in particular. The planned Bundeswehr (Federal Defense Force) needed generals, and those generals wanted to rehabilitate their comrades—which they could not have done had they been lawfully declared war criminals. To this day the Bundestag has not annulled the prohibition against applying the Radbruch formula.[36] The court's opinion in the *Mauerschützen* trial stated nonetheless that the violation of the fundamental prohibition against killing was "clearly evident even to an indoctrinated person." This is the logic that won in the end: soldiers using lethal force to guard the border should have known instinctively—that is, in their conscience—that the "natural law" that says "thou shalt not kill" superseded their purportedly legitimate orders.

But we saw in the previous chapter just how difficult it is for an ordinary citizen to stand up to the German state, and how much harder still it is for those who have sworn their loyalty to it. By allowing soldiers to be tried for following state orders, the court in effect placed the concept of the military order on trial—a concept that remains unquestioned in many other (West) German contexts. The court's judgment in effect told the East German soldiers that they should have thought for themselves before they followed orders. Yet many of the *Mauerschützen* were born in East Germany; they had never learned to see human rights other than as the East German state had taught them to see. In the trials it became clear that many had never heard of terms like "proportionate means," referring to making the severity of a punishment fit the gravity of a crime. Instead they knew terms like "group attacks" (*Gruppenangriff*)—that is, escape attempts by more than one person. And a group attack was a "crime." In such a case one *had* to shoot; the lawbreakers were not those who shot, but those who didn't. We see, then, that when the national borders of Germany changed, the borders of conscience changed as well. When the border was intact, and West Germany accepted East German sovereignty as a practical matter, it also dealt pragmatically with East Germany's efforts to guard that border with lethal force. When the Wall fell, and East German sovereignty crumbled along with it, the legitimacy of guarding that sovereignty could be retroactively denied.

The trials established that the consciences of the *Mauerschützen* should have told them that they were doing wrong. Indeed, one of the four defendants in the last trial had asked his superiors to transfer him five years after he had developed the automatic firing device. He had shown that it was possible to hear one's conscience even in East Germany. But the issue was not so simple. Where, after all, is conscience supposed to come from in an *Unrechtsstaat*? The *Mauerschützen* were in effect charged with not having had the ability to stand above themselves and see themselves from a position outside the moral perimeter of their own nation-state. Even though they were obeying the laws of their state, they were punished for not having had the foresight (or hindsight) to see that state as an *Unrechtsstaat* and then refuse to obey its orders (or to disobey them in some oblique form). They were charged with not having used their conscience, even though the criteria for using their conscience could have been developed only in dialogue with the state, as custodian and representative of a legitimate moral community, which East Germany by definition never was. As it happens, the East German state not only ordered, but even praised, the shootings on the Wall. The guards who fired at escapees did so in the conviction that they had the law on their side. How should they have developed knowledge of *Unrecht*? West German justice treated the East Germans as if they had been West Germans all along and should have conformed to West German notions of justice. They were at fault for not seeing East German law from a place transcending their status as soldiers defending their state—namely, from the vantage point of West German law. It was a judicial delegitimation that worked only because the superiority of West German politics and law were taken as already established.

After the first round of sentencing, the Federal Court of Justice (Bundesgerichtshof) found the punishment to be too harsh and made possible the often more lenient *Jugendstrafen*, or punishments for young offenders. This penalty took the form of probation rather than imprisonment. But what did probation mean in the case of these defendants? Would they really have passed a test of morality if they did not commit a crime over the next two years? Repeating the offense was almost inconceivable in any case. The state they had served was gone. No one would ever again ask them to shoot at escapees. Their conscience would remain untested. They would never have to prove that they could now resist an order to kill. Instead, they would live out normal lives, adapted to the new moral order of reunified Germany.

The sentences of East German judges form an interesting contrast. Many former judges were put on trial for the disproportionate sentences (in some cases the death penalty) that they had handed down. Many of those judges were let go. As the Berlin Regional Court stated as it justified the exculpation of two East German military judges: "It is better to exonerate unjustly [*zu Unrecht*] than to adopt the standards of an *Unrechtsstaat*." In other words, the court recognized the inviolability of *Rechtsstaat* principles, and in these cases it saw no need to invoke overriding principles of a higher law.

Another contrast emerges if we ask what happened to the leaders who had actually held up the East German regime. In these cases too the legal and moral lines were far from clear. From the beginning the East German political leaders questioned the legitimacy of West German authorities to sit in judgment over them. In reunified Germany the question of what to charge those leaders with arose. As members of the Politbureau, they could not be charged with actively committing a crime. The accused had been bureau members since the early 1980s, but the bureau had not made any decisions relating to the border since the early 1970s. Could they be charged with omission? To this charge the accused replied that they had in fact overthrown Honecker at the earliest possible moment and then opened the borders. After two failed attempts to hold the main political leaders accountable, the courts based their judgment on two inconsequential Politbureau decisions that the accused had supported.

Practical reasons also interfered with holding political leaders accountable. Of the fourteen who were tried for the deaths on the inner-German border, six were dropped from the proceedings for health reasons. The former head of East Germany, Erich Honecker, had liver cancer, and the Berlin Constitutional Court declared that putting him on trial would violate his human dignity. Honecker was released from prison that same day and joined his family in Chile, where he died the following year. (Honecker himself had prophesied, "The punishment that you obviously have in store for me will no longer reach me.") Two others were placed on probation, and the remaining six received prison sentences between three and seven and a half years. The head of the Stasi, Erich Mielke, refused any responsibility for the shootings on the Wall. He was never tried for his activities as Stasi head. Instead, he was sentenced to six years in prison (of which he served four) for having killed two police officers in 1931. He died in 2000. If the object was to hold East Germany's political leadership accountable, this record was hardly encouraging. One civil rights advocate famously summed it up: "We wanted *Gerechtigkeit* [justice], and instead we got the *Rechtsstaat*."[37]

Not everyone approved of the legal strategy adopted in the *Mauerschützen* trials. Some argued that West Germany itself had been guilty of omission because it had entered into agreements with East Germany without making the abolition of the orders to shoot a precondition of any future talks. In a sense West Germany too had known about the killings on the Wall but had chosen to ignore them. How could it now judge others who had felt similarly powerless? At one point even Mikhail Gorbachev stepped in to say that the trials violated the Unification Treaty. He said that there were neither juridical nor moral grounds for sentencing Honecker's successor, Egon Krenz, and his colleagues. Instead, he said, it was about "politics, politics, and politics again, burdened by the demons of the past."[38] Some complained that the impending legal judgments, and the unclear legal situation, interfered with the work of harmonious and peaceful *Aufarbeitung*. There was no "safe space" in which East Germans could come to terms with their past and in which "victims" and "perpetrators" could meet and make their peace, if such a peace was possible.

But the point that emerged most clearly from the trials was the asymmetry of the power of the declared *Rechtsstaat* to deconstruct its former adversary, the *Unrechtsstaat*. The alteration of the manslaughter rules offers a telling footnote. Under German law, *Totschlag* (manslaughter) carries a statute of limitations of twenty years, whereas murder charges are not subject to any limitations. However, the *Verjährungsfrist* (statute of limitations) for the East German border guards begins with the day of German reunification, October 3, 1990, since no "objective investigations" were possible before that date. The *Rechtsstaat* superimposes itself on the *Unrechtsstaat* in this instance by discounting the time itself that people spent under that other regime. Under law, as in the installation I saw before the Brandenburg Gate, East Germany can be treated as if it did not exist. As in the case of the Stasi, West German justice could lay bare, and even rewrite, the innermost workings of East German sovereignty. The compromises and elisions of the victor's justice, however, such as the ambivalent legal status of the Radbruch formula and the opportunistic rehiring of border guards I will describe below, did not become comparably transparent.

Despite all efforts of the law to craft a seamless narrative of responsibility, history gets written in many cultural spaces. I conclude with three episodes that show the messiness of that writing. In 1962, when some East German border guards were shooting at a 14-year-old who was trying to cross the border, West German customs officials tried to help the wounded adolescent. As they tied a rope around him to drag him out of the river, some of the West German customs officials fired at the East German border guards, wounding one of them fatally. The dead guard, Peter Goering, was instantly declared a hero and martyr, and the East German government named streets, schools, and plazas after him. To this day there is one street in a former East German village named after him, to the consternation of the post-reunification German authorities.

After the fall of the Wall, the West German secret service quickly hired several thousand former East German border guards and placed them along the new German border. Reporters for a news magazine discovered this fact and interviewed people who had crossed the once deadly border, those who guarded it then and now, and the West Germans who had hired those guards. In televised interviews border crossers expressed astonishment and deep dismay that members of the Stasi, who used to patrol the border with guns, were now again checking the papers of those who wanted to cross into Germany. The border guards, however, claimed they had a clean conscience. One said, speaking for others, that their work had been deemed acceptable by the East German state, that following reunification their qualifications had been tested by the new regime, and that now their work for the new Germany was equally acceptable. If there had been a problem, he said, someone would have told them. As dependable servants of the state, they had apparently adapted to their new employer. One of the officials who had hired them said that employing former border guards was a necessity, since the German borders themselves had expanded so much. One simply could not dismiss so many qualified guards.

In March 2001 the European Court of Human Rights in Strasbourg rejected the claim by Egon Krenz that he had been unjustly sentenced for the deaths on the Wall. At the same time, those judges became the first international tribunal to praise the German judiciary for how it had handled the East German past. The defendants had claimed that the sentences had violated the *Rückwirkungsverbot*, the prohibition against punishing people for deeds that were not punishable at the time they were committed. The court rejected this argument. The judges unanimously decided that the East German regime had violated the human right to life and the freedom of its citizens, both of which were *Unrecht* even in East German law. The orders of those politically responsible for the bloodshed—in particular, the deaths caused by automatic firing mechanisms and the order to shoot—could therefore not have been justly obeyed.

Abortion

Abortion policy was another crucial place where ethical differences between East and West Germany became visible, and stayed visible, although one would never guess it from the final report of the Enquete Kommission on Law and Ethics in Modern Medicine (EK), where I did my internship. That report mentions the former East Germany only a handful of times, once in a chronology of institutional developments in the field of assisted reproduction and a couple of times in a section on the extent and causes of childlessness. As I have shown elsewhere,[39] the declining population of Germany is of great importance to lawmakers, and it is little wonder that the only significant mention of East Germany should be made in this context. The report states that the number of couples who are *unable* to conceive (*ungewollt kinderlos*) is below 10 percent in West Germany and below 5 percent in East Germany.[40] The report also notes that the number of couples who are *unwilling* to conceive (*gewollt kinderlos*) is on the rise in both parts of Germany. It finds the causes of unwanted childlessness (and, by implication, of wanted childlessness as well) in different social structures, which in turn enable different approaches to life planning.

It was customary in East Germany for women to combine motherhood and professional life, but this was not the case in West Germany. When West German women in their twenties and thirties chose to pursue professional achievement, the probability that they would conceive fell drastically. Consequently, the percentage of West German women between thirty and thirty-nine who were childless was approximately triple that of women of the same age in East Germany. The proportion of childless women in the West German states ranged from 22 to 28 percent, and in the West German city-states of Hamburg, Bremen, and West Berlin it ranged from 32 to 39 percent. In East German states the proportion of childless women ranged from 7 to 9 percent, and in

East Berlin it was 13 percent.⁴¹ The report does not offer causal explanations, but a further statistic offers a clue. While the percentages of childless women with no completed professional education in East and West Germany were similar—between 10 and 15 percent—the differences became more pronounced with increasing levels of higher education. Thirty-seven percent of West German women with university degrees were childless, but only 8 percent of East German women who had completed university had no children. The Enquete Kommission cites a report of the Bundeszentrale für gesundheitliche Aufklärung (Central Federal Office for Health Education), which states that marriage and children have lost their former significance among women with high levels of education.

While the EK report looks to East Germany's higher birth rates with a longing eye, the differences in abortion policies between the two states are ignored completely. Here it is as though East Germany had never existed, and as though the differences were never significant. But what were the differences that were erased so completely in these official documents? How did abortion policy and practice differ in the two Germanys? And what might these selective removals from the field of vision tell us about the ways in which the reconstituted state wants to be seen, and wants others to see it?

In Germany, Paragraph 218 of the Penal Code (Strafgesetzbuch) regulates abortions. The paragraph entered the Reichsstrafgesetzbuch in 1871, when Bismarck formally unified the numerous independent states into a German Reich with a common law. The law criminalized abortion: both the abortion provider and the woman undergoing an abortion could be punished with prison terms of up to five years. Under National Socialism, abortion policy became a two-pronged tool of population politics. While legislation strictly prohibited abortions and even introduced the death penalty for "perpetrators who continue to inhibit the life force of the German *Volk*," it simultaneously forced the sterilization of "inferior" women and mandated mass abortions among eastern Europeans. After the war the law was relaxed to permit abortions for German women who had been raped by occupying soldiers. Whereas West Germany soon readopted the old Paragraph 218 (without the Nazi-era death penalty), the eastern part of Germany permitted abortions for medical reasons (i.e., if the pregnancy endangered the health of the mother) and for eugenic reasons (i.e., if the fetus would be born with defects).

In 1972 East Germany adopted the *Fristenlösung*, according to which a woman may abort within twelve weeks of conception.⁴² East Germans regarded the law as encoding a commitment to the responsibility and autonomy of women as equal participants in society. Abortions took place in state hospitals and were free of charge for the woman. Following the 1972 decriminalization, the number of abortions in East Germany fluctuated, but there was a steady decline, with the lowest numbers recorded for 1989. East German physicians claim that steadily improving living conditions and social support for single mothers were largely responsible for their decisions not to abort.[43]

In 1974 the West German Bundestag also adopted the *Fristenlösung*, but the conservative Bundestag parties CDU and CSU, along with the five states they governed, appealed to the Federal Constitutional Court for a judgment on the law's constitutionality. The court in 1975 declared the *Fristenlösung* unconstitutional on the grounds that unconstrained abortions are inconsistent with the unborn person's basic right to life. In 1976 the *Indikationsregelung*, according to which abortions were unpunishable if one of four conditions was met, went into effect in West Germany.[44] The law also mandated counseling in many cases.

In the process of reunification, these different legal solutions collided and sparked heated discussion. East German women were not willing to trade their liberal *Fristenlösung* for West German law. Although conservative West Germans tried to write the *Indikationslösung* with mandatory counseling into the Unification Treaty of 1990, East German resistance won out for the moment; women in the East could continue to have abortions without satisfying the indications that were required in the West. The treaty specified that East German abortion law would continue to apply in the former East German territory, but that the legislature would eventually implement a single legal solution, applicable to all of Germany, by the end of 1992.

The compromise law that the Bundestag passed in 1992 contained a *Fristenlösung*, coupled with mandatory counseling. The goal of the new law was to reduce the number of abortions by encouraging women to make an autonomous decision in favor of pregnancy. This is the explicit goal of mandatory counseling. The central aim of Paragraph 218 survives in the counseling provision, which (as described in chapter 4) requires women to justify to the state their noncompliance with the state's interest in reproduction. The current abortion law, which the Bundestag enacted in 1995, protects unborn life more strongly, but recognizes at the same time that a pregnancy cannot be enforced against the will of the mother, but only in collaboration with her. Moreover, the state no longer covers the costs associated with the abortion.

The new regulations have affected East German women the most.

They now need to get formal permission before they can get an abortion. While abortions in East Germany took place in state hospitals, getting an abortion in a hospital now has become nearly unaffordable for East Germans, given their region's high unemployment rates and stagnant economy. West German legislators, when writing the law for West Germany, had assumed that local physicians would perform abortions, but there are too few of those in the former East to guarantee easy access. One unintended consequence is that East German women increasingly resort to sterilization. According to one source, the number of sterilizations in the East German town of Güstrow in the first quarter of 1993 nearly equaled the number of births. While mandatory health insurance does not pay for abortion or for contraceptives, it does pay for sterilization. In this way, reunification inadvertently introduced new social structures and class distinctions, to the disadvantage of East Germans and their social achievements.

East Germany in the Enquete Kommission *Recht und Ethik*

An observer of one of the early meetings of the EK in 2000 would have witnessed a strange scene: at each member's place on the table was a sealed envelope containing anonymous accusations against one of the EK's expert members, Ernst Luther, and on those grounds asking for his removal from the commission. The letter was addressed to the president of the Bundestag, who in turn had passed it on to the commission for action. The accused knew neither the source of the allegations nor their precise content, but from the context we may speculate that they concerned his past East German political affiliations. It was unclear whether the allegations had come up in the routine security check conducted on all members of parliament and commission experts, or whether someone had gone out of his way to accuse Luther from outside. Without denying the accusations—for how could he deny what was left unspoken?—Luther showed that he resented the manner in which they were brought against him. It was a scene out of Franz Kafka: a totalitarian regime keeping its subjects in line by secretly and anonymously spreading rumors about them, a state whose purposes were invisible but whose effects were felt all the more pervasively. Only this time the setting was the reunified, democratic German state, and the accused was a former victim of totalitarianism.

As it turned out, the EK debated the allegations and, after hearing speakers from each of the four political parties, decided against expelling Luther, its most senior member. But what might his crime have been? Certainly his life in those days could not have been more innocently domestic. I interviewed the then 70-year-old professor in his Halle home in 2002. He spoke with pride of his five children and his ten grandchildren. His wife interrupted our conversation several times to coordinate their evening plans, and she made me a sandwich and gave me a bottle of the local *Schwarzbier* to have for dinner on the train back to Berlin.

His professional path, as I learned from our conversation, had also been littered with accolades and accomplishments, and none seemed in any sense politically questionable. Luther had written a doctoral dissertation on the medical anthropology of Victor von Weizsäcker in a philosophy department in 1961, and he wrote his habilitation (the second dissertation usually required for a professorship) on the historical development of the physician's code of ethics. Soon thereafter, in the 1970s, he turned to the fields of medical ethics and bioethics that were then emerging, and he made professional contact with leaders of the field at the Hastings Center in the United States, in the United Kingdom, and in Moscow. In the 1980s he was appointed to the first chair in medical ethics in Germany at the University of Halle. Throughout his years of professional service, some of his primary concerns had been the relationships between individual health and social responsibility and between citizens and the state in the field of health care.

Although retired from the Institute for History and Ethics of Medicine (the mandatory retirement age for academics is 65 in Germany), Luther was still active in his field. He told me that he continued to give talks and lectures in an educational institution in nearby Erfurt and was still publishing articles and reviews in various places. In addition, he had expanded his focus beyond the confines of professional medical ethics and was now running a monthly discussion forum, founded by his wife, that was designed to bring philosophical concerns to a group of local seniors who faithfully attended the events.

In reflecting on the attack on Luther, I was reminded of the times when his contributions to the collective text of the EK reports were discarded, especially when he tried to contrast current German law with alternative solutions that had existed in East Germany. One memorable scene I witnessed occurred when Luther had written several paragraphs outlining the East German abortion law, which had differed substantially from its West German counterpart, and wanted to present it as a viable alternative to the scheme that existed in reunified Germany. Two commission members from the former West Germany had called his contribution "nonsense" and dismissed it summarily.

The rejection and exclusion of East German academics, ethicists, and scientists was often more costly in personal terms than these episodes had been for Luther, and many of my East German interviewees had stories to tell about their former colleagues' professional and personal demise. Luther himself told me that friends and acquaintances of his were treated so badly that some said "if one doesn't want me, then . . ." Luther did not finish the sentence, but he did not leave room for the words to sink in either. He continued:

> Unfortunately, I am in the same situation. From my file you will see that Herr X, who is now the director of this institute—which after ten years was finally renamed again to "History and Ethics of Medicine"—was brought to Halle by me. He held a talk, and was then called to fill the position . . . and I still have the card, from 1991, wishing me a Happy New Year . . . but *as soon* as he was offered the job, he wanted to have nothing to do with me. . . . I was able to go on, but others could not handle [such disappointments and betrayals], and they left science altogether. . . . In short: before the *Wende* there was collegiality, and afterward there was competition. And competition must be eliminated.

Bioethics and the East German Public Sphere

How did reunification affect the discourse on bioethics more generally? To see this, let us first step back to the point when bioethics got its start in the two Germanys. In the 1970s, West Germany began to debate the potential benefits and dangers of genetic research. In the 1980s, and even more so in the 1990s, in response to growing international competition, scientific research became increasingly market-oriented,[45] creating a perceived gap between scientists' search for profitability and the public's demand for a science addressing social concerns. Basic science was protected by the constitutional guarantee of freedom of inquiry, but the Greens in particular argued for ethical principles to govern a science that they saw as increasingly instrumental and applied. The parliamentary Enquete Kommission on Opportunities and Risks of Gene Technology that I discussed in chapter 1 constituted a partial answer to these demands for a socially responsible science. The relationship between science and the public, moreover, was continually mediated by science journalists.

In East Germany, history unfolded differently. East Germany also set out to build a better society, with better people in it, but the East German program was overtly one of social engineering, designed to replace the capitalist system with a different, more equitable social order. Science was going to play a major role in improving lives and living conditions. East Germans were less concerned about certain negative outcomes of science because, for them, science was as good as the social system in which it was practiced. The dangers did not lie in science itself, or in its applications, but in science's alliance with the corruptions of capitalism. Socialism, in contrast, would free up the positive potential of science.

Orthodox Marxism-Leninism held that individuals were wholly shaped by society, and until the mid-1960s, the official Lysenkoism propagated by the Soviet Union denied the existence of genes altogether. When official dogma finally crumbled, a configuration emerged that was different from what existed in West Germany. The political authorities were delegitimized among scientists by their previous official denial of the existence of genes. Officials thus retreated as scientists, artists, and literary figures carved out a space in which interested nonscientists could communicate directly with geneticists about the biological, environmental, and social implications of genetic research.[46] In these cautious dialogues, all participants were equally welcome to contribute their perspectives on the social consequences of the new gene technologies.

These discussions became progressively more ingrown. After the war, many people moved from the East to the West. In 1961, after more than two million people had left, East Germany built a physical wall around itself, and it began to restrict scientific travel shortly thereafter. East German scientists were able to visit conferences in the West until the late 1960s, but inter-German contacts gradually decreased as East Germany secluded itself more completely. Now the professional societies became a kind of second home for many East German scientists. It was as if East German science as a whole became domesticated, removed from the internationally open public sphere. According to several scientists I interviewed, one result of this was that individuals tended to form closer interpersonal connections—with both scientists and nonscientists—than many of their Western counterparts. Transparency between science and the public grew as transparency between East and West diminished. Questions arising in the context of new developments in biomedicine could be discussed openly in professional gatherings. In some sense these meetings fashioned a collective future, if only in the participants' imaginations.

These widely watched discussions occurred in several forums. In

1970 the geneticist Erhard Geissler inaugurated the Kühlungsborn Kolloquia. He wanted to bring together scientists, doctors, philosophers, social scientists, writers, and artists to discuss biological weapons and the risks of molecular medicine. By the early 1970s East German scientists recognized that genetics raised social problems that geneticists were not equipped to solve alone. Persistent in the discussions were questions about the individual's autonomy and the dialectic between the individual and society.

In 1979 Geissler speculated that it was now possible in principle to conduct experiments on the human genome. He said that this recognition opened the door to thinking about whether one might improve human beings genetically. These eugenic speculations existed in dialogue with a widely read novel, *Krabat oder die Verwandlung der Welt (Krabat, or the Transformation of the World)*, by the East German author Jurij Brezan, which personifies the conflict between scientific rashness and cautionary voices through the ages. In an interview Brezan warned of technological hubris, and he wrote that he was afraid of biologists.

To sharpen my own perspective on these events, I spoke to Rainer Hohlfeld, a West German biologist turned social scientist working at the Berlin-Brandenburg Academy of Sciences. Throughout the 1970s and 1980s, Hohlfeld had observed the developments of East German science through East German publications, and he had authored numerous scholarly works on the different social structures of biology in the two Germanys. He told me that in East Germany there were futuristic visions like those of Lee Silver in the United States,[47] but there was always strong opposition to those visions as well. Opponents used the argument of human dignity even though, in contrast to West Germany, human dignity was not anchored in the East German constitution.

The discussions that these two public figures, Geissler and Brezan, began in 1979 continued into the 1980s in *Sinn und Form*, the East German literary and cultural magazine. Because East Germany had no public press in which journalists could comment on the topics of the day, literary authors took over this role, as they did in Czechoslovakia, Poland, and the Soviet Union. Those authors occupied positions from which they could address new developments in the sciences critically. Hohlfeld told me that in East Germany, authors, and artists more generally, had a very special function: they in effect had the task of coauthoring a vision of what sort of society East Germany was going to be. The state, of course, hoped that those authors would amplify the official socialist ideology, and those who strayed too far from it were censured. But there was a large gray area in which authors could voice their concerns, and in these cases they became legitimate and valued participants in debates about new developments in the life sciences.

The *Gaterslebener Begegnungen*, or "Gatersleben Encounters," grew out of the *Sinn und Form* discussions. According to Anna Wobus, a prominent biologist who organized the Gatersleben Encounters and the successor Gatersleben Talks, the idea in moving the meetings to Gatersleben was that the concerns that were in the air should be discussed with a larger public. In 1986 the Institut für Pflanzengenetik und Kulturpflanzenforschung (IPK), a member institution of the East German Academy of Science, where both Wobus and her husband worked, invited the authors who had participated in the *Sinn und Form* debate to join them there. Among the early participants were mainly literary authors (Christa Wolf had a leading role), while authors like Richard Pietrass and Manfred Voltra joined later. Eventually some Marxist philosophers were asked to join, Wobus told me, so as not to raise suspicion with the authorities. As a result, the discussions were often inflected by the philosophy of Marxism-Leninism and its belief that scientific progress can bring about a better world. Here too members of the public were an integral part of the conversations from the beginning. In East Germany these bioethical forums effectively constituted what we might call a public sphere, even if this sphere was tightly circumscribed in its focus on the implications of scientific advances—what in other nations went by the name of bioethics.

The Gatersleben Encounters always took place in May or June, benign summer months in northern Germany, and in the evenings of the first and second days there would be readings by authors and outdoor displays by artists, graphic designers, and sculptors. On the afternoon of the second day there would be a round table. The meetings typically began with authors who posed questions to the natural scientists, often by reading a piece of prepared prose. Then the scientists would give scientific presentations on those topics. Next there would be reflections on the ethics of the respective positions. When I asked Wobus which ethical aspects were discussed at the meetings, she said that the central questions were those of risk and of a humane and responsible society. Wobus told me that back then, people had more or less the same concerns they still have today: they worried about environmental pollution, especially in the wake of the Chernobyl disaster in 1986; they were concerned that gene technologies would erode individuals' right not to know; and they were worried about the looming prospect of reproductive cloning and the potential selection of certain traits over others. The perennial question underlying the *Gaterslebener Begegnungen* was this: How are we, as a society, going to deal with the problems of science and medicine? The main problem that needed to be addressed was that too many people were not part of making these decisions that affect everyone. It was particularly important, therefore, that Gatersleben remained a place of face-to-face encounters, with approximately sixty nonscientist participants from all over Germany.

Every meeting in Kühlungsborn and in Gatersleben was turned into a conference volume. These volumes were very official and were carefully edited to get past the official censors, Hohlfeld told me, but the arguments become visible in the footnotes, where the controversies are to be found. What was at stake is, interestingly, not in the text itself, but in the subtext.

Some East German scientists, like the geneticist Erhard Geissler, were critical of authors who got mixed up in debates that they thought ought to be internal to science, and they voiced their concerns at the meetings. Wobus quoted Geissler, the founder who later worked at the Max Delbrück Center for Molecular Medicine, as saying that "authors ought to do their *Abitur* before they speak out." But, Wobus insisted, of course an author can make his or her unease public (*sein Unbehagen äussern*). That is why the aim of the meetings from the beginning was not to teach, but to inform. The initiators wanted to bring the viewpoint of natural science closer to the authors and artists, but they wanted to accept the perspectives of the authors and artists on developments in the biosciences as having equal standing (*gleichberechtigt*).

Wobus claimed that the meetings were under intense surveillance by Stasi spies. When I asked her how she knew, she said that in a small place like Gatersleben, everyone knew who worked for the Stasi. For this reason it was imperative that one have a few Marxist philosophers in the audience; otherwise the authorities might have cancelled the entire event. In 1988, Wobus remembered, the Stasi people sat right behind her. She felt very anxious, she said, but also somewhat secure, since Gorbachev, who had come to power in 1985, had just recently loosened the hold of socialist governments on their populations.

Official authorities generally did not interfere with what got said at Kühlungsborn or Gatersleben, but rather observed and even mediated the events. One insider I spoke to said that no one interfered because these talks were like "children playing in the sandbox," meaning that nothing significant could come of them. East Germany could afford to wait and see how opinions developed because bioethics was insignificant for the government. This evaluation is reminiscent of how James

Watson, the head of the U.S. National Institutes of Health's Human Genome Project, regarded that project's ethical, legal, and social issues (ELSI) program. The ELSI program, which originally received 3 percent of the genome project budget, was to be, Watson hinted, a sideshow to the *real* science.

But Wobus clearly saw things differently. She was proud of having really listened to and understood the public's concerns about scientific developments. She and the other scientists had tried to take the public's concerns seriously and, as scientists, had tried to answer them honestly. She told me that she did not know whether they had succeeded.

When I asked Wobus what the Gatersleben Encounters had accomplished, she conceded that they had not changed the world. But, she said, they had succeeded in raising the questions of the critical public, of the social sciences, and of the artists. The meetings had had overwhelming resonance with authors who believed in responsible science. To them, and to others as well, the personal exchange had been paramount. One might say that the meetings put a human face on the conflict between science and society. The other ever-present benefit, in Wobus's view, was the sensitization of the natural scientists. She said, "Problems were articulated where we had not seen any, but where the public had concerns."

Many of the East Germans I talked to about my research seemed explicitly concerned with the notions of the public and of the public sphere. What kind of entity had East Germany been? Was it a state? Was there a public sphere? What happened to it? According to Hohlfeld, the idea of a "victor's justice" cannot be sustained in the case of science, much less in the case of bioethics. Although there had been no separation of powers, and although the East German state had not permitted a public (sphere) to exist, Hohlfeld insisted that there *was* a public sphere in the case of bioethics. This East German public sphere was not to be found in the newspapers, which were all controlled and censored by the state, but in public events where opinions clashed in representative ways. In East Germany one had a different concept of the state and a very different kind of public sphere. The Kühlungsborn Kolloquia and the Gatersleben Encounters exemplified what Hohlfeld called a "*stellvertretende Öffentlichkeit*" (alternative public sphere).

When the Wall fell in 1989, there was an initial period of optimism in East as well as West Germany. Both sides assumed that many institutions in East Germany would have to be reformed, but not radically redefined or eliminated. Many East Germans hoped for a confederation, in which the social models of both Germanys would find room for expression and possibly collaboration. In many universities and research institutes, East Germans themselves began the work of reform.[48]

Decades of state control, however, had left East Germans insecure in the face of the certainties that West Germans brought to the task of reunification. The speed of the process quickly disabled all prospects for autonomy, or for any alternatives, and most East German scientific institutions were quickly dismantled. In some cases East German institutions developed proposals for their continued existence, and even received positive evaluations from international science teams, but the final decisions made by the German Science Council declined most of the applications. In its five-thousand-page report, the council recommended cutting about two-thirds of the East German research positions.[49] The notion of the East as a supposed *Wissenschaftswüste*, or scientific desert, remained latent throughout the evaluation process. It was only in 2004 that officials openly acknowledged that East German scientists had been dismissed too quickly. A study commissioned by the Science Administration of the Berlin Senate showed that the "integration" of East German scientists into the German science system had largely failed. In early 2004 Thomas Flierl, Berlin's official responsible for improving conditions for science (Wissenschaftssenator) and himself an East Berliner, planned a formal ceremony at the Rotes Rathaus, Berlin's city hall, to honor the dismissed East Berlin scientists for their lifetime contributions. Even this small gesture proved controversial, however, and several conservatives maintained that Flierl was trying to manufacture the appearance of injustice where none had occurred; some insisted that the dismissals had been an issue of quality, not of prejudice against East Germany.[50]

One West German evaluator summed up the process of reordering East German science when he said that "only the re-integration of East German scientists into the international system of scientific production . . . will enable their reintegration into *the one* world of science."[51] Statements like these demonstrate the structural impossibility of taking the work of East German scientists seriously. In fact, the Unification Treaty did not plan for the structural reform of East German institutions, but only for the "adaptation of science and research to the common research structure of the Federal Republic."[52] It did not help that the evaluating committees were overwhelmingly composed of West Germans. The one world of science was going to be the international world in which West Germans played a defining role.

Hohlfeld told me, with some pride, that after the border opened, his observations turned out to be essentially correct. Leading East German scientific institutions, such as the IPK itself, generally adapted to West German conditions very rapidly and very competently. Within two years East German institutions were at the same level as their Western counterparts, much as Hohlfeld had expected. This alone should have disproved the idea of an East German *Wissenschaftswüste*.

The dismissal of East German scientists on the basis of their isolation and domestication through their close alliance with the state suggests another reason for the difficulty of incorporating them. Almost by definition, scientists working in a socialist state could not buy into the Mertonian norms of universalism, communalism, and disinterestedness.[53] Holders of the Mertonian ethos would be bound to reject as "not science" any knowledge produced through avowedly nonuniversalist means. Work by Jens Lachmund on West Berlin ecologists supports the idea that enclosure in Berlin was less of a problem than the East Berlin socialist ethos.[54] Lachmund shows that these ecologists developed a similarly structurally isolated position, but were nevertheless taken seriously as producers of robust knowledge.

With the end of East Germany, new pressure for competitiveness and short-term successes entered from the West, and it affected both scientific practice and medical care. The standards by which scientific work was measured were no longer industrial applications, nor even the education of young scientists, but publications in high-impact journals and the associated need to be first in the race for scientific discovery. Perhaps the greatest losses in this transition to a Western research model were the possibilities of cooperation within science. One glaring example was the lost connections to Eastern European scientists; another was the sacrifice of East-West connections that had been built during this era.

A similar fate befell the Gatersleben Encounters. Although Wobus told me that the number of attendees doubled once the Wall fell, Hohlfeld said that the number in fact remained constant, and even decreased.[55] He told me that the meetings had been the defining events for the IPK, and that the organizer tried to portray them as a model of successful post-Wall integration, when in fact they represented the subordination of a logic of solidarity to the more forceful logic of capitalism. According to Hohlfeld, the audience's composition changed as well. Before the Wall fell, the meetings had been genuinely critical encounters between artists and researchers: debates were open and wide-ranging, and opposing positions were taken seriously. After the fall of the Wall, the meetings became events where scientists justified their work without much of an ear for reciprocal public concerns—in line with the prevalent Western construction of the public as scientifically illiterate and needing better "understanding of science." The artists who had fulfilled such an important role as co-producers of socially robust knowledge in East Germany were no longer welcomed as equal voices in bioethical debates. The public sphere around science, he said, was constricted when the Wall fell. The values of science became less transparent. Ethical relations between science and society were framed in narrower terms, and art diminished in significance as a critical voice from the people. In some sense the IPK adapted to the West after the fall of the Wall. The scientists agreed to dissolve the forum in 2003.

Coda—A Very Private Place

But where did the East German public sphere disappear to? What answer do I take away to my original question: What happened to East Germany and to the public moral order that bound its citizens to their state? Over the past few years, memories of East Germany have been normalized to some degree. Several prominent films have portrayed life in East Germany from the perspective of East Germans. The recent cultural outpouring from East German authors and filmmakers suggests that as the public story of East Germany got written out of German history, its achievements and yearnings were transposed to the private sphere. In public, there could be room for only one vision of law and ethics; in private, many colorful stories of solidarity continued to coexist. In private, one can still express some of the hopes that public life no longer sustains.

There is, for example, the film *Sonnenallee*, which is named after a street that linked the West Berlin district of Neukölln and the East Berlin district of Treptow. Under Hitler the street was named Braunauer Strasse, after Hitler's birthplace, Braunau am Inn, but after the war it reacquired its original name. The Wall divided the street, and its shorter end—in German, *das kürzere Ende*, a metaphor for the shorter end of the stick—was located in the East. In an early scene an East Berlin housewife and mother finds a West German passport dropped by a day visitor, an older woman. The attractive, family-loving East German woman spends the rest of the day trying to make herself look as old, bitter, and unattractive as the West German woman in the passport photograph. After a long time she succeeds and goes to the border guards. At the last moment, with the guard demanding to see her papers before the clock strikes and the gate closes, she opts for her family and rushes back home. Much of the film consists of the coming-of-age story of several teenagers and a young man's first love. The young man watches helplessly as the girl he half-openly admires becomes attracted to another young man who regularly visits from West Berlin, drives a large car, knows the latest pop albums, and in general seems much more worldly and suave than the poor East Berlin youths. But it turns out that the West Berliner drives large cars because he parks them for a fancy West Berlin hotel, and that he is more subservient to his boss's orders than the East Berliners are to their authorities. The film seems to be saying to West Germans: No matter what material advantages you have, we were not inferior in our social lives, and we were also having fun. Needless to say, the East German eventually gets the girl.

Better known outside Germany is the film *Goodbye Lenin!*, which has won numerous international awards. This 2002 film, directed by Wolfgang Becker, documents the life of an East Berlin family through the year of the *Wende*. The film begins with a retrospective that shows a happy family in 1978. Two children watch Sigmund Jähn, the first German cosmonaut, on his way into outer space. In the background the Stasi informs the mother that her husband has not returned from an official trip to West Berlin. She sees her world falling apart, and her son Alex, who narrates the film, tells us that then, to overcome her pain, as a good communist, she married the socialist fatherland.

When the mother, on her way to attend the official celebrations of East Germany's fortieth anniversary, sees her idealized state using force against her demonstrating son, she has a heart attack and falls into a coma. When she wakes up again after eight months, she has noticed nothing of the "victory" of the West over the East and the subsequent invasion of capitalism into her socialist world. Alex, warned to spare his bedridden mother any excitement, resurrects East Germany in their small apartment. As Alex searches all over Berlin for the items that made up an East Berlin household, we see just how radical the transformation has been—many everyday items can be found only in dumpsters or in abandoned East Berlin apartments. The East German ethos has shifted. Those who are not young enough or mobile enough to find employment in the West are now unemployed. The former "Heroes of Labor" (*Helden der Arbeit*) have become aggressively self-interested or alcoholics, and their former solidarity is gone.

Although Alex goes to great lengths to protect the idealized version of the state his mother clings to, she manages to go outside, where she sees for herself the profound transformations in her neighborhood: West German cars and stores have displaced what was there before. She watches the statue of Lenin being carried away through the street, and as her eyes meet the statue's, she seems to say goodbye to the ideals she once held dear. In a television show he tapes for his mother, Alex "explains" that Erich Honecker has opened the Wall to let in the West Germans desperately trying to escape from unemployment, extreme self-interestedness, and the rise of right-wing ideologies at home. In the process he designs an idealized East Germany, the way he might have wished it to have been: a generous society that values tolerance, solidarity, and peaceful coexistence. Throughout, Alex molds his representation of society to suit the desires of the person who needs him most. In one scene we see him sitting in front of a sign that says "The Human Being Is at the Center of Socialist Society."

The mother seems willing to believe that the fantasies she has dreamed have come true, and she makes a confession to her children. Their father did not abandon the family for another woman: the couple had in fact planned the escape together, but when it was the mother's turn to apply for an exit visa to follow with the children, her courage had failed her. She could not stand the thought that her children might be taken from her. To spare her children the painful truth, she had hidden the dozens of letters that their father had written over the next three years, before he remarried. In her agony the mother had then denied her feelings and turned to the work of improving the state by participating in social functions meant to further solidarity. Like Alex, she had gone about crafting a fiction for her loved ones at home.

After his mother's confession, both Alex's sister and his girlfriend urge him to reveal his deceptions to her as well. Alex never does. Instead, he tries to make fact and fiction converge by filming a third and final news show in which he makes Honecker step down and cosmonaut Sigmund Jähn replace him. On the show, Alex's childhood idol is made to say:

> Socialism means taking a step toward the other, living with the other; not merely to dream of a better world, but to make it reality. Many are searching for an alternative to the struggle for survival within the capitalist system. Not everyone wants to participate in a world in which career and consumption come first. Not everyone is made for the elbow mentality. Those people want a different life.

While Alex is away at one point, his girlfriend reveals the truth to his mother. The lie at the center of the family eventually comes to light for the mother; but Alex believes until the end that the world he created for his mother has remained intact. It is as if, for him, the truth of human connection wins out in private, interpersonal communication, even if this means that public reality, mediated through television, can remain mired in lies. Alex's last words in the film complete the transition of East Germany to a site of private memory: "East Germany was a country that never existed in this way in reality. It is a country that in my memory will always be connected with my mother."

6 Stem Cells, Interrupted

Ethical Imports at Last

On December 23, 2002, while walking home late in the evening, I saw newspaper sellers on the street offering the next morning's edition of the Berlin daily *Tagesspiegel*. Tomorrow's news was here today. The main headline, central, bold, and above the fold, announced that human embryonic stem cells could now be imported into Germany. A few days earlier, without announcing its decision to the public, the newly appointed Central Ethics Commission for Stem Cell Research (Zentrale Ethikkommission für Stammzellforschung, or ZES) had approved the application of the Bonn neurologist Oliver Brüstle to do so. Brüstle was quoted as being relieved that after two years of waiting, he would now be able to import the embryonic stem cells that the Haifa gynecologist Joseph Itskovitz-Eldor had cultivated. Brüstle would receive three fingernail-sized containers with different cell lines in them. He hoped to grow these cells in culture and use them to develop therapies for degenerative neurological disorders like Parkinson's or multiple sclerosis. He added that the long wait had seriously dampened the enthusiasm for the project in his Institute for Neuropathology in Bonn.[1]

CHAPTER SIX

After a year and a half of studying the ethical debates in parliament and in the public sphere over just this importation, I received the long-anticipated news almost like a Christmas present. I was no longer the anthropologist of a virtual object, no longer "waiting for stem cells." With the arrival of those elusive and dangerous bearers of pluripotency, I would now have "real" physical entities to trace through German culture. It remained uncertain, of course, whether these potent cells would enter the country before I would leave it again several months later.

The tiny opening in the law that legislators had carefully crafted to allow a particular and strictly circumscribed kind of stem cell to pass through had finally yielded a result. The dam had been breached, and many feared a torrent that only the most carefully considered of laws would be able to manage. To avoid this outcome, the undifferentiated cells had been precisely classified and their promising potential carefully controlled.

Yet the torrent did not come. Parliament and the media had mobilized great resources and great emotion in orchestrating the stem cell debate, and a new ethics commission had been created with no other mandate than to receive and judge applications for stem cell research. By the summer of 2003, one year after the Stem Cell Law was implemented, the Central Ethics Commission and its affiliated agency, the Robert Koch Institute (RKI), had approved a total of three applications for importing stem cells: first, in December 2002, that of Oliver Brüstle at the University of Bonn; second, in January 2003, that of Jürgen Hescheler at the University of Köln; and third, in March 2003, that of Wolfgang Franz at the Grosshadern Clinic, near Munich. Compared with earlier expectations, those three applications represented a mere trickle. Brüstle, who had been under police protection during the parliamentary debate in early 2002, and who had told his story to numerous reporters over the previous two years, twice refused to be interviewed by me. I did speak with the other two researchers, however.

In this chapter I investigate how the German Stem Cell Law (*Stammzellgesetz*), which was passed with significant input from the parliamentary ethics commission and, to a lesser extent, from the chancellor's ethics council, has been applied in practice. I begin by describing the purpose and structure of the law itself, and I show the paradoxes and ambiguities that are built into it. I then describe the ethics commission that was appointed to oversee the application process through which researchers receive permission to import human embryonic stem cell lines into Germany. I show how ambiguities in the law itself were erased in order to provide an orderly, and ethical, narrative of German research that would be credible to both Germans and others. Finally I analyze the experiences of two stem cell researchers in trying to follow this law, and I show how the law and its execution brought German stem cell research to a near halt.

"No Embryo Shall Die for German Research"

On January 30, 2002, the Bundestag passed a bill that prohibited the importation of stem cells into Germany as a matter of principle, but allowed for certain exceptions. The plenary debate was celebrated as a *"Sternstunde des Parlaments,"* or a moment in which parliament decisively affirmed its ordering capacity and its relevance in directing the course of scientific research. The debate stood as a monument to an ethical state, to responsible science, and to democratic decision making. Margot von Renesse, who was then head of the parliamentary Enquete Kommission on Law and Ethics in Modern Medicine (EK), was one of the bill's main authors. Renesse stated publicly and privately that "no embryo shall die for German research." The first paragraph of the Stem Cell Law formulates the law's purpose accordingly: "The goal is fundamentally [*grundsätzlich*] to prohibit the importation and use of embryonic stem cells; to prevent the derivation of such cells, or the production of embryos for their derivation, in Germany or on behalf of German researchers; and to determine the conditions under which the importation and use of these cells may be permitted for research as an exception."

These imperatives required three forms of implementation: First, one needed to make the process of determining which cells were eligible for importation completely transparent. Second, one needed to specify the ethical criteria that would warrant the exceptional permission to import stem cells in good conscience. And third, one needed to specify what precisely one meant by "German research" and "on behalf of German researchers." This last point was by no means trivial, given that so much research involves international collaborations and that German researchers themselves are highly mobile and active outside German borders. As I have shown in the preceding chapters, each of these concepts (*transparency*, *conscience*, and *Germany*) came with specific cultural trappings. Let us now look at how the Stem Cell Law encodes these three concepts and at the new paradoxes and ambiguities it expressed and generated.

Ethics Becomes Law

The Stem Cell Law is brief, and its contents are easily summarized. The early paragraphs specify that the law applies to the importation and use of embryonic stem cells, and they define the terms stem cells, embryonic stem cells, embryonic stem cell lines, embryo, and importation. The law is built around a central contradiction: it categorically prohibits the importation and use of embryonic stem cells, but then specifies exceptions to the prohibition and formal criteria with which to test for those exceptions.

The law lays out a two-step process by which two institutions divide responsibility for evaluating scientists' applications to import stem cell lines for research. Researchers must first submit their applications to a subordinate authority designated by the Department of Health. The Department of Health chose the Robert Koch Institute (RKI), the highest federal office for disease control and prevention, one of whose mandates is to prepare scientific studies in support of health policy decisions. The RKI ensures that the application is complete. To be complete, the application must demonstrate, first, that the stem cell lines were produced before January 1, 2002. Second, the source embryos must have been produced in the context of reproductive medicine, and it must be certain that they will not be implanted, but for reasons that have nothing to do with the embryos themselves. In other words, the embryos must not have been selected out because of a perceived flaw or defect. Indeed, no selection must have taken place.[2] Third, no monetary compensation must have been given or offered in exchange for the embryos. Further, the import may be prohibited if the derivation of the cells occurred in obvious violation of the fundamental principles of German legal order. The fact that the stem cells were derived from human embryos, the law states, is not in itself reason enough to block their entry into Germany. Every import and, indeed, every use of stem cells requires separate permission. Each application must document the stem cells to be imported and used, or otherwise show that they are identical to those listed in an officially recognized and authorized registry.

Once the application is complete, the RKI must notify the applicant and at the same time forward the confidential application to the ZES for ethical evaluation. This formally independent commission, which is located at the RKI, is made up of five physicians and natural scientists as well as four philosophers and theologians (plus nine designated alternate members who have voting rights when "their" regular commission member is absent).

The nine voting members of the ZES then decide whether the application is ethically acceptable (*ethisch vertretbar*) according to three criteria that the law lays out. First of all, the research must serve goals of high priority in the area of basic research or in the extension of medical knowledge in the realms of diagnostic, preventive, or therapeutic procedures in humans. Second, the applicant must have explored the research question as far as possible using animal studies. Third, the research question must be answerable only through stem cell research. In other words, for stem cell research to count as ethical, it must be of high priority, qualified scientists must do it, and there must be no alternatives to using human embryonic stem cells.

From the day the application is complete and the ZES has submitted its evaluation, the RKI has two months to decide whether to permit the import. If the authority deviates from the commission's judgment, it needs to give its reasons in writing. If it fails to come to a decision in time, then the import counts as approved. Violations of the law are treated as criminal offenses. Those who import or use embryonic stem cells without proper permission from the RKI (or attempt to do so) can be punished with up to three years in prison and a fine. Those who make false claims on their applications can be fined up to 50,000 euros. These punishments are approximately equivalent to those for having an unauthorized abortion. Finally, the law mandates that the executive submit a report on its experiences with implementing the law to parliament every two years.

Converting Ethics into Reason

In July 2003, a year after the Stem Cell Law went into effect, I called two administrators at the Ministry of Education and Research (BMBF) who, one of my EK co-workers had told me, had been instrumental in writing the actual legal text. One of them was on vacation, but I reached the other, a lawyer. He appeared confused by my call and asked me several times to repeat my name and my reason for calling. He apparently did not receive many calls asking him to explain the mechanics of legislative drafting. His answers to my questions were brisk and brief, even curt, and he kept insisting that he could tell me nothing interesting. He clearly wanted to get the conversation over with as quickly as possible, but his mandate as a public servant prohibited him from simply cutting off a citizen in search of information about the workings of government.

When I asked about the rationale of the law, he told me that the logic was very simple: the goal was to permit the research that was prohibited by the Embryo Protection Law, but to do so without changing that law itself. In other words, the goal was to amend the Embryo Protection Law and thereby to liberalize it somewhat. It was to find the lowest common denominator and to avoid rejecting any research if possible. When I pointed out that the Stem Cell Law explicitly prohibits stem cell research, my interlocutor saw the dilemma: "One can certainly call this schizophrenia, or a double moral standard for research."

I asked him what role the BMBF had played in the process of turning the bill into law, and he told me that it was the same role they always played: that of manual craftsmen (*Handwerker*). Extending the metaphor, he added, "Our job is to make sure that whatever it is that politics wants to do is cast into reasonable [*vernünftige*] legal forms." Thus is public reason made—through the craftwork of the law. One of my interviewees, Margot von Renesse, later contradicted this administrator's modest claims to legal authorship and authorization by telling me that she herself—retired judge, head of the EK, and first author of the successful bill—had, together with three others, played the crucial part in writing the law.

I then asked the administrator whether, in his opinion, the 2002 stem cell bill had been translated completely into the final Stem Cell Law. He responded that the translation had not been 100 percent complete, but 90 percent. Asked what was missing, he explained that the calls for transparency had not been heeded entirely. The newly appointed ZES that the law mandated would be housed at the RKI, and therefore was not a public commission. As I later ascertained, the commission does in fact meet behind closed doors and does not make its inner workings public. While the process of deciding the architecture of the law was as transparent as lawmakers could make it, the law was written, and is implemented and executed, in comparative seclusion. Once again, transparency is about selective vision and about specific judgments regarding what should be seen and what cannot be shown. The official I spoke to excused this lapse, however, by reminding me that transparency would be ensured in two ways: every two years the ZES would produce a public report on its activities, and the executive branch of government would separately report on its experiences in implementing the law. Both reports would be due at the end of 2003. Apparently, the public display of reasoning was felt to be key, consistent with the principle of *Nachvollziehbarkeit*, but that display could take place in writing.

When I asked my lawyer-interlocutor what the BMBF was currently working on, he told me that it was preparing a law to regulate genetic testing. When I mentioned that no such bill had yet been voted on in parliament, he told me that such a law would be inevitable. He then referred to his workload and brought our ten-minute conversation to an end.

This phone call to the executive branch fitted with impressions conveyed by the NER. According to a friend who attended a summer school that the NER and its French counterpart sponsored in September 2004, bioethics was presented there almost as a way of finding loopholes or ambiguities in the law and exploiting them. According to my friend, the NER promoted an ethics of do-ability. In the interpretation of the Stem Cell Law, it seemed, the stylistic conflicts between the EK and the NER (described in chapter 1) were resolved in the NER's favor.

Reading the Law

The Stem Cell Law is the outcome of broad public reflection on, and objections to, particular forms of scientific research, so let us look at its practical effects on researchers. The law aims to make stem cell research transparent and conscientious and to mark that research as characteristically and properly German. We will see, however, that all three concepts—transparency, conscientiousness, and "German research"—are impossible to pin down or define precisely.

The law contains numerous paradoxes and ambiguities, many of which have been noted by scientists, lawyers, and philosophers. The definitions in its Paragraph 3, for example, are ambiguously phrased. What does it mean, some of my interviewees asked, to define as an embryo "every human totipotent cell that is capable of dividing and developing into an individual under the *conditions that are further necessary* for this process" (my emphasis)? As one scientist explained to me, under the right conditions, *any cell* can be made to develop into an individual. Another scientist told me that totipotency can only be shown empirically. It therefore makes little sense to define certain cells as totipotent, and off-limits for research, if one is not permitted to test them for their totipotency. Moreover, I once heard Jens Reich, a former East German scientist and almost-politician as well as an NER member, say at an NER hearing, "We should not make totipotency into a fetish. It is an operative concept that is meant to tell us to what extent something can develop into an organism. It's certainly not advisable to bring this into a juridical, definitional form." In other words, for scientists, totipotency is a condition that they attribute to cells through their analysis and investigation, not a condition of being or not being that inheres in cells, as the law implies. Reich also said that he found it problematic to hold the ethical discussion on the basis of the present state of biology. He said that one needed to base ethical discussions on something more durable. Reich seemed to imply that the attempt to tie the law to a rapidly progressing science would almost inevitably reproduce the perception of the "ethics lag" that I discussed in chapter 1.

The Cutoff Date—An Unenforceable Line

Paragraph 4 of the Stem Cell Law categorically prohibits the importation of human embryonic stem cells for research, but then lays out a formal procedure for importing them in exceptional cases, and in Section 2 institutes a cutoff date for the importation of stem cell lines. The idea of a cutoff date leans on George W. Bush's solution of providing U.S. federal funding for research on only those human embryonic stem cell lines that existed before his speech of August 9, 2001.[3] Whereas the American policy pertains only to federally funded research, the innovation in Germany was to turn Bush's directive into a law that is applicable to all stem cell research, regardless of the source of funding. When I asked Ulrike Riedel, a lawyer who formerly worked for the Department of Health and was a member of the EK I worked at, by what right one could prohibit the importation of something that was not the product of criminal activity, she agreed that there was a contradiction. After puzzling for a while, she tried to remember what she knew of the law's genesis. Eventually she told me that stem cell imports carried the danger that foreign researchers would kill more embryos in the future. As she rephrased this statement several times in our conversation, each iteration strengthened the causal connection, until she stated matter-of-factly that "the importation . . . practically *causes* the killing of embryos." She went on to say that even the indirect endangerment of a legal good (in this case, the embryos' human dignity and right to life) may prompt a legislative response. And where fundamental principles are concerned, the state may act even when no individual's rights are in danger. She elaborated that the killing of embryos elsewhere in the world might, "in the minds of people, and of society," turn the embryo into an instrumentalizable thing, with which one can then do as one pleases. In order to prevent this from happening, she argued, one may restrict such basic freedoms as that of research. When I reminded her of a complaint scientists frequently made, that the already existing cell lines were in all probability infected with animal viruses and therefore unusable for creating therapies, she told me that stem cell research would not make progress toward therapeutic applications anyway.

The cutoff date, she further told me, made the lawyers sitting on the ZES superfluous. Since the criteria for ethical research (the strictest of which is the cutoff date) are defined so clearly and merely need to be applied to the research proposals, it almost does not matter who sits on the commission. This view was consistent with Anna Wobus's remark to me that her current work on the ZES was very different from the ethical deliberations she had organized at Gatersleben. In this new context, she was not eliciting ethical concerns from the public, but merely enforcing principles that the law had already set down. As we will see, however, scientists themselves assumed that the commission consisted of lawyers, since those were the people the scientists had the most contact with when they submitted their proposals to the RKI. In fact, one stem cell researcher I interviewed had been preoccupied mainly with the legal questions of importation during the application process. The RKI lawyers can prolong the decision-making process until they consider the application complete, and an insider told me that the ethics commission uses the RKI lawyers as intermediaries to help delay approval when scientific aspects of the application are unclear to it.

Not everyone saw the value of the cutoff date with equal clarity. Though precise in wording, it may be unenforceable in practice. During a March 2002 joint session of the Gesundheitsausschuss and the Forschungsausschuss (the Bundestag subcommittees on Health and on Research, respectively, that formally authored the Stem Cell Law), one committee member asked an invited expert whether it could be decisively proved that a particular stem cell (not a cell line) was produced before the cutoff date and received an affirmative answer. My EK coworkers, who were also present, expressed doubts: "She can't seriously mean that!" Other scientists of my acquaintance have suggested that some countries might produce new stem cell lines and then falsely claim that they existed before a given date so that they could profit from exporting these "unethical" lines to wealthy researchers. Others have said that having a deadline in a law is a bad idea anyway because it can be altered all too easily.

Prohibited yet Permitted

One scientist told me in an interview that he found the idea of institutionalized exceptions to a moral prohibition intolerable. The former German minister of health, Andrea Fischer, said in a public forum that despite such contradictions, the Stem Cell Law would not be revisited because a working consensus had been achieved to support it. Laws are generally left unchanged even when contradictions are glaringly obvious to outsiders, and even when they are in conflict with subsequent legislation. The contradictions themselves are seen as hard-earned compromises whose value lies in the inclusion of as many voices as possible. The cultural value of consistency is trumped in practice by the achievement of a consensus. Not surprisingly, then, later attempts by the Berlin Senator for Science, Brigitte Zypries, to reopen the issue and adapt the law to changing scientific knowledge for a long time came to nothing. Zypries claimed that the Stem Cell Law had become outdated and needed to be liberalized to ensure the continuity of promising research, but opponents of that research brought the discussion to a halt. It was only in early 2008 that parliament voted to move the original 2002 cutoff date to May 1, 2007, thereby increasing the number of stem cell lines available to German researchers from just over 20 to about 500. Critics referred to this move as turning the law into an "ethical sand dune" (*ethische Wanderdüne*)—a dubious structure that lacks a firm foundation and shifts with the winds. Once a cutoff date becomes alterable for the expediency of research, their argument went, then the entire purpose of the law—preventing German scientists from destroying embryos for research—is undermined.

Ethical German *Research*

No embryo shall die for German research—but what is German research? What does "German" mean in a field in which international collaborations are often vital for scientific experiments? Is German-ness tied to the nationality of the researcher, or to the territory where the research physically takes place? It is not difficult to imagine a German researcher who picks up the phone in Germany and asks for stem cells for which embryos may be illicitly destroyed elsewhere. Or a German who is part of a multinational research team and uses collectively gathered data from research on embryos, not all of which was gathered according to German ethical standards. What happens when German researchers are board members of multinational corporations conducting stem cell research through foreign subsidiaries? What if German researchers are asked to review stem cell research proposals by colleagues in other countries? What if federally funded researchers spend time in foreign laboratories where research takes place that German law prohibits? Despite the categorical language of the law, it is not clear in principle where the boundaries around "German research" will be drawn. Only through actual practices in the lab and elsewhere will those boundaries become clear.

Shortly after the law went into effect, the Deutsche Forschungsgemeinschaft (DFG) commissioned two legal expert statements (*Rechtsgutachten*) to interpret the Stem Cell Law with respect to the meaning of "German research."[4] One concluded that Germans may conduct research on embryonic stem cells outside of German borders. Thus German researchers evidently are allowed to be potent outside Germany, but not to be potentiated by stem cells within their own country. Unless, that is, they are *Beamte*—civil servants committed to upholding the state—a category that in Germany includes almost all university professors. *Beamte* are bound by national law even when outside that law's formal jurisdiction. In other words, Germans may go and do "unethical research" elsewhere as private individuals, but not as public employees. Those who want to dissent from the German consensus on stem cell research are permitted to become un-German upon leaving the country. Only when they directly represent, or embody, the state are they required to carry the territorial law in their own persons, making it operative in other sovereign locations as well. But this, too, is a contested question. The second expert statement imagined many more possible scenarios and came to more differentiated evaluations, which in some cases contradicted the first statement.

One of the arguments for permitting stem cell research was to turn Germany into a "state-of-the-art location for science and economy" (*Wissenschaftsstandort*; *Wirtschaftsstandort*). In this respect too Germany was conceived as a unified territory marked by special attractions, including its standards of ethical research. Yet making Germany a place where science thrives creates new kinds of border crossers, and ethical cross-dressers too. In spring 2003 a conflict erupted in Europe when the research commissioner of the EU, Philippe Busquin, demanded that European funds be spent on research projects on human embryonic stem cells. Germany had prohibited that research at home, but was the largest single financial contributor to the EU, which meant that "German" money would be used to finance research that German scientists were forbidden to do. The issue was not resolved in Germany's favor: the Sixth Framework Program regulating EU research permitted research with spare embryos from IVF, which were not allowed to be created in Germany.

The ZES and the RKI Reconfigure Science and Ethics

Since the Stem Cell Law mandates that scientists will form the majority of the ZES, the commission that determines the ethical acceptability of stem cell imports, one may assume that the commission will be research-friendly. Indeed, by 2003 the commission had reviewed three research proposals and approved all of them. Of the three applicants, two had already imported human embryonic stem cells from the United States before parliament decided on the conditions under which such imports would be ethically permissible. In talking to them, I did not get the sense that they had consciously acted unethically, but rather that they had gone about their work within the legal framework as it had existed in their minds before the Stem Cell Law was passed. What struck me, however, was how they conceived the division of labor between science and ethics. Although both scientists told me that they valued having their work accepted and appreciated by society, both also downplayed their own expertise and agency in determining what counts as "ethical." For these actors, then, a sense of morality (i.e., conscience) was not in operation prior to practice, but was the result of an externalized process of reflection that the actors themselves did not need to be, or even feel, part of.

In September 2002 I called the RKI to clarify how the RKI and the ZES in fact divide between themselves the labor of deciding on applications to import stem cells. How were the bureaucratic competencies of the federal authority separated from the normative work of the ethics commission? The woman in charge of administrative affairs for the ZES told me that the RKI first checks to see whether the formal conditions of Paragraphs 4, 5, and 6 of the Stem Cell Law are fulfilled. If they are, the application counts as complete and is passed on to the ZES, which then checks to see whether the substantive criteria of Paragraph 5 are met. In other words, the RKI first ensures that the application fully addresses all the requirements regarding the lineage of the imported cells, and the ZES then judges whether the answers to the law's substantive demands justify underwriting the research as ethical.

This particular separation of form and substance is significant. It means that the ethical criteria, those subject to assessment by the ZES, are those of Paragraph 5. That paragraph, we recall, demands that the research aims be of high priority (*hochrangig*), that the scientists doing it be qualified, and that there be no alternatives to it (*alternativlos*). As one of the law's authors told me, these terms are empty categories to be filled in later. She said that she was aware that "high priority" and "absence of alternatives" are flexible concepts that can be given many meanings. The point, however, was to set the bar high enough so that not everyone would be able to import stem cells, but at the same time low enough for qualified researchers to continue their work.

Over the course of my ethnographic research, the ethical questions that had sparked so much debate (the embryo's moral status and the threat of a *Dammbruch*) were effectively reconfigured into two sets of formal criteria. In the pre-legislative debates, the public had been concerned that embryos might be killed for research and that research might soon expand to encompass higher forms of life. The law, however, translated those overriding moral questions into formal administrative criteria. At the same time, matters previously intrinsic to science and accessible mainly to the scientific community alone (whether the research is "good enough" to justify its constitutional protection against regulation), now became reframed as "ethical criteria." The split procedure that the RKI and the ZES follow implies that ethical acceptability is now equated with scientific merit. The ontological status of the embryo that had sparked so much debate became supplemented in part by scientific criteria so that, in effect, doing stem cell research ethically (once the importation criteria were met) came to mean doing stem cell research scientifically. The realm of science had opened up to allow the realm of ethics in, while the protection of human dignity had become a matter of bureaucratic interest for lawyers to check and cross off. What was meant to be a law about protecting human dignity became, in effect, a law for regulating the ethics of basic research—although, to be sure, research at the frontiers of defining human dignity. Seen this way, perhaps the job of ethics had been to take the public concerns and form from them a collective conscience that could be, and was, then written into law. Lawyers are thus the agents of collective conscience, expressed and solidified through the law.

As the lens of ethics shifted its focus from human dignity to the justification for research, other changes also took place. One significant shift occurred in the meaning of "high priority." During the time of public debate, scientists pressing for a liberal law had claimed that therapies would arrive quickly and almost automatically, but the law as enacted emphasizes that knowledge gains in "basic research," far removed from eventual therapies, are enough to demonstrate "high priority." A 2001 newspaper article portrayed stem cell researcher Jürgen Hescheler as wanting to "import stem cells to cure heart disease," implying that stem cells would soon be available in therapeutic form. By contrast, this same scientist's (confidential) research proposal to the ZES reads that the research "might potentially have medium- to long-term relevance for the expansion of medical knowledge in the development of diagnostic, preventive, or therapeutic applications in humans." In an interview Hescheler told me he believed that therapies would arrive eventually, but only in the distant future.

Another significant shift occurred in the meaning of "no alternative." During the public debates, scientists had emphasized that only human embryonic stem cells could do some jobs that adult or animal stem cells were unsuited for. The law, however, no longer mentions adult cells, and merely requires that research to have been successfully performed on animal cells. While public debates and media coverage had suggested that embryonic stem cell research alone would lead to the production of spare organs and new therapies, that rhetoric was quickly toned down, and the submitted research proposals look little different from any other proposals for basic research. While the public and consumers of mass media were left with the impression that Germany had successfully navigated its way to permitting particularly promising research that was also exceptionally ethical, science in the laboratory is progressing pretty much as it had before, with the exception that some researchers now have to get their proposals past a state-approved ethics commission.

The meanings of "high priority and "absence of alternatives" were not absolutely clear to the ZES members themselves, and an insider told me that the ZES and the NER were planning a joint conference to discuss the two concepts. I heard that at one point the ZES even considered visiting scientific laboratories, presumably to get a firsthand impression of how scientists experience and interpret the concepts.

Nor has the enactment of the law calmed all the doubts of legal professionals. The law professor and NER member Jochen Taupitz once said in a public forum that the Stem Cell Law was unconstitutional, since it retroactively gave legal protection to something that no longer existed. After all, one imports only stem cell lines, not the embryos whose destruction the law intended to prevent. Moreover, the cutoff date itself was enough of a limit to research, he said. Article 5 of the Basic Law specifically guarantees the freedom of scientific research. To require stem cell scientists to prove that their research goals were of "high priority" went beyond what the constitution permits legislators to enact.

Laws, we see, can incorporate all kinds of contradictions, as long as they have been opened up for debate at some point. After the debate they are black-boxed and become self-referential with respect to their legitimacy. So long as one can demonstrate that the proper procedure was followed, it is enough to establish that there once was a democratic decision. This subsequent black-boxing is what makes papers, protocols, and reports so important: they are the archives of transparency. The stages of the decision-making process are retained for posterity so that, in principle, anyone can access them again, even if no one (save the occasional social scientist) ever does this. If the procedure has been stamped as legitimate, then its outcome and further consequences must be legitimate as well. In this understanding of the law, it becomes almost irrelevant whether its implementation actually calls forth any public involvement or oversight.

Inside the ZES

The ZES is officially the place where the ethics of stem cell research are defined and deliberated. To better understand the institutional location and the inner workings of the ZES, I called Anna Wobus, one of the organizers of the Gatersleben Encounters and a member of the ZES. From her Gatersleben lab, Wobus said that she could give me information on the three approved stem cell imports, but (in a move reminiscent of what the experts at the Dresden conference said to the citizens) that I might just as well read about them on the Internet. As a *mündiger* citizen, I could simply look up what I wanted to know. She could not talk to me about the applications then under review because those, like the commission's deliberations on them, were still strictly confidential.

Wobus suggested that I might learn more about the ZES's functioning from the head of the commission, Ludwig Siep, a professor of philosophy at the University of Münster. I followed her suggestion and went to Münster, a quiet university town just over three hours from Berlin by train. Siep was a healthy-looking, tall, tanned, white-haired man with a ready smile. His office was spacious, with very tall ceilings, large windows, and an enormous desk with some shelves behind it. When the six o'clock church bells interrupted our conversation, it was easy to imagine those evening bells harmonizing the daily rhythm of this small university community, distracting it from its worldly worries for a few minutes.

Siep was appointed to the ZES, he said, because he had served on numerous other ethics commissions. After writing his first and second dissertations in philosophy on the German idealists Fichte and Hegel, Siep was drawn to the natural sciences, the field in which his wife worked. When he came to Münster, he began teaching "practical philosophy." At that time the subfield of "practical philosophy" was still shaped by approaches derived from philosophers up to Hegel, and it only later underwent a turn toward the present. When a position opened on the ethics commission for clinical research at Münster's medical faculty, those who appointed Siep thought that his background in what they viewed as "practical ethics" suited the position, not knowing the historical orientation of his field. Thus, ironically, it was not his professional background that brought him a position in bioethics, but the appointment that led him to the field of bioethics. These early activities made him eligible for further positions on other ethics commissions, and he became so well known that his name came up almost automatically when the Department of Health was looking for members for the ZES.

Our conversation then turned to the puzzling fact that the ZES had no jurists, who were otherwise considered indispensable on ethics commissions. As Siep realized, this conspicuous omission presented a problem, since it made the supposedly independent ethics commission dependent on the RKI for all juridical questions. Siep mused that the reason was probably that the RKI, and not the ZES, acted as the ultimate permission-granting authority. An application to import stem cell lines must be addressed to the RKI, to which the ZES, in turn, is attached as a formally independent ethics commission. The ZES then generates an independent evaluation of the ethics of the application and makes a recommendation to the RKI, which the RKI must follow or else explain its contrary decision in writing. The idea, Siep thought, was that the RKI itself, as a federal authority at the highest administrative level, had its own jurists: competent administrative lawyers who in many cases had once served in various federal departments. The RKI, Siep said, covered the juridical aspects of imports, while the commission covered the ethical aspects. This division of labor, in effect, made ethics subservient to the law.

The ZES had avoided this problem in some sense because a highly positioned lawyer from the Department of Health was present at all commission meetings. According to Siep, she interrupted the discussions when they seemed to contradict the text of the Stem Cell Law. Since the commission members then occasionally pitted their own understanding of the text against hers, the commission in effect engaged with legal expert knowledge at almost every meeting. Siep added that this lawyer almost thought of herself as a member of the commission.

The composition of the commission—four ethicists from the fields of philosophy and theology and five natural scientists from the fields of medicine and biology—followed a certain rationale. One wanted to represent the "relevant disciplines," and one wanted to have a certain balance of power among them. The aim was to keep ethicists from dominating the natural sciences. Siep was aware that this composition made the commission vulnerable to two charges, coming from opposite sides. On the one hand, there were those who charged that here were ethicists who used vague terms such as "high priority" to make claims about biological facts and obstruct progress in science and medicine; on the other hand, there were those who claimed that the commission consisted of stem cell researchers plus a few yea-saying ethicists who allowed any research proposal to pass using flexible concepts such as "high priority" and "without alternatives" (*hochrangig* and *alternativlos*).

The balance of power also played out in more subtle ways. The law specifies that importation of stem cells is ethical if the research is of high priority and if it cannot be conducted on other cells. To determine whether those conditions are met, one needs the testimony and judgment of competent natural scientists. One cannot decide these questions purely as an ethicist; on the contrary, Siep said, the other commission members constantly needed the scientist members to explain to them what the applicants wanted to do, what it meant for research more generally, and whether the research was of high enough priority to justify the import. Ethics was caught, in effect, between the hammer of science and the anvil of law.

How then did Siep interpret his own mandate? At one point I asked him about the meaning of conscience, and he told me that in the ZES there was no need to use his conscience, since the law prescribes so closely what one has to test for. He could use his conscience in private, but not when holding an office (*Amt*). He told me that it was a tactic of the RKI to keep a research proposal afloat for as long as possible before rejecting it, first, because prohibiting research is a serious infraction of a basic right, and second, because the commission's decisions could be contested in court. If there was any chance that an applicant could improve his application, or withdraw it, then this would be preferable to ruling against it.

Siep mentioned that after a year and a half, there had been only five applications to import stem cells, so that everyone assumed that no one was doing stem cell research in Germany.[5] There had not even been enough requests to justify adding personnel to the ZES office, as Siep had wanted to do. Ironically, then, the issue that had justified intragovernmental competition, protracted debate, the crafting of a law, and the appointment of a dedicated ethics commission was no longer significant enough to justify an infrastructure to help support it.

Why had there been so few applications? Unlike Jürgen Hescheler and Wolfgang Franz (see below), Siep did not connect the trickle of applications to the stringency of the requirements. Siep said that he was himself surprised by it because the DFG had been so adamant that it had applications from scientists who needed to conduct research and should not be slowed down by an overly restrictive law. Siep had expected perhaps twenty proposals to start with. He was also surprised that researchers had not gone ahead and imported human stem cells before the Stem Cell Law was written, since "the Embryo Protection Law had not prohibited the imports." (Of course, two researchers had indeed gone ahead, and they were heavily criticized for it.) Siep said that the volume of research in the United States, where (in summer 2004) sixty applications for federal funding were under review, far exceeded that in Germany, even though the restrictions were almost equivalent in both cases. But, he conceded, "of course America has a completely different research landscape."

The Stem Cell Law went into effect on July 1, 2002, and the ZES approved the first application to import stem cells in December of that year. Why did it take so long to approve that first application? Siep said that applicants often did not know how to write their applications. In order to help them with the procedure, the commission produced an instruction sheet that is now available on the Internet. The most important threshold criterion was that the application should be complete; only then could the commission properly evaluate it. Communication back and forth had been necessary because the scientists did not understand what they had to submit. They focused on the potential of their research, Siep said, but failed to state where the stem cells would come from or whether they had certifiably existed before January 1, 2002. Nor did they state that no one had paid for them, that no pressure had been used in taking then, and that they had originated in reproductive medicine, without selection. As mandated by the law, the RKI must make sure that all these initial criteria relevant to the ontology of the cell line are addressed. Only then can the ZES test the ethical criteria.

Siep saw two extreme interpretations of the "absence of alternatives" that researchers had to demonstrate. On the one hand, some said that there was no alternative way to understand the workings of embryonic stem cells other than working with them.[6] On the other hand, some said that before using embryonic stem cells, all other options had to be explored, and those options could be extended ad infinitum. Yet, Siep mused, if the law had intended either extreme, then it would have been phrased differently. Therefore the law's intention had to be somewhere in the middle.

When I mentioned to Siep that at least two stem cell scientists I had interviewed were frustrated with the overly legalistic initial communications from the RKI, he was sympathetic. Siep too was bothered that the RKI could not find out why German researchers were so reluctant to apply for ethical approval. When asked about this, a woman from the DFG had told the commission that young researchers were hesitant because the future of stem cell research in Germany seemed so uncertain. Siep had offered to simplify the application procedure, in case this was what was holding them back. When this offer was leaked to the newspapers, there was a small scandal, since to outsiders it appeared as if the very person who was in charge of controlling the tiny hole in the law that allowed fundamentally "unethical" research to proceed in an ethical way was trying to enlarge that hole, with uncertain consequences for the ethics of German research. One article openly questioned whether Siep's views made him unfit to run the ethics commission, to which Siep replied that he was perfectly capable of publicly enforcing a law as head of a commission while privately disagreeing with the law as a philosopher.[7] Siep was, in effect, living the dichotomy between the public and private uses of reason that Kant had written about so forcefully two centuries earlier. For Siep, it was a problem that some people equated "ethically problematic" (*ethisch problematisch*) research with "ethically bad" (*ethisch schlecht*) research. When a question is "ethically problematic," he told me, it means that one has to weigh the pros and the cons and reflect on the ethical dimensions. The only research that was "ethically unproblematic" (*ethisch unproblematisch*), Siep joked, was investigating how grass grows.

In sum, the law that aimed to make German stem cell research ethical produced a version of transparency, of conscience, and of Germany itself. The Stem Cell Law mandates "transparency" in the sense that it requires one to follow its reasoning in detail; it aims in this sense at making the ethics of stem cell research *nachvollziehbar*. The law also mandates *Nachvollziehbarkeit* in that everyone can follow the origins, destinations, and purposes of the stem cell lines. But in this law transparency came up against its limits. It confused and frustrated researchers, and it produced conflicting legal expert statements. It hardly achieved the desired *Rechtsklarheit*, or clarity of law. Criteria of collective conscience were written into the law, but were implemented as a formal checklist; associated questions about the limits of scientific research were relegated to a new ethics commission, which defers to scientists and lawyers in interpreting its ethical mandate. Germany, in the meantime, was defined as a place with the highest ethical standards, but to avoid the uncertainties and delays of the law, it appeared that scientists would do better to leave Germany.

In the remainder of this chapter I move into the stem cell scientists' world as they fit their work to these contradictions and ambiguities. I show how they inhabit the spaces that law and ethics both create and constrain.

Jürgen Hescheler

In August 2003 I went to Köln to visit Jürgen Hescheler, the second researcher who had received permission from the RKI to import human embryonic stem cells. While I waited in his office, I chatted with his secretary. She was somewhat surprised that a researcher should be interested in the social aspects of science, since in her experience scientists were solitary fighters (*Einzelkämpfer*). My project became more plausible to her as she began to think of interesting "social aspects," but she declined to name any.

Hescheler called to say he would be fifteen minutes late. When he arrived we went into his darkened office. It was pleasantly decorated, with a wall of books and an ailing palm tree. On the walls were Japanese prints and other pictures. When I told him of my anthropological interests, he was reminded of a conference in Heidelberg the previous year where Jewish and Christian theologians had argued vehemently against one another. This encounter had opened his eyes to cultural differences as well as to the fact that even within specific religions there were no unified opinions that could place themselves in concentrated opposition to science. He gave me the name of a Jewish theologian in case I wanted to follow up on the debate. I had been to a conference on Jewish bioethics in New York City the previous year, where I had confirmed Hescheler's impressions for myself. I had met several Jewish bioethicists who headed the ethics commissions of biotech companies that had made international news for conducting controversial experiments in stem cell research and cloning, among them Geron and Advanced Cell Technologies. In the evenings the conference participants had sat down in a circle with excerpts from the Torah, and I had observed with interest how the discussion leaders inferred the permissibility of such cutting-edge experiments from ancient texts.

Hescheler had always been first. He had been enthusiastic about science from an early age, and as an adolescent he and a friend had won first place in a national math contest. While still studying medicine, he began to work on a dissertation on what regulates the heart and makes each of its cells beat in synchrony. Again he proved precocious, and he received invitations to international conferences to present his doctoral work, which brought together the previously disparate fields of electrophysiology and biochemistry. Another formative moment was his success in making visible, again for the first time, how single protein molecules on a membrane open and close to let ions through, creating so-called ion channels. This discovery led to his being the first to find a signal transmission cascade that activates ion channels. These breakthrough moments, along with the stimulating international atmosphere of the pioneering laboratories he worked in, shaped his fascination with science. To this day, he said, science meant to him being a discoverer, the first to do something new.

His overriding research question had been how cells, which are closed systems surrounded by a membrane, are directed to perform certain functions. What activates the signal transmission cascades that end up governing cell functions? How can cells, he was asking, communicate with one another so that each of them ends up doing something meaningful (*etwas Sinnvolles*) and everything harmonizes?

This question about functional control led to the question of how a cell system develops in the first place. Hescheler recounted his good fortune in meeting Anna Wobus at a conference in 1989, just after the fall of the Wall. In East German Gatersleben, Wobus had been able to differentiate mouse embryonic stem cells into heart cells, but she got stuck when she could not show how those cells functioned physiologically. Wobus's and Hescheler's areas of competence complemented each other, much as many post-Wall enthusiasts hoped the East and the West would now fit together to form a whole. Right after the Wall came down, Wobus went to Berlin to collaborate with Hescheler, and this East-West team conducted the very first functional measurements on heart cells derived from mouse embryonic stem cells. Together, they were the first to show that those cells were, in functional terms, truly behaving like heart cells.

Five years before James Thomson of the University of Wisconsin at Madison first isolated human embryonic stem cells, Hescheler predicted that those cells would make medical testing more accurate in the future, when one would be able to test medications directly on human stem cells. Animal rights advocates liked the idea. Hescheler regretted that no one had listened to him seriously back then, since the ethical debates could have started, and been resolved, much sooner.

When Hescheler wanted to focus more directly on embryonic stem cells, he encountered resistance from his supervisor and from other scientists, who claimed that he would never produce functioning heart cells that he would be able to control (i.e., make contract in unison), but that opposition did not deter him. Support for his work came from animal rights advocates, who saw that tests on stem cells might make some animal testing of medical or commercial products superfluous. He told me that the research leading to a single medication might kill millions of animals.

While Hescheler had always worked on both heart cells and nerve cells, he now specialized on heart cells because the heart was the best understood organ, and thus medical success with heart cells would most rapidly translate into applications in transplant medicine. He wanted to find out what it was that could turn a completely unspecialized and inactive cell into a highly specialized heart cell. And what made heart cells beat in unison? To answer these questions, Hescheler needed to find out whether a single heart cell already beat in any kind of rhythm, or whether rhythm developed when a critical number of cells came together. In other words, does a specific rhythm inhere in each and every cell of the heart, or is rhythm the result of coordinated action that emerges in a group of cells?

In trying to bring a common rhythm into a cell culture, Hescheler discovered that even the first cell carries in it the *principle* of rhythm, but that it is not yet perfected. As more cells are added, rhythm is built up, and even at a very early stage, the heart cells begin to beat in synchrony; from then on they always act as a coordinated group.

Some time before Brüstle's application to import stem cell lines in accordance with the Stem Cell Law, Hescheler had ordered and received stem cells from Thomson, after explaining that he had experience in investigating their physiology. Here too he had been first. Lawyers whom he had consulted had claimed at the time that the Embryo Protection Law did not explicitly prohibit the importation itself.

The first delivery of American stem cells, for which Hescheler received much negative publicity in the German media, arrived a mere two days later, but the cells turned out to be damaged. He then negotiated an agreement with the same Israeli researcher, Joseph Itskovitz-Eldor, with whom Brüstle collaborated. This German-Israeli collaboration was put on hold, however, while Germany sorted out its standards of ethical research. After Hescheler had done all the necessary paperwork and had received the ethical stamp of approval from the RKI, Itskovitz-Eldor changed his mind about the collaboration, to Hescheler's great disappointment. Apparently Israel had caught up during the almost two-year debate in Germany and no longer required Hescheler's expert knowledge to proceed with its research. Hescheler emphasized that his enormous lead in the field had disappeared and that the loss of time had disadvantaged him. Israel still collaborated with Brüstle, Hescheler speculated, because the Israelis had no experience of their own in working with neuronal cells, which were Brüstle's specialty. Now Hescheler worked with American stem cells, which meant that all rights to commercial use of his results rested in the United States.

In Hescheler's view embryonic stem cells still held the greatest scientific promise. Adult stem cells could not be transdifferentiated (transformed from one cell type into another without first being turned back into pluripotent stem cells) in the same ways, since during development a cell places what Hescheler called "bookmarks" in the genetic text. To transdifferentiate an adult stem cell, one would have to go back and reread the already marked-up genome of the cell.

The national regulation of research that is fundamentally and almost inherently international presents problems, Hescheler said. The institutional and financial ties that make certain kinds of research possible in the first place, as well as the publication of scientific results, often cross national borders. In Hescheler's view this meant that Germany could not simply stand back and claim to have a different ethics, much less a better one, than other nations.

The one time that Hescheler was second, it was only by accident. Hescheler told me that he had submitted his research application to the RKI at the same time as Brüstle, and that it was sheer coincidence that his Bonn colleague's proposal was handled first. It may not have been a complete coincidence, however: Oliver Brüstle had set off the German stem cell debate when in 2001 he was the first to apply for DFG funding to import human embryonic stem cells.[8] The scientific community was keenly aware of how long Brüstle had been waiting for approval. Hescheler knew that the RKI had no experience in handling this kind of scientific research proposal, and he recounted how the RKI lawyers kept calling him with questions. Through this interaction with the lawyers, his application slowly became complete and therefore eligible for ethical evaluation. The ZES then discussed his and Brüstle's proposals together, approved them both, and formulated a 20- to 30-page response that specified in detail what research was permitted, and under what conditions. Although the commission was wary of prohibiting research, it could nevertheless be extremely intrusive. Brüstle, for example, was not permitted to develop heart cells, while Hescheler was not permitted to culture nerve cells, even though that had been one of his earlier specialties. The ZES, in other words, decided that Brüstle was an expert on neuronal cells, while it made Hescheler a specialist on heart cells alone. Research was ethical only in the field in which one possessed absolute competence, while more general or exploratory research became unethical.

Clarification questions arose where the language that scientists used deviated from the language that the law used. It was hard to explain to the lawyers, Hescheler remembered, what a cell is in the first place. Is a completely undifferentiated stem cell the same cell after it has altered somewhat? Is it still part of the same cell line? After he had explained the biology to them, the lawyers concluded that the originally imported cell was the cell researchers could work with, and that everything that scientists derived from that cell counted as that same cell, even if the products were genetically altered.

This conclusion seemed to me to resonate with German juridical definitions of embryos as continuous and identical with later persons and as containing the potential to become those later persons. I asked whether one might say that the lawyers treated the early embryo as identical with the later adult person while, in insisting that a cell line was identical with its genetically altered offspring cells, they were at the same time treating a parent as identical with her child. Hescheler agreed that there was a paradox and said that in the end, the problem was that lawyers had very simple ideas while biology was incredibly complex. In Hescheler's view complexity existed more in the categories of nature than in those of society. The law and its clumsy categories, he said, simply could not capture the fact that any alteration in one of the 40,000 genes of a cell would produce an entirely different system.

Perhaps it was this awareness of biological complexity that led Hescheler to admit that he did not see embryonic stem cells as persons, but as very early cell lines that one could develop into human tissues. He recalled the time when he had worked with hospital patients for whom no therapies existed and said that helping them outweighed respect for the blastocyst from which stem cells were derived. In theory, one needs only one blastocyst in order to develop a stem cell line that might cure thousands. Hescheler did not understand the scruples of those who were unable to see things from that perspective.

When Hescheler listed the ethical preconditions that he had to fulfill before importing the cell lines, he said that the parents had to permit the release of the blastocyst and that the cells had to be certifiably free of diseases—details that the law does not mention. Hescheler also thought that the existing cell lines would be unusable for therapies because they were grown on mouse feeder cells that could potentially transmit viruses from the animal cells to the human cells. He suggested that new stem cell lines should be created in Germany and then stored in a public institution where researchers could access them.

In spite of all his reservations, Hescheler saw the Stem Cell Law as necessary because the debate had been so heated that only a clear procedure could calm things down again. The law made clear how things would proceed; it set limits and provided procedures; it tamed the scientists who were perceived as unruly. The law as written seemed to him the only compromise possible under the circumstances.

At the end of our three-hour interview, Hescheler offered to show me his laboratory. He led me through each room as if I were a student to whom even the most basic things about research needed to be explained and would be fascinating, and he tried visibly to portray everything precisely and thoroughly. Besides the complete division of laboratory labor (which, Hescheler told me, represented the stages that all members of the lab had to pass through), I noticed the lab diaries on the tables in which every step of every experiment was meticulously noted. Hescheler said these diaries were juridical documents that could be subpoenaed in case an experiment could not be replicated elsewhere. Although the steps appeared to be recorded precisely, it struck me at once that one needed a lot of preexisting knowledge in order to make sense of the chaotic-seeming numbers, graphs, and abbreviations. I remarked that nonreplicability could clearly have other causes as well. Hescheler said that researchers presume a certain familiarity with the procedure, and they accord a certain amount of trust to one another.

Throughout our walk through the laboratory, and later through the basement where the lab animals were kept, I noticed signs with instructions, warnings, and reminders: "Please don't forget to shut this door." "Keep the temperature at this setting." "Bring dead animals into the basement and do not leave them in the refrigerator." On many of the doors were cartoons. In combination with the overstuffed shelves and the messy worktables, these things gave the impression of a children's playroom in which all the toys were labeled and the rules for using them written out. The experimental life seemed like a site of highly scripted play.

Wolfgang Franz

In August 2003 I went to Munich to visit Wolfgang Franz, the third scientist whom the RKI had officially permitted to import embryonic stem cells for research. Franz was a youthful, dark-haired man in his early forties who seemed concerned that I might be a journalist interested in his prelegal importation of stem cells. He seemed reassured when I told him that it would take several years for the results of my own research to be published. He had studied medicine in Munich and then spent three years at the Salk Institute in San Diego learning molecular biology. When he returned to Germany the Gene Technology Law had just gone into effect, and he was worried about his academic future. However, the Max Planck Institute for Biochemistry in Martinsried, near Munich, offered him a position, and then, three years later, he went to

Heidelberg to help expand a cardiology clinic there. Later, in Lübeck, he began working on mouse embryonic stem cells to study the differentiation of heart muscle cells. When he succeeded in deriving heart muscle cells in mice, he became interested in applying his work to human cells. He had many discussions with ethicists, he told me, who seemed to concur that the Embryo Protection Law did not prohibit the importation of embryonic stem cells. Like Jürgen Hescheler, he went ahead and imported them before the legislature closed the apparent "gap in the law" (*Gesetzeslücke*). He seemed not to think that he had done anything wrong, since the absence of a prohibition, to him, implied permission. Even back then he had, in effect, fulfilled the current criteria of the ZES in that his previous work had plausibly suggested that experiments on human embryonic cells would be successful.

At present, he had in hand an approved application for DFG funding, the positive evaluation of the ZES, and a Material Transfer Agreement from Israel. His new stem cells could now arrive, and he hoped to begin his research on them in the coming weeks. One of his colleagues had already been to Israel, where he had first learned how to handle the stem cells and then conducted, with permission, preliminary experiments. Franz's first shipment of cells had come from the company WiCell in Madison, Wisconsin, which had forced him to give up all rights in his findings. The cells he now hoped to work with were from a firm in Israel, with which he had signed a *joint* intellectual property agreement.

When I asked where he kept the cells he had originally imported, he said that they were in an institute at his present university, but that a commissioner (*Komissar*) was responsible for handling them. They were not stored in his own laboratory, Franz said, because then nothing would have kept him from doing research on them. It needed to be visibly obvious that the researcher was physically removed from conducting potentially dangerous experiments. It was as if being a scientist meant being incapable of self-restraint.

Franz was frustrated by the poor communication on the part of the ZES and the RKI. He had followed the specifications of the law and had demonstrated the presence of the three ethical criteria of high priority, sufficient qualification through prior work, and absence of alternatives. When he mentioned the high priority of his research, he pointed to a thick file folder that contained his correspondence with the RKI. After adding up the dates of this complicated back-and-forth, he said that he had spent some eight months writing letters and answering questions from the commission.

Although the ZES technically had only two months to make its decision once the application was complete, Franz explained to me that the commission had used the RKI as a front to relay technical questions back to the researcher. If the RKI asked those questions, the implication was that the proposal was not yet complete and that the 2-month clock of the ZES had not yet started ticking. Franz said that there had been at least three calls for clarification or more information from the ZES, all of which had cost valuable time. In this way, Franz inferred, the ZES benefited from a poorly regulated process. It took away the pressure to work very hard at quickly understanding complicated scientific matters.

Like Siep and Hescheler, Franz thought that the reason there was so little stem cell research in Germany was because the hurdles were set too high. Pointing to another large folder, he told me that this was the material necessary for getting his application through. The ethics commission was not the only obstacle to his research; he also had to apply for financial backing from the DFG, and he had to negotiate an agreement with the research unit sending the cells—each of which also involved significant labor. Finally, Franz was frustrated that the restrictions left so few stem cell lines to work with. How many stem cell lines were there, he asked, that were left over from reproductive treatments from before January 1, 2002, that were donated voluntarily (i.e., without compensation), and that had no other chance for life?

In our interview Franz explained how he had proved scientifically to the ZES what "high priority" and "absence of alternatives" meant in the context of his work. The ZES then had to confirm that the preconditions necessary for the importation in fact existed. In doing so, the ZES equated research under those formal conditions with "ethical research." In that sense "ethical" became a category as empty as "high priority" and "without alternatives." Whereas before the Stem Cell Law research proceeded in accordance with scientists' autonomous understanding of that term, there is now a new category of "ethically unproblematic research," produced through the practice of the ZES in lengthy and often frustrating communication with researchers.

: : :

A transmutation of sorts occurs when the law demands that German research open itself up to the state's gaze. The Stem Cell Law hybridized the space in which scientific research takes place. It is now a space where research is "ethically unproblematic," as science becomes relabeled as ethics (through the power of the ZES). In its zeal to protect human life and human dignity, the state now controls the motivations and practices of scientists. The ZES becomes the public conscience of the scientific community. It ensures that German research will proceed transparently and ethically. But, given that scientists (and to some extent lawyers) dominate the discussion at the ZES, and given the members' feelings of powerlessness, it is difficult to ascribe ethical agency to this new voice of German conscience. Rather, it seems to serve as a site where the forms of reasoning in which the state has acquired a new interest will proceed not in the private thoughts of scientists, but in the public play of meeting bureaucratic requirements. The "moral law within" will in this way be extracted and made publicly available, but only in accordance with the particular configuration of transparency within the closed meetings of the ZES.

It seems that the German state knows how to deal with potential threats to the ethics enshrined in the Basic Law only by morally subjugating them and bringing them into line. To bring the scientists, as potential transgressors, into line, it needs to craft a common space with them, but, as scientists point out, it does not really succeed. German researchers not only have to characterize stem cells and cut their ties to human life and potential personhood, and to the inviolable dignity that these imply. They are also suddenly asked to turn their internal practices inside out and reveal them to the state. Now science itself is transformed as the state asks the scientists to display a clear conscience. In being required by law to display their qualifications and motivations, the researchers even become potential criminals. They can be fined and imprisoned for transgressions. The Basic Law's guarantee of free inquiry is in some sense abridged, as scientists are asked to renew their contract with the state.

When I looked at the letters of approval that the RKI had published by summer 2003, I was struck by their uniformity. Entire paragraphs of the two- or two-and-a-half-page letters closely resembled one another, if they were not completely identical. Such standardization is probably not surprising in a governmental office that develops rules and procedures it can follow, but these similarities in ethical evaluation somehow seemed inconsistent with the excitement that human embryonic stem cells had caused, the uniqueness of treatment that their unique capabilities seem to call for, and the depths of ethical reflection that had preceded the law. The careful consideration of individual cases had given way to a routinized process that checks off technical/ethical criteria. Although differences of opinion are doubtless tucked away in the "subtexts" of those voluminous files of correspondence, the public is presented with a clear example of ethical research, approved by a state-appointed ethics commission.

Conclusion

In a short story entitled "Funes, The Memorious," Jorge Luis Borges imagines a young man who, after a horseback-riding accident, finds himself physically hopelessly crippled, but equipped with perfect memory. To pass the time, the bedridden Funes requests a book from the narrator, an estranged childhood friend, who has just begun the methodical study of Latin. The disbelieving narrator, somewhat tauntingly, lends him a volume of Pliny's *Historia Naturalis*, along with a Latin dictionary. When urgent news calls the narrator away, he visits Funes one evening to retrieve the books. Funes greets him with an enumeration, in Latin, of the mnemonic feats that Pliny recounts. The two men spend the night talking, in darkness.

Following his debilitating accident, Funes's infallible memory began to record every perception with flawless accuracy and to retain it indefinitely. At first he found his immobility a small price to pay for this newfound mental capacity. His life now took place within, with unprecedented clarity and brightness. Remembering everything meant that he could perceive the minutest differences where others perceived only sameness. His mind was filled with endless contiguous memories of change. But it was a disconcerting brightness that so clearly illuminated each moment. Each day the sight of his hands shocked him anew because he could watch himself aging.

Remembering a day meant reconstructing the entire day, a task which itself took a day. He closed his eyes or turned his face toward dark walls whenever possible to avoid creating further impressions.

Funes could not help noticing the imprecision of the categories into which we sort things in the world, and he invented new words and systems of enumeration to capture the utter particularity of each thing at every moment. Dissatisfied with how many words it took to say a number, he attached each number to a single, uniquely identifying verbal expression, saying "Maximo Perez" for seven thousand thirteen, or "The Train" for seven thousand fourteen. Before he gave up a few days later, he had reached 24,000, with the new words for each number etched into his mind. At the same time Funes tried to devise a usable mental catalogue for all the images in his memory, but gave up again because of the scale of the task. Cataloguing his childhood memories alone would have preoccupied him for the rest of his life.

The narrator points out to Funes the futility of such attempts at particularizing. Order, he notes, depends on classification and categorization, which in turn depend on generalization. Both thinking and acting meaningfully in the world demand giving some perceptions priority over others. In turn, selective perception and selective remembering of the past make the future possible. When the night comes to an end, the implications of the conversation also dawn on the narrator, and they leave him as paralyzed as Funes. Aware that each word and every gesture would live on in the other's "implacable memory," the narrator feels "benumbed by the fear of multiplying superfluous gestures." Funes dies shortly thereafter, of pulmonary congestion.

Reading Borges, Reading Germany

Borges, as usual, offers us a philosophical point dressed in the form of a short story. His tale of the physically crippled but mentally hyper-invigorated Funes is about the conditions for meaningful recollection of the past and the possibilities for purposeful action in the future. Funes's ultimate failure lies in his inability to turn sensory impressions into meaningful patterns. He is incapable of doing the work of erasure and selection that is necessary to separate a text from its context. He is unable to generate order—even numbers lose for him the sequential and relational qualities that make them invaluable for human communication—and he fails in his attempts to make sense of ordinary experience. His fate reminds us of the plight of Germany, consigned by history to making moral sense of the Holocaust over and over again.

At times it seems that Germany, too, is caught in the trap of an implacable memory where the Holocaust is concerned. No detail can be forgotten; events must be remembered and recorded to the extent it is humanly possible. Artworks spring up all over the country that try to recollect the name of every victim. Plaques memorialize every site where something tragic happened to someone. Commemorative rituals mark almost every date, and these rituals constitute a kind of living historiography. November 9, for example, encompasses layers of history: it was the day of the declaration of the Weimar Republic, of Hitler's first failed attempt to seize power, and of the *Kristallnacht*. When the Berlin Wall fell on just that fateful day, the weight of memory almost grew too much. Instead of celebrating this first successful German revolution, the official observance of reunification was displaced onto October 3, the date of the signing of the Unification Treaty, which the West German government had chosen to be almost purposefully lacking in historical resonance. In erasing November 9 from the official calendar, the memory of the East Germans' victory over their state was perhaps subordinated to Germany's need for a less ambiguous commemorative moment; yet November 9 remains imprinted on the minds of citizens, and the state needs a show of force to keep those memories from spilling into unruly protests.

Transparency—Text and Context

The German state is committed to the political principle of transparency. It seeks to make its inner workings visible and demonstrates that commitment, both materially and symbolically, in the glass architecture that has sprung up everywhere in Berlin's government quarter. In the Pretext I described a couple standing near the Reichstag, itself the primary icon of German democracy, in front of nineteen upright and contiguous glass plates on which the basic rights of the German Basic Law are etched. What these observers saw was a function of where they stood. The words of the law became visible only when the viewers looked at the text from a particular standpoint and at a particular angle that made the surrounding glass fade into background. Only then did the text stand out from its context. The physical self-positioning required for this act of sense-making can be seen as a metaphor for how transparency functions in contemporary Germany. When the state makes a display of its transparency, it does not reveal all there is to be seen. As I have demonstrated, the state is selective about what it shows. Like the couple reading the letter of the law, its citizens are asked to stand in a particular place and to view its workings from angles that the state itself prescribes.

German culture, as I also noted in the Pretext, emphasizes texts, textuality, and reading. Making meaning depends on the reader's capacity to distinguish that which will have meaning—that is, the text—from that which will be considered either a source of explanation or irrelevant detail—that is, the context. Separating text from context, I have suggested, involves a fundamental, and ultimately arbitrary, distinction, and it requires significant erasures. Bringing something into relief, in sculptural terms, either means removing extraneous material until the relief alone stands out clearly, or it means cutting and carving the figure into the material, accentuating it as different from the unworked surface. It is the choice of what will be relevant and irrelevant that produces both text and context.

As we have seen, either operation involves labor. Producing transparency, in the case of the citizen conference on genetic testing at the German Hygiene Museum, meant displaying particular features of the place, the process, and the participants while de-emphasizing others. Not coincidentally, the event itself took place in a location known, since its first display of the Transparent Human (*Gläserner Mensch*), for its commitment to making visible the inner workings of science and of bodies as part of the pedagogical enterprise of making good, well-educated (*gebildete*) citizens. The citizen conference on genetic testing also involved training laypersons to see the right things and, in the process, to become capable citizens. It meant turning them into proficient readers of expert claims so that they could see themselves as a vital organ of a representative democracy committed to progress through science and technology. We saw that it took many preparatory weekends, and the guidance of an *Universalmoderator*, before the citizens could look responsibly into the workings of expertise. And, in an ironic twist, the citizens themselves became transparent as objects of scientific observation, so that the state could construct a working model of democracy to suit its future policy purposes.

In Germany, as we have seen, something is considered transparent when one sees its "logic" clearly: that is, when the state's workings appear *nachvollziehbar* to its citizens. One can disagree with official reasoning and voice differing opinions, but the result of a visible process is accepted as having been arrived at democratically. In that sense transparency carries its own logic of legitimacy. From transparency, the logic goes, democracy will follow. In this Habermasian scenario, pure reason will flourish in the public sphere, provided only that the state makes available a fully visible, and accessible, meeting place.

And yet this kind of *Nachvollziehbarkeit* was often difficult to accomplish and required its own forms of training and concealment. To produce a meaningful ethical text, much of the actual work of crafting ethical principles had to be hidden from view, some of it in parliamentary archives accessible only to some intrepid future researcher, some available only through the eyes of the never fully integrated anthropological field-worker. We saw how the glossary of the EK reports marked medical terms and unmarked legal ones, thereby emphasizing the agency of science while rendering the law invisible. We saw how large amounts of work, like the section on organ transplantation in the final EK report, were excluded in deference to more state-of-the-art analysis done elsewhere, thereby reinscribing the narrative of ethics lagging behind science. We also saw how attempts to include references to East Germany were rejected, thereby ratifying the view that the failed state had nothing relevant to teach about bioethics.

Even after a law has been adopted, there is ongoing contestation over what its text means, and the context of its interpretation becomes, in effect, part of its meaning. In the interactions between the RKI and the ZES, for example, the place of the law in the ZES, and therefore the relation between law and ethics, were negotiated through practice—partly in invisible interactions between the applicants and the authorities and partly through contingent advice giving by the RKI's legal representative to the ZES. The meanings of *hochrangig* and *alternativlos*, which came to define ethical stem cell research, were in this way the outcomes of practical decisions, arrived at on the spot—decisions that could not be generalized in a principled way from one situation to the next.

Text and context, transparency and opacity, we see then, are in a dynamic relationship. This relationship further implies that transparency and democracy are linked not by causality, or necessity, but rather dialectically, through culture. Making something transparent means making it easier to see through to the workings of something otherwise opaque. But what one sees more clearly depends on what is being made transparent. In Germany, the state makes itself transparent in the service of democracy, perhaps more self-consciously so than in any other contemporary European nation. But it is only because state and citizens are *already* committed to certain prior notions of democracy—such as notions of *Nachvollziehbarkeit* or of lawfulness—that more transparency, in their eyes, produces more democracy.

In some sense the state's performance of transparency even deprives citizens of agency, though it is designed to do the opposite. The state displays itself not with the goal of making its transparent workings open to critique, but with the purpose that citizens understand, and adopt, its governing logic. Something is most *nachvollziehbar* when a particular reading practically forces itself on the readers; that is, when there is the least interpretive choice. Transparency is most persuasive when only one interpretation remains possible. If, on the other hand, reading a public text involves agency on the interpreter's part, then not everyone can instantly follow a given reading, and transparency breaks down. Like Peter Eisenman's Holocaust memorial, which seeks to constrain and discipline memory much as it disciplines the possible paths of visitors, transparency in government acts to shut down alternative readings.

Potentialities—Setting Limits as an Ethical Act

The discourse of bioethics in Germany is partly about how to ensure that scientists conduct ethical research and how to prevent unethical research from taking place. As we have seen in the case of human embryonic stem cell research, these imperatives involve two main points of contestation. First, there is disagreement about the relation of limits to ethics. Can ethical principles be laid down transparently, in advance, for all to accept and follow? Or is scientific research a fundamentally transgressive process that constitutes ethical limits by violating our ethical sensibilities? Second, under what circumstances is the individual justified in violating collective norms? When must laws be obeyed, and when may they be broken?

The act of setting limits occupies a special place in the German imagination. In following political discussions in Germany, I often encountered the sense that definite limits must be set and clear boundaries drawn around what one can and cannot ethically do. When Oliver Brüstle applied for federal funding to import human embryonic stem cells, for example, he was fully aware that, due to a "gap in the law," their importation was not explicitly prohibited. In interviews, however, he emphasized that he wanted *Rechtssicherheit* and *Rechtsklarheit* (legal security and clarity) before he would import the cells. He placed the topic on the national agenda by provocatively applying for federal funding, thereby forcing the state to take a firm and decisive stance on the controversial issue. *Rechtssicherheit*, in the actors' view, does not mean that the law must be written in a particular way, mandating one or another course of action, but only *that* the issue must be regulated, at least in theory, even if uncertainties and ambiguities persist in practice. It is less important what the limit is than that a limit is set in the first place; even a self-contradictory legal mandate is better than no mandate at all.

In Germany the act of setting limits often creates a space for reflecting on when limits may and may not be ethically crossed. The relation between law and ethics, while acknowledged to be intimate, is open to contestation. Being ethical can mean setting limits, but it can also mean knowing when to break limits, or when to set new ones. One is ethical so long as one engages with limits. At a panel discussion in Berlin in January 2003, I heard Heribert Kentenich, a physician working in the field of reproductive medicine, argue for embryo research by saying, "We have *always* followed the logic of breaking norms, and of developing new norms, and of breaking them all over again." His discussion partner, former minister of health Andrea Fischer of the Green Party, drew the opposite conclusion. To her, Kentenich's point was trivial and not even worth making, and she emphasized the opposite: "At the same time as it has broken them, society has *always* set itself boundaries." Kentenich denounced the Stem Cell Law as schizophrenic because it pinned the ethical acceptability of stem cell importation on the cell lines' having been produced elsewhere and elsewhen. Fischer conceded the legal schizophrenia, but said that a similar compromise law, Paragraph 218 on abortion, was similarly internally schizophrenic, but also wise in bringing together a previously divided society. She argued that the Stem Cell Law had captured a comparable consensus in a pluralistic society. It was as if, for Kentenich, breaking limits (again and again) was the act that constituted ethical social progress, while for Fischer, the act of setting oneself limits was constitutive of social order.

The latter view is rooted in Kantian philosophy. For Kant, self-limitation equaled demonstrating personal freedom and moral autonomy. The view that setting limits is an ethical act almost in itself has acquired special currency in Germany, and its persistence today is perhaps a reaction to seeing what can happen when the state does not set such limits for itself and for others. Perhaps Germans distrust individual autonomy precisely because individuals feel subordinate to collective "reason" and social "order" as well as to the law that embodies these collective judgments.

At a meeting of the Nationaler Ethikrat (NER), Horst Dreier, a law professor and NER member, stated what must be axiomatic to most Germans and, indeed, to most believers in the rule of law: "A legal order is obligated to be internally coherent." (*Eine Rechtsordnung ist zur inneren Kohärenz verpflichtet.*) But for some, the Basic Law provides at best an ambiguous answer to the question of embryo research. In his Berliner Rede of 2001, the German president Johannes Rau had implored scientists to restrict their research to this side of the Rubicon, without crossing the boundary between ethically acceptable and unacceptable research. Ernst-Ludwig Winnacker, the head of the DFG, publicly contradicted Rau. Winnacker complained that researchers were almost forced to have a bad conscience if they as much as mentioned the words "freedom of research," even though that freedom was also a constitutional right. According to Winnacker, there was no research on this side of the Rubicon; like Kentenich, he saw research as a history of broken taboos. Compared with the taboos that Darwin and Freud had violated, he said, the stem cell debate was negligible.

To others, it remained unclear how precisely to ground the limits that researchers set for themselves. Jens Reich, the politically active scientist and NER member, once conceded in a public forum that there was no moral reason to do embryo research, but *also* no moral reason to stop it. What, then, could be so dangerous about getting on this supposedly slippery slope? Other scientists asked similar questions: If we counter all the objections that the public has against stem cell research or cloning by technical means, will there *still* be arguments against such research? If we remove all the obstacles to moving ahead with cloning, for example, will we then move ahead in fact? Reich once asked whether we will have good reasons to prohibit cloning by law when all the present technical and scientific difficulties have been removed. After all, he said, the only way in which a clone is different from a twin is that the former has been produced through a technical process.

Some seemed to answer the question by applying a version of the precautionary principle: if one doesn't know the potential consequences, then it is better to err on the side of caution. In an interview, lawyer and EK member Ulrike Riedel explained the logic of protecting stem cells. She told me that we did not yet know what status they had: "Perhaps biology will at some point tell us that they are in fact human beings, and then it will be better that they are protected already." In her view it was better to protect something unnecessarily than to leave (potential) human beings to their fate in the hands of researchers. Riedel in effect recommended drawing a circle of safety around entities that might deserve human dignity in the future. By making the future inviolable, she was making potentiality itself inviolable.

Still others want to leave moral decisions unequivocally up to individuals. We recall that, in response to the Law for Aerial Security, which briefly permitted pilots to shoot down planes on military orders when they presented a danger to the public, Berlin lawyer and bestselling author Bernhard Schlink argued that responsibility for firing should rest not with a military superior far removed from the action, but rather with the individual who makes such localized moral judgments in dialogue with his or her inner self. But Schlink's argument obviously did not carry weight with the Federal Constitutional Court that invalidated that section of the law, nor with the press or the courts that assessed Wolfgang Daschner's controversial orders in the Gäfgen torture case.

Whereas Riedel, the policy actor, sought to use politics for the purpose of strategic moralization, and Schlink, the author, argued for individual exceptions to collective norms, Reich, the philosopher-scientist, asked how any collective could be morally bound. On what grounds can we write laws that society must obey? What counts as justification for legislating something "we believe" is unethical? On one occasion I heard Reich say, "I know *for myself* that this research is wrong; I have religious reasons. But I cannot make my personal religion into a greater law. I am looking for an *argument*." Reich implied that religious beliefs constrain the individual, but that only reason can ground laws that will bind a collective. Or perhaps it worked the other way around: when the collective feels constrained in a particular way, we call it reason; when the individual feels compelled in a way that may go against the collective, then we call it (mere) belief.

In any case, it seems that in Germany only the law itself can offer the necessary protection. There is deep distrust of individual judgments. That is why the first article of the Basic Law enshrines Germany's highest ethical principle, the inviolability of human dignity. And those individuals who, through their personal actions, come anywhere near to breaking that fundamental taboo, such as stem cell researchers and pregnant mothers seeking abortions, are under a particular compulsion to demonstrate to the state's satisfaction that their reasons are truly *reasons of conscience*, not arbitrary acts of noncompliance.

Talk of potentiality permeated the German stem cell debates. The embryo is defined as a potential human being, and the stem cell is a potential cure for the genetically predisposed but asymptomatic person, who is defined as a potential patient. And yet confusion over potentialities surrounded the definition of stem cells in the Stem Cell Law. Lawmakers and scientists disagreed on legally defining attributes of cells that could be verified only empirically. Stem cells could have become anything. They were immortal, and in their immortality they were full of potential. They could have turned into persons. They could have become cures for diseases. But for now stem cells remain trapped in suspended animation, where law and science maintain them. Law, in restricting research, prevents viable stem cell lines from coming into existence, let alone becoming cures; science, in tapping their curative potential, prevents them from becoming full-fledged persons. In its state of perpetual suspension, the stem cell almost expresses the very ambivalence that animates so much of contemporary German culture. At times the ethical discussion of the embryo's potentiality seems to parallel an underlying, implicit discussion of Germany's potentiality as a place where change and growth can happen in an ethical way.

In the stem cell debates, recurrent references to the future determined options for action in the present. The talk of the unborn evoked and invoked future generations—those not yet alive. In thus constructing a new generation, the debate raised the question of who exactly this new generation was going to be. How and when and where would it succeed existing generations? The stem cell debate thus encodes a "generational conflict." Future generations have ambivalent resonances in Germany, a country with a declining birth rate and a shrinking "native" population, where many voice concern over the problems of multiculturalism that follow failures of integrating and assimilating immigrants. Perhaps part of the ambivalence stems from uncertainty over whose future Germany is really trying to secure: that of a decaying, aging *Volk*; a displacing immigrant nation; or a genuinely invigorated *Bevölkerung* that would incorporate the best of both past and future.[1]

Taboo—Dammbruch

A dominant German image of discontinuity, which also encodes a continuity, is the *Dammbruch*. It too rests on selective visions and perceptions. It is about rhetorics and images of radical disjuncture: the *Dammbruch* itself, the Rubicon, the *Sonderweg*; the schizophrenic laws that both forbid and enable; the ideas of *Technikfolgenabschätzung* (technology impact assessment), of *Rechtssicherheit*, and the threat of horror scenarios; the idea of all or nothing; the fear of what may happen if one frees oneself from all those paralyzing "German" qualities of overregulation and the obsession with order.

The *Dammbruch* metaphor rests on particular constructions worth making explicit. It assumes, first, that there is a clear and present danger, a body of water that comes surging from an uncontrolled direction. Second, it assumes that there is a dam, constructed for a particular purpose by particular people, resisting the onrushing water. Third, it assumes that there is a given territory that will be flooded if the dam is breached and that it can be protected by fortifying the human-made barrier. The fear of a *Dammbruch* expresses the concern that if the fortified wall of law and ethics is breached, then unwanted waters will come rushing through and submerge any ground that society holds dear. To ward off the destructive flood, strong laws must cover every inch of contested ground. The dam will impede destructive movement and allow progress to flow only at a carefully regulated speed.

Dammbruch arguments are about vigilance. They imply that if one does not watch out at all times, then fearful things will happen. Psychotherapists have said that "what we fear the most has already happened."[2] The German state knows perhaps better than most what humans (and states) are capable of if one stops paying attention to cracks and leaks.

In some ways the *Dammbruch* is the counterimage of potentiality and pluripotency. The potentialities that stem cells promise became, for many months of public debate, an imagined vehicle for German innovation and self-reinvention. The *Dammbruch* inverts the promise and turns potentiality into a threat. While proliferating stem cells might produce a world in which all is possible and good, in the world of the threatened *Dammbruch*, too, anything is possible, but much of it is damning. Utopia meets its counterpart in dystopia.

The *Dammbruch* argument is about an ultimate kind of sense-making, and its critics are as outspoken as its defenders. It is about the making of a single history, a single set of norms, and a single way of guarding against moral disorder or chaos. For some German observers, however, those certainties disappear on closer inspection. The theologian and NER member Richard Schröder put it to me in this way in an interview:

> The *Dammbruch* argument is a prognosis. . . . It rests on the assumption that, in the past, if we had been careful from the beginning, *this* would not have happened. It is an interpretation of the Nazi era: if we don't say "no" at a particular point in time, then everything else will necessarily follow. For the installation of a dictatorship this is true to some extent. When they can arrest any opponent, then it is too late to overthrow them. . . . But we are not talking about putting up a dictatorship; we are talking about ethical standards. And these standards are treated like the democracy in the Weimar Republic: if one doesn't watch out, then they disappear irrecoverably. . . . One reason why this is nonsense is that every generation critically looks at the preceding one . . . so that the next generation creates *hypermoral*

standards. . . . Every generation undergoes a process of self-reinvention. . . . Another reason [why this is nonsense] is that the *Dammbruch* argument, when applied to ethical considerations, implies that we are the last people who can still think clearly. The next generation, if any danger arises, will abandon reason and allow the most horrendous things to happen, unless *we* now prevent *them* from doing this. Of course it's true that we learn mainly from catastrophes, but if someone produces a malformed clone, then you can be sure that people will *not* allow cloning to move forward. On the contrary! . . . The *Dammbruch* argument is a prognosis, but one that is derived from the wrong materials. Our experience shows that any local tendencies to desensitization [*Verrohung*] . . . are more than counterbalanced by sensibilities that increase over the course of our cultural development.

A media theorist at the Technical University of Berlin, Norbert Bolz, said that it was not necessary to invent horror scenarios, since skepticism and self-criticism are built into the German political system. According to him the problem was not that people were not critical enough of new technologies, but rather that there were too many people in important political positions who had made it their profession to develop horror visions.

Interestingly, the *Dammbruch* seems to happen only in debates, or in people's heads. Some people tried (in some cases successfully) to make the potential *Dammbruch* disappear by simply framing the issue differently. Like the couple near the Reichstag trying to find a place from which to read the text of the law, some tried to find a vantage point from which the text of the law might be better read. When the NER publicly presented its statement on preimplantation genetic diagnosis of embryos, for example, one of its members, a lawyer, tried to resolve a widely perceived disjuncture between the permissive abortion law and the restrictive Embryo Protection Law. To assuage concerns that an untenable divergence might put nascent life at risk of easy access by researchers, he offered as a starting point the parents' reproductive freedom. Parents had a right to *their own* child, he suggested, and "the disharmony between the abortion law and the Embryo Protection Law can thus be resolved. . . . There will be no *Dammbruch*." A different framing of the danger had made the danger disappear.

Proponents of stem cell research adopted the *Dammbruch* argument by inverting it. In arguing for legalizing the research he wanted to do,

Oliver Brüstle offered the following rationale, countering the direction of causality that the *Dammbruch* argument assumes: "If we don't do stem cell research now, then we will have to be consistent later on and deny ourselves the therapies from stem cell research as well." In highlighting the schizophrenic aspect of the Stem Cell Law, Brüstle also drew attention to the fact that ethical decisions are not isolated, but have ramifications in multiple directions.

Law and Memory—Recht und Unrecht

The Holocaust is Germany's ever-present text and context. It is the text of the past to be read, reread, interpreted, and explained, and it is also the context that serves as an explanation for political and moral action in the present. Germans had the Holocaust *because* they were what they were; Germans are what they are today *because* they had the Holocaust, and must never let it happen again. Germans increasingly seek to master the Holocaust by cultivating the memory of it. The *Historikerstreit* of the 1980s centered on just how to tell that story and how to fix it in memory so that its lessons would not be forgotten. Degussa, the company whose subsidiary had supplied the poison gas Zyklon B, would be memorialized for its infamous role in the Holocaust on the very memorial to the Holocaust it helped make resistant to water and weathering. Even Nazi architecture is sometimes proudly displayed, with new contexts giving it new meaning, as in the case of the monumental Berlin stadium that Hitler had built for the 1936 Olympics and that was reconstructed for the 2006 soccer World Cup. As the Hamburg architect Volkwin Marg, whose firm oversaw the $283 million project, put it, "You can't overcome history by destroying it. We have to overcome our role in history by demonstrating it."[3]

But West Germany's highly developed culture of memorialization met its greatest challenge on a political field, when, after 1989, it had to accommodate its East German counterpart. As Jeffrey Herf points out, forty years of East German memorialization had resulted in divided memories and distinct pasts.[4] How could these ever be incorporated into a single history?

The West German way of reading the Holocaust won out in all significant public respects. In the course of reunification, the law of the *Rechtsstaat* worked to turn the East German state into an *Unrechtsstaat*, a state in which there was no law to be recognized as lawful. The Stasi became the defining image of how the East German state had operated, and the trials of the *Mauerschützen* (border guards under military orders to shoot) defined relations between that state and its citizens as they *should* have existed, but so evidently had not. East Germany ceased to be a text; archived away in the extensive documentation of two Enquete Kommissions, it was no longer obligatory reading for the present. Instead it became a context: a collection of now irrelevant political life that could at most explain East Germans' moral decline and ultimate failure. In the conflict between Marianne Birthler and Robert Leicht over Chancellor Kohl's Stasi files, we saw how a West German sensibility prevailed over an East German one when it came to investigating the full extent of the Stasi's powerful reach in order to understand its abuses of power and prevent their recurrence.

At the same time, East German public life, and its ethics and bioethics, were entirely erased. As we have seen, these erasures left permanent traces, for example, in the Enquete Kommission on Law and Ethics in Modern Medicine's refusal to accommodate comparisons of German abortion policy with that of East Germany. Such comparisons, it seemed, might improperly suggest that the state that had been relabeled an *Unrechtsstaat* could provide an alternative way of thinking about bioethical topics. We have also seen how, in the course of reunification, the Gatersleben Encounters changed as one way of making science accountable eclipsed another. In the years after the fall of the Wall, some participants felt, scientists no longer entered into a dialogue in which both sides anticipated a change in perspective; instead the biannual meetings became a place where scientists explained their doings to a public redefined as "lay" and as deficient in its understanding of science.

East Germany had operated, more than just rhetorically, as a tightly bound collective to which everyone contributed and from which almost everyone benefited. East German bioethics had been part of a wider biopolitics, a forceful solidarity that engaged citizens in public forms of collective caretaking. Health care had been universally available, women were officially equal to men in public life, and abortions were paid for by the state. The regime shaped subjects who were, and to some extent still are, engaged in weaving a dense tapestry of collective life. The ongoing private activities of the retired ethics professor Ernst Luther as a pillar of his local moral community resonate with the activities of Alex's mother as she seeks to further solidarity from her bedroom in the film *Goodbye Lenin!* With the takeover of the former East German public sphere, however, public solidarity has either disappeared or become entirely privatized. The tourists who travel by boat through the parts of the former East Berlin where the reunified government has now taken shape can hear the words that I heard so many times while I worked at the Enquete Kommission on Law and Ethics in Modern Medicine: "*Aber das ist jetzt alles Vergangenheit.*"

: : :

Two generations after the end of World War II, Germany remains preoccupied with learning the lessons of the Holocaust. As we have seen, there is no clear consensus on how this learning should proceed—either in general or in the specific context of fashioning bioethics. The admonition "Never Again!" that I frequently heard and as often read expresses a determination never again to start a war from German soil, reducing the nation to ruin, just as it means that never again will a murderous dictatorship arise in Germany. World War II and the Holocaust remain "moral resources" for Germany's younger generations, but as the unruly demonstrations of November 9 dramatize, German youth are not united in how to interpret their meaning. References to the Holocaust both permit and disallow certain systemic changes and adaptations; similarly, fifteen years after reunification, Germans from the former East and West said they had grown further apart in their sense of national identity.

I have argued throughout this book that progress through *Bildung* is central to contemporary German culture, but that in attempting to master memory, it will still be necessary for Germans to traverse many highly contested pathways of sense-making. The process of making moral sense of stem cells, and in that context finding grounds for ethical research, has illustrated some of these perplexities.

A recent German newspaper article contained this statement: "If we would learn the lessons of the Holocaust, we would not need ethics commissions." Perhaps so, but for now, at least, it appears that Germany is not yet ready to stop learning. Accordingly, bioethics in Germany is, as we have seen, a place of engagement, even, at times, of schizophrenia; it is not a reflection of a complete *Aufarbeitung* of the past, nor of a resulting moral complacency.

Acknowledgments

In writing this book I have had many supporters. I am deeply indebted to a great number of people in Germany who not only made Berlin a fascinating research site but also made the city feel like a home to me. It is impossible to name everyone, but I thank particularly the parliamentary Enquete Kommission for allowing me to attend its meetings in early 2002 and the members of the secretariat for their patience in answering my questions. Cornelia Beek's generosity and conscientiousness greatly facilitated my work. Ingo Härtel provided many insights into otherwise obscure proceedings, and became a friend. I also thank my interviewees—scientists, physicians, politicians, philosophers, ethicists, genetic counselors, and nonprofessionals—for the time and care they took in sharing their views with me, even when the topics and questions I was interested in must have seemed dry and obvious to them.

Throughout my research in Germany, and on subsequent visits, I have been affiliated with the Institute for European Ethnology at Berlin's Humboldt University. I thank Wolfgang Kaschuba for opening the institute's doors for me and for providing a workspace. I also thank Stefan Beck and Michi Knecht for integrating me into their growing intellectual community and for inviting me to teach several courses in medical anthropology and in science studies with them. Their friendship and intellectual curiosity have occasioned many stimulating discussions on anthropology, science, and cultural difference that have pushed my thinking forward.

Material support is also essential to anthropological work, and I was fortunate to have many generous supporters. A fellowship from the Social Science Research Council funded my first year of research in Germany. When it became clear that I would stay for a second year, the Institute for Science and Technology Research (IWT) at the University of Bielefeld generously admitted me as a temporary stipendiary research fellow into its Graduiertenkolleg Auf dem Weg in die Wissensgesellschaft. I thank the members, and especially Peter Weingart and Alfons Bora, for engaging with my work and for allowing me to present parts of it in a weekly colloquium. I wrote a draft of this book while I was in residence in the Program on Science, Technology and Society (STS) at Harvard University's John F. Kennedy School of Government. An NSF training grant (Reframing Rights: Constitutional Implications of Technological Change, NSF Award Number SES-9906834) and a Charlotte W. Newcombe Fellowship supported me during those years.

I thank my editor at the University of Chicago Press, Karen Merikangas Darling, for her unfailing support and for seeing the book through to publication. Two anonymous reviewers made many helpful suggestions that greatly improved the clarity and concision of the final manuscript. Dominic Boyer, who also reviewed the book, commented generously on several drafts. His careful reading and his knowledge of German social history were invaluable in revising and strengthening my arguments. Andreas Glaeser provided a final set of helpful comments. I also thank Norma Sims Roche, my copyeditor, for a masterful job of flagging errors and inconsistencies. It was humbling to see how much she improved what I had considered a well-edited manuscript. All remaining flaws are of course my own.

Several readers have helped me make this book what it is. Collectively, these friends and colleagues were generous enough to make my concerns their own, and they gave me more of themselves than I could ever have asked for. I thank John Borneman for his detailed reading of my work. His expertise on East and West Germany, and on the reunification of moral orders, is evident throughout this work. Alexander Görsdorf and Frances Chen read the entire manuscript, and I thank them for their many thoughtful comments. I thank Dörte Bemme for her intense intellectual engagement. Throughout much of the early writing process, she brought her powerful ethnographic sensibilities and empathic ability to bear on my thinking and writing. I also thank Anne

Dippel for bringing a well-trained eye for historical detail to her reading of the manuscript and for the rigor and sympathy with which she debated and supported many of my observations and interpretations. James Pile has been a model of intellectual rigor and deep theoretical engagement. His friendship, and his generosity and enthusiasm in sharing his insights and his stories from the field, have been inspiring.

I most especially thank Sheila Jasanoff, who for years talked through my material with me almost daily. On countless occasions we discussed my observations, and many of the ideas contained in this book originated in our ongoing dialogue. I cannot thank her enough for her friendship and for sustaining a deeper form of intellectual engagement than I could ever have hoped for.

∴ ∴ ∴

Some of the material in this book has appeared in print before. Parts of chapter 3 were published in "Knowledge Rites and the Right Not to Know," in *Political and Legal Anthropology Review* 30(2): 269–87. Portions of chapter 5 were published as "The Politics of Transparency and Surveillance in Post-Reunification Germany," in *Surveillance and Society* 8(4): 396–412. And parts of chapter 6 have been published in "Converting Ethics into Reason: German Stem Cell Policy between Science and the Law," in *Science as Culture* 17(4): 363–75.

Notes

PRETEXT

1. The Paul-Löbe-Haus is named after the Social Democrat Paul Löbe (1875–1967), who served as the last president of the Weimar Reichstag. Löbe had helped draft the Weimar constitution, which made him a symbol of Weimar democracy, as well as Germany's current 1949 constitution. The Paul-Löbe-Haus contains 550 offices for representatives, 19 conference rooms, and about 450 offices for parliamentary commissions.

2. Unless otherwise noted, all translations are my own.

3. This difference may have to do with the different ways in which democracy was established in the two countries: through acts of men rebelling against a distant power in the United States, and through victorious outside forces in postwar Germany.

4. Katrin Gerlof, "Nichts zu verbergen," in *Deutscher Bundestag—Blickpunkt*, March 2002.

5. See Harvey's *Paris, Capital of Modernity* for complex and richly textured analyses of the changes that modernity brought to that city.

6. For an analysis of what Deborah Ascher Barnstone calls the "ideology of transparency" in the construction of parliament buildings, see her book *The Transparent State*, in which she challenges the official idea that a particular form of architecture can bring about democracy.

7. There had been some debate over the fact that a foreign architect would be putting a new cap on the German parliament, thereby seemingly imposing transparency on German politics from the outside. Early misgivings were quickly forgot-

ten, however, and Germans now find the glass dome attractive, as millions of visitors over the years attest. The Reichstag dome has become a recognizable logo for Berlin and appears on subway decorations and even in beer ads around the city. A travel article in Britain's *The Independent* listed it at the head of Berlin's top fifty attractions.

8. Numerous brochures intended for Bundestag visitors make these connections in more or less explicit form.

9. For a study similar in spirit, see Paul Rabinow's *French Modern*, which analyzes discourses of colonialism, urbanism, and ethics to trace perceptions of the social environment in France from 1830 to 1940.

10. The most famous of these ruins is perhaps the Kaiser Wilhelm Memorial Church, near Bahnhof Zoologischer Garten. Built around 1900, its 113-meter tower was the tallest in Berlin. Having sustained heavy damage during World War II, the 68-meter-tall ruin was left standing as a war memorial and surrounded by modernist structures that inform visitors about the war.

11. There was a seven-year struggle over whether the inscriptions should contain a standardized text or whether the party affiliations of the mainly Social Democratic and Communist representatives should be included. See Endlich and Lutz, *Gedenken und Lernen an Historischen Orten*.

12. A friend of a friend, a native of another EU country, told me she would have to take 600 hours of cultural training before she was eligible for the full welfare benefits of German citizenship.

13. Laura Nader coined the term "studying up" in her influential 1969 article "Up the Anthropologist—Perspectives Gained from Studying Up," a call to study those with power rather than the powerless. More recently, Ulf Hannerz used the term "studying sideways" in his 2004 book *Foreign News*.

14. I will use the term "bioethics" throughout in the sense it has acquired in the United States—namely, to refer to the ethical analysis of social questions arising from developments in the life sciences.

15. Both the EK and the NER underwent changes in the years after I left Germany, and in 2008 they were succeeded by the German Ethics Council (Deutscher Ethikrat), which incorporated features of both of its predecessors. Although formally succeeding the NER, the German Ethics Council held its first meeting in the Reichstag building. Its twenty-six members are appointed for four-year terms by both parliament and the executive.

16. Stem cells come in many variations, and most uses are noncontroversial. In the remainder of this book, I will use the terms "stem cells," "stem cell research," and "stem cell debate" to refer to the controversial human embryonic stem cells, research on human embryonic stem cells, and the debate over human embryonic stem cells. This usage is consistent with the terms commonly used by the people I studied and by popular media in Germany, and even with the wording of the Stem Cell Law.

CHAPTER ONE

1. Gerth and Mills, *From Max Weber: Essays in Sociology*, 196–264.
2. See, for example, Jürgen Habermas's influential booklet *Die Zukunft*

der menschlichen Natur, in which he warns that the rise of choices available in creating and enhancing life might lead to a liberal variant of eugenics.

3. One example is a series of billboards drawing on questions submitted to the website www.1000fragen.de, which the organization Aktion Mensch had set up. Sample questions, provocatively illustrated, included "Do embryos have an expiration date?" "What happens if my child wants designer-parents?" "Do *I* want to know everything, or does my insurance company?" "Would *you* have passed a pre-natal check?" (http://www.1000fragen.de/projekt/anzeigen/index.php?sid=5897dddc3193fe176e1bb8d9beb6cbcb, accessed April 25, 2012.)

4. For an early and concise statement of this relationship, see Edwin J. Holman's 1972 article "The Time Lag between Medicine and Law."

5. Other nations, of course, share this sense, and similar situations arise there. Here I show how concerns over legislating research acquired a special texture in Germany, where concerns over research ethics perhaps resonate more deeply than in most other national contexts.

6. See, for example, Kipphoff, "Verweile Nicht, du bist so schön!"

7. See, for example, Leicht, "Vom Reichstag fallen die Hüllen; Aufbruch zur Neuen Republik."

8. See Leinemann, "Das Neue ist die Größe."

9. See Steven Rosenberg, "Merkel faces tough EU challenge," BBC News, Berlin, January 19, 2007, http://news.bbc.co.uk/2/hi/europe/6281323.stm (accessed April 25, 2012).

10. See Charles Maier's *The Unmasterable Past: History, Holocaust, and German National Identity*, for a nuanced discussion of the German *Historikerstreit*, a debate over how to write Germany's difficult history.

11. For a reflection on the significance of this dedication in light of the struggle over German identity, see Sperling, "Managing Potential Selves."

12. According to one PDS member I interviewed, parliament had routinely rejected proposals that his party had made. According to this interviewee, these rejections were an example of discrimination against the successor to East Germany's Socialist party.

13. This symbolic power would become actual in the summer of 2005, when President Horst Köhler dissolved parliament after Chancellor Schröder had called a vote of no confidence.

14. The word "legislator" (*Gesetzgeber*, literally "lawgiver") is used in the singular, which connotes the perceived unity of parliament.

15. See Bimber, *The Politics of Expertise in Congress*, for a detailed discussion of the Office of Technology Assessment.

16. According to Germany's national pay scale, B3 is the salary class for an Oberst (colonel), a high-ranking staff officer commanding an entire army or air force regiment, or a Kapitän zur See (captain at sea) in the navy.

17. The Benda Commission was named after its president, Ernst Benda, a former president of the Federal Constitutional Court in Karlsruhe. It comprised scientists, physicians, theologians, philosophers, and jurists who were tasked with analyzing the ethical and legal aspects of reproductive medicine, embryology, and gene therapy and to recommend legislative measures. In its

1985 report the commission recommended the prohibition of all embryo research, excepting only research that directly benefited the embryo itself or that served well-defined medical goals of high priority (*hochrangig*).

18. Beck-Gernsheim is quoted in "Eher ein Zustimmungsgremium" ("More of an Approval Commission"), *Tageszeitung*, May 3, 2001.

19. Minister of Health Andrea Fischer of the Green Party resigned in January 2001 because of the BSE scandal. Within 24 hours, Schröder nominated Ulla Schmidt (SPD) to be her successor. The ethics commission under Fischer had a reputation for caution about controversial developments in medicine, such as preimplantation genetic testing of IVF embryos, and it had been working on a bill to regulate and restrict gene technologies. Schmidt halted this endeavor, dissolving the commission altogether in March of 2001 and making some radical personnel changes in her ministry. According to her, the chancellor's Nationaler Ethikrat was now the more appropriate forum for exploring the implications of modern medicine.

20. Others who declined the chancellor's call were Ernst-Wolfgang Böckenförde (a former judge on Germany's Federal Constitutional Court), Frank Schirrmacher (a well-known science editor for the *Frankfurter Allgemeine Zeitung*), and Hans Peter Stihl (a former president of the German Chamber for Industry and Trade).

21. Another NER member gave me a different explanation for Kollek's inclusion. According to Wolfgang Van den Daele, Kollek represented a conciliatory gesture toward parliament, since she had been a staff member in the earlier Enquete Kommission on Chancen und Risiken der Gentechnologie. Van den Daele himself had served on that commission as an expert member.

22. The German president Roman Herzog had held the initial Berliner Rede in 1997, in which he made the now famous remark, "*Durch Deutschland muss ein Ruck gehen.*" (A push must go through Germany.) He was referring to a widely felt need for optimism, vision, and courage to implement social changes that would prepare Germany for the twenty-first century. In subsequent years foreign dignitaries were invited to give the Berliner Rede, until Johannes Rau became president and used the institution of the Rede to deliver an annual address of his own to the nation.

23. "Sie (die Ethikräte) sollen die Konflikte nicht entscheiden, sondern sie so klar wie möglich präsentieren und dabei Demagogie, falschen Schlüssen und vorschnellen Schuldzuschreibungen vorbeugen. Mehr nicht. In Ethikräten sitzen keine Spezialisten für moralische Antworten, die gibt es nicht. Die Antwort muss jeder/jede schließlich selber finden—wie bei der Abtreibung." Van den Daele is quoted in "Wir können mit Grenzen umgehen," *Die Welt*, May 30, 2001.

24. In my final discussion with Cornelia Beek, the head of the EK secretariat, she told me that EK members had gone on several international trips, to the United States, to Great Britain, and to Iceland. The function of these trips had been to gather *authentic* information, to reformulate the questions asked, to widen members' horizons, and to make contact with people who were engaged in similar struggles over research ethics in their own countries. The EK's overall goal was to shore up its legitimacy by enrolling allies and building a comprehensive conceptual map. Of course people in countries with less

restrictive laws were not so interested in a country as restrictive as Germany. The British, for example, had little interest in getting to know the German solutions to genetic testing. In Cornelia's telling, no one had any interest in looking at restrictions, since this meant looking back rather than forward. The Icelandic protesters against DeCode's genetic database, on the other hand, were very interested in the Germans' experience.

25. I discuss this shift and others in greater detail in Sperling, "Converting Ethics into Reason."

26. The NER inspired the media's ironic imagination. Shortly after the council's appointment, the prominent German weekly *Die Zeit* began a column called "Nationaler Ethikrat," in which it humorously offered "ethics advice" for people who were in the news, or offered ironic reflections on life's deep questions. One might contrast such lampooning of a national ethics council with a *New York Times Magazine* column called "The Ethicist," in which Randy Cohen resolved readers' dilemmas for 12 years before Ariel Kaminer succeeded him. Readers' questions frequently revolved around the ethics of money (such as the technicalities of just compensation, or the mechanics of fair exchange), and irony is not a marker of this column. Cohen's undertone was always that laws had to be followed, that money is serious and real, and that a market ethics of reciprocity should govern interpersonal relationships.

CHAPTER TWO

1. Although politicians are important social actors, there is little ethnographic work on politicians in action. One exception is Fenno, *Watching Politicians*. While Fenno's account provides in-depth journalistic portraits of high-level politicians, I want to understand the production of political order rather than the behavior of individual political actors.

2. The case was controversial in Germany because it suggested that some lives are not worth living, an implication that people with disabilities particularly resisted. In the United States the legal concept of "wrongful birth" captures similar cases.

3. While the commission may begin its work when twenty-five percent of the members are present, votes require a fifty percent presence. Since textual changes involve constant votes, technically, any commission member could have halted work on the text at any time.

4. To this day, former secretariat members organize an annual summer barbecue in a Berlin park, which scheduling conflicts have unfortunately kept me from attending most years.

5. In German usage the greater degree of intimacy and lesser degree of formal respect of the "*Du*" is "offered" by the higher-status person to the lower-status person.

6. At the NER those relations became fraught in other ways, however, when romantic relations crossed hierarchies. Some persons now were subordinate to one partner of a couple while the other partner was subordinate to them.

7. See Latour and Woolgar, *Laboratory Life*, for an analysis of how material inscriptions in scientific laboratories are transformed into scientific papers.

8. I sometimes got the sense that commission members felt the pride of discovery when they found the latest piece of relevant news and could bring it to the communal table. It was almost as if they were thereby sharing in the scientific discovery itself.

9. I witnessed numerous contexts in which a vote on a text made that text immutable, even though all those who had voted were still in the room and could have changed their minds. I never witnessed the reopening of an issue that had been closed by a vote.

10. The term *Flickenteppich* (or *Flickenteppiche* in the plural), when used to refer to legal matters, denoted lack of rational order and had a derogatory meaning.

11. For a strong and sustained critique of the universalizability of *Homo economicus*, see the oeuvre of Marshall Sahlins, in particular "The Sadness of Sweetness."

12. The book, a 700-page interdisciplinary commentary on German transplantation law, was edited by the law professor Wolfram Höfling. It included contributions from other lawyers, physicians, and social scientists as well as a comprehensive collection of original documents related to the practice of organ transplantation in Germany.

13. Michael Polanyi, in his *Personal Knowledge*, coined the term "tacit knowledge" to draw attention to the idea that unconscious and unarticulated operations underlie many cognitive processes and behaviors.

14. Commission meeting, January 28, 2002.

15. Commission meeting, February 4, 2002.

16. See Latour, *Science in Action*, for an elaboration of this argument.

17. The Deutscher Ethikrat now occupies the NER's former website, www.ethikrat.org. Archived documents of the NER and the EK are available at http://www.ethikrat.org/archiv/nationaler-ethikrat (accessed April 25, 2012) and at http://www.ethikrat.org/archiv/enquetekommissionen (accessed April 25, 2012), respectively.

18. See Leinemann, "Das Neue ist die Größe."

19. Ironically, breakthroughs in reproductive technologies such as the cloning of Dolly the sheep have tended to render the male superfluous, not the female.

20. In his *Outline of a Theory of Practice*, the French sociologist Pierre Bourdieu draws on research in Algeria to show that in rule-bound societies there are what he called "rules for breaking the rules," which remain implicit. In distinction, I am pointing out the move of German bureaucracy to offer one set of rules and then to make explicit another set of rules for how to follow the first one.

CHAPTER THREE

1. For an interesting set of observations on the German government's efforts to link transparency to democracy through the symbolic use of glass in its official architecture, see Ascher Barnstone's *The Transparent State*.

2. For juridical treatments of what the role of transparency in legislation

and politics ought to be, see Bröhmer's *Transparenz als Verfassungsprinzip* and Steffani's early critique, *Parlamentarismus ohne Transparenz.*

3. Franz Boas, *Anthropology and Modern Life.*

4. Marshall Sahlins, in *Islands of History*, compressed Boas's principle into the line "the seeing eye is the organ of tradition."

5. Strathern, "The Tyranny of Transparency."

6. Foucault, *Discipline and Punish*; and Ezrahi, *The Descent of Icarus.*

7. One of the central figures in Shapin and Schaffer's seminal *Leviathan and the Airpump* is the gentleman witness who "modestly" observes experiments and thus helps produce disinterested knowledge of the natural world that all can agree on.

8. See Jasanoff, *States of Knowledge*, for an important collection of essays analyzing the relationship between scientific knowledge and political power through Jasanoff's concept of co-production.

9. Stölb, "Undurchsichtige Tricks mit der Gläsernen Kuh." The tracking system requires cow owners to register and unregister their animals promptly upon birth, death, or change in ownership.

10. Comaroff and Comaroff, "Transparent Fictions; or, the Conspiracies of a Liberal Imagination: An Afterword."

11. A slightly modified version of this brochure is available at http://www.bundestag.de/kulturundgeschichte/architektur/reichstag/architektur/index.jsp (accessed April 25, 2012).

12. Foucault, *Discipline and Punish*, 207.

13. Ibid.

14. Ibid.

15. Ibid.

16. Foucault, *Discipline and Punish*, 317.

17. For a thorough critique of where political practice does or does not live up to an ideal(ized) version of transparency, see Eckert's *Transparenz im Gesetzgebungsprozess*. See especially pages 131–49 for Eckert's discussion of the plenary debate.

18. For a summary review of such claims, see Brodwin, "Genetics, Identity, and the Anthropology of Essentialism."

19. For an ethnographic account of the interpersonal dilemmas raised by genetic testing, see Konrad, "From Secrets of Life to the Life of Secrets."

20. These concerns have given rise to a growing academic literature that explores the potential consequences of increasingly understanding ourselves through our genes. For exemplary and representative treatments, see Wolff, "Eugenik und genetische Beratung"; Duster, *Backdoor to Eugenics*; Paul, "Eugenic Anxieties, Social Realities, and Political Choices"; and Novas and Rose, "Genetic Risk and the Birth of the Somatic Individual."

21. Schiller, "Brief an Friedrich Christian von Augustenburg, 13. Juli 1793."

22. Schiller, "Brief an Friedrich Christian von Augustenburg, 13. Juli 1793," 262–68.

23. See Dumont's *German Ideology*, particularly pages 69–144, as well as Boyer's *Spirit and System*, especially pages 46–70.

24. Klaus Vogel, "Editorial," in Bürgerkonferenz: Streitfall Gendiagnostik Newsletter, 06/2001.

25. For images of the building and further information on history and events, see the website of the Deutsches Hygiene Museum Dresden, http://www.dhmd.de/neu/index.php?id=12 (accessed April 25, 2012)

26. Roth, "Menschenökonomie," 46.

27. Quoted in Roth, "Menschenökonomie," 51.

28. For some of the strategies that Lingner and others used to present hygiene as both a personal and a national problem, see Nikolow, "Der statistische Blick auf Krankheit und Gesundheit."

29. Galison, in his "Aufbau/Bauhaus," offers a complex treatment of how Bauhaus and logical positivism in the 1920s and 1930s articulated mutually constitutive visions of modernism that emphasized what Galison calls "transparent construction."

30. For images of the factory and the motors, see Roth, "Menschenökonomie," 58–59, 63.

31. Roth, "Menschenökonomie," 53.

32. Images such as the one I describe were not uncommon at the time, and Kahn himself authored numerous other anatomical and physiological illustrations. The image I describe is his best known. It can be seen at the website of the U.S. National Library of Medicine, National Institutes of Heath, http://nlm.nih.gov/exhibition/dreamanatomy/da_g_IV-A-01.html (accessed April 25, 2012).

33. Roth, "Menschenökonomie," 43.

34. Roth, "Menschenökonomie," 41.

35. This is the percentage mentioned in the evaluating report, Zimmer, *Begleitende Evaluation*, 11. Public descriptions of the conference mentioned 255 responses, which is a standard response rate.

36. In Schicktanz and Naumann (editors), *Bürgerkonferenz: Streitfall Gendiagnostik*.

37. Zimmer, *Begleitende Evaluation*, 11.

38. On how these concerns play out in national debates on immigration and stem cell research, see Sperling, "Managing Potential Selves."

39. Zimmer, *Begleitende Evaluation*, 12.

40. See Habermas, *Theory of Communicative Action*.

41. Zimmer, *Begleitende Evaluation*, 29.

42. The term *mündig* (competent) is used in legal language to denote a person of legal age and full competency. To German ears the term, when used in the context of citizenship, resonates with Kant's essay "What is Enlightenment?," which defined enlightenment (*Aufklärung*) as "man's release from his self-incurred tutelage [*Unmündigkeit*]."

43. According to the evaluation report, the participants also criticized the unnecessary repetition resulting from the experts reading well-known statements. Just as the experts had received the participants' questions in advance, the participants too had received their responses in advance. Because it was not clear to everyone whether the participants would have sufficient time to read everything beforehand, the organizers asked the experts to deliver their statements orally in abbreviated form. The experts, however, opted to read

their statements out loud in their entirety. See Zimmer, *Begleitende Evaluation*, 23.

44. The participants complained that a real dialogue with the experts never occurred. Instead, they noted, the physical seating arrangement exacerbated the polarization. See Zimmer, *Begleitende Evaluation*, 22.

45. Zimmer, *Begleitende Evaluation*, 10.
46. Zimmer, *Begleitende Evaluation*, 38.
47. Zimmer, *Begleitende Evaluation*, 37.
48. Zimmer, *Begleitende Evaluation*, 40.
49. Zimmer, *Begleitende Evaluation*, 20.
50. Zimmer, *Begleitende Evaluation*, 45.
51. Zimmer, *Begleitende Evaluation*, 42.
52. Zimmer, *Begleitende Evaluation*, 41.
53. Zimmer, *Begleitende Evaluation*, 18.
54. Zimmer, *Begleitende Evaluation*, 19.
55. Zimmer, *Begleitende Evaluation*, 33. There has been some disagreement about how to evaluate the citizen conference. A professor of sociology from the University of Bielefeld, who accompanied the conference as a researcher, claimed that it was absolutely necessary to produce audio and video recordings of every step of the participants' deliberations. The conference conveners, on the other hand, insisted that "the procedure itself had primacy" (*das Verfahren selbst hat Vorrang*) and that "the citizens have to decide" (*die Bürger müssen entscheiden*) the extent to which they would be documented.

56. Zimmer, *Begleitende Evaluation*, 24.

57. See Duden, *Die Gene im Kopf—der Fötus im Bauch*; and Samerski, *Die verrechnete Hoffnung*, on similar claims. Petersen, in "The Best Experts," reports on an empirical study aimed at understanding how patients diagnosed with genetic diseases learn about, live with, and manage their conditions. He finds that living with such information presents unique challenges due to the inevitable and heritable nature of genetic illness.

58. An example that has been widely cited, including at the citizen conference, is the drastic decline in Germany in births of children with trisomy 21, which causes Down's syndrome.

59. Michael Sandel also makes this argument in "The Case against Perfection."

60. For an emblematic statement, see Arno Orzessek's "Perfektion ist ein Verbrechen," *Süddeutsche Zeitung*, November 28, 2001. The article closes with a quote from Baudrillard that supposedly summarizes the citizen opinion's main concern: "Crime is never perfect, but perfection is always a crime."

61. Zimmer, *Begleitende Evaluation*, c.

62. The evaluation report concedes that the resonance of the conference in both the national and local press was shallow.

63. Zimmer, *Begleitende Evaluation*, 48.
64. Zimmer, *Begleitende Evaluation*, c.
65. Zimmer, *Begleitende Evaluation*, 34.

66. A comparable sense of revelation accompanied the announcement of the human genome's decoding in 2000. Many claimed that we knew some-

thing "essential" about human nature now that we had made the human genome "transparent."

CHAPTER FOUR

1. Germany specifically opted to call its constituting document a (provisional) *Grundgesetz*, or Basic Law, rather than a (permanent) *Verfassung*, or constitution, because it thought the division of the nation would be temporary, and that a reunified Germany would then give itself a *Verfassung*. Over the decades, however, the Basic Law came to be seen as a stable and permanent foundation of German democracy, and it was not altered significantly after reunification.

2. See, for example, Arendt, *Eichmann in Jerusalem*.

3. The following pages lean on pages 296–299 of Martin Morlok's comprehensive commentary on Article 4 of the German Basic Law.

4. Bainton, *Here I Stand: A Life of Martin Luther*, 142–44.

5. Hobbes, *Leviathan*, chapter 7, page 48.

6. Foucault, *Madness and Civilization*, 213.

7. The U.S. constitution does not grant freedom of conscience as such. While the German Basic Law protects individuals from the state by safeguarding their inner core of conscience and dignity, the U.S. constitution spells out a number of freedoms and rights that empower citizens to protest infringements on their ability to believe, speak, and associate freely for political or religious purposes.

8. Today the legal protection afforded to both faith and conscience is stronger in Germany, or at least more explicit, than in a number of other European states.

9. Morlok, "Glaubens-, Gewissens- und Bekenntnisfreiheit, Kriegsdienstverweigerung," 302.

10. See Hobbes, *Leviathan*; see also *Oxford English Dictionary*, 2nd ed., s.v. "conscience."

11. In German: "Zwei Dinge erfüllen das Gemüt mit immer neuer und zunehmender Bewunderung und Ehrfurcht, je öfter und anhaltender sich das Nachdenken damit beschäftigt: Der bestirnte Himmel über mir, und das moralische Gesetz in mir."

12. Kant, *Metaphysik der Sitten*, 96–97.

13. Kant, *Metaphysik der Sitten*, 400–401.

14. As I showed in the previous chapter, *Nachvollziehbarkeit*—the ability to follow the reasoning in support of a collective decision—is part of the German understanding of transparency.

15. Deleuze, *Kant's Critical Philosophy*, x.

16. Media examples include the newspaper headlines "Heute sind wir alle Spanier" (after the March 2004 terrorist attacks in Madrid); "Seveso ist überall" (after the dioxin leak in Italy); "Tschernobyl ist überall" (after the nuclear reactor meltdown in Ukraine); "Wir sind alle Amerikaner" (after the attacks of September 11, 2001).

17. In German, the two words *Pflicht* and *Pflege* are also linked etymologically.

18. See primarily Habermas, *Theory of Communicative Action*.
19. Kant, "Was ist Aufklärung?," 7–20.
20. Foucault, "What is Critique?," 23–61.
21. For more information, see the website of INFO-SERVICE Beamte/Öffentlicher Diens, http://www.beamten-informationen.de (accessed April 25, 2012).
22. Max Weber, "Politik als Beruf," in Gerth and Mills, *From Max Weber*, 77–128.
23. The paradigmatic account of these "desk-criminals" (*Schreibtischtäter*) is Arendt's *Eichmann in Jerusalem*.
24. Stödter, *Deutschlands Rechtslage*.
25. Heyland, *Das Widerstandsrecht des Volkes*, 70.
26. Heyland, *Das Widerstandsrecht des Volkes*, 81.
27. Heyland, *Das Widerstandsrecht des Volkes*, 82–83.
28. BGHZ (Bundesgerichtshof für Zivilsachen) 13, p. 296.
29. This information can be found on a website commemorating the East German uprising of June 17, 1953, which the Bundeszentrale für Politische Bildung (Federal Central Authority for Political *Bildung*) maintains: http://www.17juni53.de/chronik/5312.html (accessed April 25, 2012). See also Klaus-Detlev Godau-Schüttke, "Von der Entnazifizierung zur Renazifizierung der Justiz in Westdeutschland," in an on-line magazine hosted by the Max Planck Institute for European Legal History; available at http://www.rewi.hu-berlin.de/online/fhi/articles/0106goau-schuettke.htm (accessed April 25, 2012).
30. See Peine and Heinlein, *Einführung in das Recht des Berufsbeamtentums*.
31. Details are available on the archive page of www.waechterpreis.edu (accessed April 14, 2012).
32. See, for example, *Die Zeit*, "Ein bisschen Folter gibt es nicht," November 25, 2004, www.zeit.de/2004/49/Folter (accessed April 14. 2012).
33. Representative statements taken from readers' letters read:

> "It is always depressing to note how little influence legal, historical, or humanitarian arguments have. Pointing out that torture in all its forms is absolutely prohibited nationally and internationally is not a legalistic nicety. . . . All torture, including 'a few slaps in the face,' is prohibited. It has taken centuries for humanity to recognize the inhumane and destructive essence of all torture. . . . It is especially necessary for us Germans to look back on the years of 1933–1945 to see what barbarism results from making humanitarian norms relative. . . . I wonder if members of the police are not planning to use this case to touch/violate [*antasten*: resonates with *unantastbar* (inviolable)] the first articles of the Basic Law and of the UN Anti-torture Convention."

> "This case is not about emotions; it is about the foundations of our society, and about fundamental questions of our *Menschenbild*."

"The dignity of the victim cannot be weighed against the dignity of the perpetrator."

"According to Christian teachings the human being does not have a *Wert* [price, value], but a *Würde* [dignity]. And of dignity there is no more or less."

34. Article 1 of the UN Convention against Torture states in part that "torture" encompasses any action causing physical or psychological (in German: *seelische*) pain or suffering to obtain a confession.

35. An article on torture from the *New York Times* reads in part: "Mr. Goss offered a strong defense of what he called 'professional interrogation,' saying that it had resulted in 'documented successes' in averting attacks and capturing terrorist suspects, and he asserted that the procedures currently used by the C.I.A. did not amount to torture. 'Torture is not productive,' Mr. Goss said. 'It is not professional interrogation; we don't do torture.'" Douglas Jehl, "C.I.A. Chief Defends Interrogation Policy and Disavows Torture," *New York Times*, March 17, 2005.

36. Schlink, "An der Grenze des Rechts," 34ff.

37. As Schlink points out, the law holds the inherent danger that, by analogy, the armed forces might soon see justification for bombarding trains or school buses if those vehicles are controlled by terrorists and carrying explosives toward densely populated areas.

38. Schlink's view stands in stark contrast to that of the American pilot of the *Enola Gay*, who, on orders from President Harry Truman, dropped the first nuclear bomb on Hiroshima in 1945, instantly vaporizing an estimated 80,000 Japanese. In an interview, the pilot, Paul Warfield Tibbets, famously claimed that "it was just another mission."

39. The judgment is available at http://www.bverfg.de/entscheidungen/rs20060215_1bvr035705.html (accessed April 24, 2012.)

40. The situation was somewhat muddled because Daschner did not think he was breaking the law. My main point, however, remains intact.

41. Similar information is available at the website of the agency Pro Familia, http://www.profamilia.de/erwachsene/ungewollt-schwanger/schwangerschaftskonfliktberatung.html (accessed April 15, 2012).

42. In January 1998 Pope John Paul II sent a letter to the German bishops asking them to stop signing the post-counseling certificates that enable women to have abortions. Many bishops protested these orders, but the German Catholic Church nevertheless adopted them.

43. The question of confidence had been raised in parliament three times before. Willy Brand in 1972 and Helmut Kohl in 1982 used the vote of no confidence (*Misstrauensvotum*) to dissolve parliament and call for new elections. Helmut Schmidt, in 1982, survived a first vote of no confidence, but lost power seven months later, when parliament voted him out of office with a vote of no confidence on a different issue.

44. In an interesting German twist, the voter does not really vote for a representative, and it is only in very rare cases that a representative is directly

elected. More commonly, one votes for one of the political parties, and the party then decides which persons it sends to parliament. This system explains why the party can demand a degree of loyalty from its members—it was the party that put the individual politicians into power in the first place—and it goes some way toward explaining how the collective of the party can become the reference point of each legislator's individual conscience.

45. This statement resonates with a sentence that I heard Jens Reich repeat several times in public contexts: "I *know* it is wrong, but we must find *reasons*." (Ich *weiss* dass es falsch ist, aber wir müssen *Gründe* finden.)

46. Kass, "Wisdom of Repugnance," 20.

CHAPTER FIVE

1. The plenary protocol is available on the website of the German Bundestag, http://dip.bundestag.de/btp/13/13044.asc (accessed August 18, 2008).

2. Herf, *Divided Memory*.

3. Herf, *Divided Memory*, xi.

4. The Nazis had imprisoned several of East Germany's later leaders, including Erich Honecker.

5. See, for example, Borneman, *Belonging in the Two Berlins*.

6. Borneman, *Settling Accounts*.

7. Borneman, *Settling Accounts*, 7.

8. Röper, "Allzeit verschiedene Deutsche."

9. Niederberghaus, "Abbau West."

10. Unfortunately, when I asked him, Berth had no comparable numbers for young West Germans. It would have been interesting to compare their view of the progress and prospects of reunification with those of young East Germans.

11. One of the amusing exceptions to the perception of East German deficiency was the widely quoted survey finding that East German women made for more pleasurable bedmates. The rumored reason was that they had grown up without the performance pressures that surrounded West German women and had therefore maintained a more "natural" and relaxed approach to their own bodies and to ways of using them.

12. See, for example, Langdon Winner, "Do Artifacts Have Politics," in *The Whale and the Reactor* (Chicago: University of Chicago Press, 1986), pp. 19–39. See also Bruno Latour, "Der Berliner Schlüssel," in Bruno Latour, *Der Berliner Schlüssel* (Berlin: Akademie Verlag, 1995), pp. 37–51.

13. Peter Eisenman on German television, ZDF, November 13, 2003. The video can be watched at http://www.zdf.de/ZDFmediathek/inhalt/25/0,4070,2168345-0,00.html (accessed April 4, 2006).

14. Borneman, *Settling Accounts*.

15. See Anderson, "Census, Map, Museum."

16. The head of the authority had a term of five years, with one possible reelection, and the first "special official" (*Sonderbeauftragter*), Gauck, became a "federal official" (*Bundesbeauftragter*). In 2000, after Gauck had served for ten years, the civil rights advocate Marianne Birthler succeeded him.

17. Over the years the Stasi File Law has also changed several times. In

the first change, in February 1994, the Gauck authority received permission to use information from East Germany's central citizen registry (*Zentrales Einwohnerregister*) to aid its investigations. A second change made it mandatory for all public offices to notify the Gauck authority of any Stasi files or copies thereof in their possession. The third change, in 1998, stated that activities of the Stasi would not be reported if they took place before 1976.

18. Funder, *Stasiland*, 268–69.

19. The documentary film is entitled *Das Ministerium für Staatssicherheit—Alltag einer Behörde*. In it the perpetrators themselves tell of the Stasi techniques for monitoring and coercing individuals.

20. See Macrakis, *Seduced by Secrets*, for a close look at the spy tools that the Stasi used.

21. See Schulzki-Haddouti, *Im Netz der Inneren Sicherheit*, for an overview of the novel surveillance technologies used in the name of national security.

22. These critical voices were published at http://www.politik.de/forum/innenpolitik/17729-image.html. When I checked the site again in late 2011, it had been taken off-line to fix security issues. It is still off-line, but now announcing the arrival, in a few months, of a new portal for citizen participation on-line.

23. More information about the Mauermuseum is available at the museum's website: www.mauermuseum.de/frame-index-mauer.html (accessed April 25, 2012).

24. According to some estimates, more than five thousand East Germans successfully crossed the Berlin Wall alone.

25. Gauck, *Die Stasi-Akten*, 40

26. Gauck, *Die Stasi-Akten*, 92ff.

27. Wagner himself had been key in saving many of the Stasi files from looting and destruction. He worked for the file authority and was offered the job of directing work in the central Stasi archive—a job he declined. He has been unemployed since 1997, when it came to light that the Stasi had recruited him as an unofficial collaborator—*inoffizieller Mitarbeiter*, or IM—in 1981. Wagner's 2001 book, *Das Stasi-Syndrom*, adds a different and often marginalized perspective to the tone that dominates the Stasi file discourse.

28. Wagner, *Das Stasi-Syndrom*, 175ff.

29. Gauck, *Die Stasi-Akten*, 38.

30. Gauck, *Die Stasi-Akten*, 55.

31. Gauck, *Die Stasi-Akten*, 131.

32. Knabe, *Die Unterwanderte Republik*.

33. In 1998 the Stasi File Authority discovered in its archive a secret Stasi order dating from 1970 that showed that the East German state had systematically tried to hide deaths and injuries on the border. In one 1986 case the Stasi pretended that a man had successfully escaped, when in fact border guards had killed him. In a 1966 case the Stasi made the deaths of two children appear to be swimming accidents.

34. The Federal Constitutional Court wrote in 1996: "It is an invalid reason that places the prohibition against leaving the GDR over the right to life of human beings by permitting the premeditated killing of unarmed escapees, because it obviously and unbearably violates elementary commandments of

justice and human rights that *Völkerrecht* protects. This violation is so pronounced in this case that it violates the legal convictions that rest on the value and dignity of human beings and that are common to all peoples; in such cases the positive law must give way to justice (the so-called Radbruch formula)."

35. See "Ehemalige DDR-Politiker über Verfassungsurteil empört," Associated Press, November 12, 1996; see also Heribert Prantl, "Schreibtischtäter heissen Täter und sind es auch," *Süddeutsche Zeitung*, November 13, 1996.

36. Uwe Wesel, "Juristische Konstruktionen," *Tageszeitung*, August 26, 1997.

37. The courts also sentenced fifteen members of East Germany's military leadership, ten of whom went to prison. The commander of the border troops and vice minister of defense, Klaus-Dieter Baumgarten, received the longest sentence: six years, of which he spent three in prison before being pardoned. Baumgarten has since published a book in which he defends the Stasi. West German courts prosecuted numerous members of the SED (Socialist Unity Party); there were about 65,000 court cases against 100,000 accused. Only 5,000 were sentenced, among them Erich Honecker's successor, Egon Krenz.

38. Mikhail Gorbachev, quoted in "Haftstrafen fast einhellig begrüsst," *Süddeutsche Zeitung*, August 26, 1997.

39. Sperling, "Managing Potential Selves."

40. Enquete Kommission, *Schlussbericht*, 58.

41. Enquete Kommission, *Schlussbericht*, 59.

42. Abortion has always been a controversial topic—the 1972 *Fristenlösung* was the first and only law passed in East Germany to which opposition was voiced. See Aresin, "Schwangerschaftsabbruch in der DDR," 92.

43. Although East German women aborted almost twice as often as West German women, their birth rates were approximately 50 percent higher.

44. These conditions were medical indication (danger to the health or life of the mother), eugenic indication (health problems for the fetus), ethical/criminal indication (the pregnancy resulted from rape, incest, or another crime), and social indication (the birth of the child will cause social distress—e.g., poverty—for the mother).

45. See, for example, Gottweis, *Governing Molecules*.

46. Bielka and Hohlfeld, "Biomedizin," 79–142.

47. Lee Silver is the author of the well-known and popular book *Remaking Eden*. There Silver suggests that the developments that the opponents of genetic manipulation of humans fear are just around the corner or already here in slightly disguised form.

48. Bielka and Hohlfeld, "Biomedizin," 111–13.

49. Bielka and Hohlfeld, "Biomedizin," 13–15.

50. Regina Köhler, "Flierl will Ostberliner Wissenschaftler rehabilitieren," *Die Welt*, February 12, 2004.

51. Quoted in Bielka and Hohlfeld, "Biomedizin," 115.

52. Quoted in Bielka and Hohlfeld, "Biomedizin," 114.

53. Merton, "The Normative Structure of Science."

54. Lachmund, "Knowing the Urban Wasteland."

55. According to Wobus, on the other hand, Hohlfeld first attended the Gatersleben Encounters in 1990, after the Wall had fallen. Wobus claims that

Hohlfeld had not "been there" before the Wall, implying that he is not justified in talking about that period. Hohlfeld told me that he was there in 1988. After the Wall fell, he told me, he was no longer invited because the DFG thought he was too critical.

CHAPTER SIX

1. Hartmut Wewetzer, "Stammzell-Import erstmals erlaubt," *Der Tagesspiegel*, December 24, 2002, http://www.tagesspiegel.de/politik/stammzell -import-erstmals-erlaubt-bonner-wissenschaftler-darf-embryonales-gewebe/ 375794.html (accessed April 13, 2012.)

2. Selection is a cultural taboo in Germany, and the Stem Cell Law mandates that imported stem cells must not have been derived from embryos that have been selected. That is, German researchers must not have induced others to kill embryos, nor can they have encouraged others to sort embryos into those worthy of life and those unworthy of it. During the parliamentary debate on January 30, 2002, one person said that hearing the words "embryo selection" reminded him of the ramp in the concentration camp Auschwitz, where Jews were sorted into those who were still fit for labor and those who were doomed to death.

3. For a comparative reading of this decision, see Jasanoff, *Designs on Nature*.

4. One was prepared by Albin Eser and Hans-Georg Koch, the other by two attorneys, Dahs and Müssig.

5. These numbers increased only slightly over the years. The ZES report issued at the end of 2010 mentions that a total of fifty-eight applications for importation had been submitted since the commission's inception, an average of just under seven per year. The report is available at http://www.rki.de/DE/ Content/Kommissionen/ZES/Taetigkeitsberichte/8taetigkeitsbericht.pdf?__ blob=publicationFile (accessed April 24, 2012.)

6. In Hans-Jörg Rheinberger's terms, they are "epistemic things"—they have (or supply) biological meaning only by being worked with. See Rheinberger, *Toward a History of Epistemic Things*.

7. See Christian Schwägerl, "Bock im Ethikgarten: Der Oberembryonenschützer hält nichts vom Embryonenschutz," *Frankfurter Allgemeine Zeitung*, May 16, 2003.

8. Brüstle's application was for work within a general stem cell research program that Wobus had initiated following the isolation of human embryonic stem cells in 1998.

CONCLUSION

1. See Sperling, "Managing Potential Selves."
2. Rogers, *A Shining Affliction*.
3. Quoted in Mark Landler, "In Berlin, Every Cheer Casts an Eerie Echo," *New York Times*, December 8, 2005.
4. Herf, *Divided Memory*.

Bibliography

Anderson, Benedict. "Census, Map, Museum." In *Imagined Communities: Reflections on the Origin and Spread of Nationalism*. New York: Verso, 1991.

Arendt, Hannah. *Eichmann in Jerusalem: A Report on the Banality of Evil*. New York: Viking, 1963.

Aresin, Lykke. "Schwangerschaftsabbruch in der DDR." In *Unter anderen Umständen: Zur Geschichte der Abtreibung*, edited by Gisela Staupe and Lisa Vieth, 86–95. Dresden: Deutsches Hygiene-Museum, 1996.

Ascher Barnstone, Deborah. *The Transparent State: Architecture and Politics in Postwar Germany*. London: Routledge, 2005.

Bainton, Roland. *Here I Stand: A Life of Martin Luther*. New York: Abingdon-Cokesbury, 1950.

Bartolf, Christian. *Mein Gewissen sagt nein: ausgewählte Begründungen von Kriegsdienstverweigerern*. Berlin: Wichern-Verlag, 1996.

Beier, Rosmarie, and Martin Roth, editors. *Der Gläserne Mensch: Eine Sensation*. Stuttgart: Gerd Hatje, 1990.

Bielka, Heinz, and Rainer Hohlfeld. "Biomedizin." In *Wissenschaft und Wiedervereinigung*, edited by Jürgen Kocka and Renate Mayntz, 79–142. Berlin: Akademie Verlag, 1998.

Bimber, Bruce. *The Politics of Expertise in Congress: The Rise and Fall of the Office of Technology Assessment*. Albany: SUNY Press, 1996.

Birthler, Marianne. "Jede Medizin hat Nebenwirkungen: Warum die Stasiabhörprotokolle in Sachen Helmut Kohl kein Staatsgeheimnis sein dürfen." *Die Zeit*, January 25, 2001.

Boas, Franz. *Anthropology and Modern Life*. New Brunswick: Transaction Publishers, 2004.
Borges, Jorge Luis. "Funes, the Memorious." In *Labyrinths*, 59–66. New York: New Directions Publishing, 1964.
Borneman, John. *Belonging in the Two Berlins: Kin, State, Nation*. Cambridge: Cambridge University Press, 1992.
———. *Settling Accounts: Violence, Justice, and Accountability in Postsocialist Europe*. Princeton: Princeton University Press, 1997.
Bourdieu, Pierre. *Outline of a Theory of Practice*. Cambridge: Cambridge University Press, 1997.
Boyer, Dominic. *Spirit and System: Media, Intellectuals, and the Dialectic in Modern German Culture*. Chicago: University of Chicago Press, 2005.
Brodwin, Paul. "Genetics, Identity, and the Anthropology of Essentialism." *Anthropological Quarterly* 75, no. 2 (2002): 323–30.
Bröhmer, Jürgen. *Transparenz als Verfassungsprinzip: Grundgesetz und Europäische Union*. Tübingen: Mohr Siebeck, 2004.
Buruma, Ian. *Wages of Guilt: Memories of War in Germany and Japan*. New York: Farrar, Straus & Giroux, 1994.
Comaroff, Jean, and John Comaroff. "Transparent Fictions; or, the Conspiracies of a Liberal Imagination: An Afterword." In *Transparency and Conspiracy: Ethnographies of Suspicion in the New World Order*, edited by Harry West and Todd Sanders, 287–99. Durham: Duke University Press, 2003.
Deleuze, Gilles. *Kant's Critical Philosophy: The Doctrine of the Faculties*. London: Athlone Press, 1984.
Duden, Barbara. *Die Gene im Kopf—der Fötus im Bauch: Historisches zum Frauenkörper*. Hannover: Offizin, 2002.
Dumont, Louis. *German Ideology: From France to Germany and Back*. Chicago: University of Chicago Press, 1996.
Duster, Troy. *Backdoor to Eugenics*. New York: Routledge, 1999.
Eckert, Christoph. *Transparenz im Gesetzgebungsprozess: Das Prinzip der Öffentlichkeit staatslenkender Entscheidungen zwischen Anspruch der Rechtsordnung und Realität*. Hamburg: Verlag Dr. Kovac, 2004.
Endlich, Stefanie, and Thomas Lutz. *Gedenken und Lernen an Historischen Orten*. Berlin: Landeszentrale für Politische Bildungsarbeit, 1995.
Enquete Kommission Recht und Ethik der Modernen Medizin. *Schlussbericht*. Berlin: Deutscher Bundestag Referat Öffentlichkeitsarbeit, 2002.
Ezrahi, Yaron. *The Descent of Icarus*. Cambridge, MA: Harvard University Press, 1990.
Fenno, Richard. *Watching Politicians: Essays on Participant Observation*. Berkeley: Institute of Governmental Studies, University of California, Berkeley, 1990.
Foucault, Michel. *Discipline and Punish: The Birth of the Prison*. New York: Vintage, 1995.
———. *Madness and Civilization: A History of Insanity in the Age of Reason*. New York: Vintage, 1988.
———. "What is Critique?" In *The Politics of Truth*, 23–61. New York: Semiotext(e), 1997.
Funder, Anna. *Stasiland*. Melbourne: Text Publishing, 2002.

Galison, Peter. "Aufbau/Bauhaus: Logical Positivism and Architectural Modernism." *Critical Inquiry* 16 (1990): 709–52.
Gauck, Joachim. *Die Stasi-Akten: Das Unheimliche Erbe der DDR*. Hamburg: Rowohlt Verlag, 1991.
Gerth, H. H., and C. Wright Mills, editors. *From Max Weber: Essays in Sociology*. London: Routledge, 1991.
Gottweis, Herbert. *Governing Molecules*. Cambridge, MA: MIT Press, 1998.
Habermas, Jürgen. *The Theory of Communicative Action*. Boston: Beacon Press, 1984.
———. *Die Zukunft der menschlichen Natur: Auf dem Weg zu einer liberalen Eugenik?* Frankfurt: Suhrkamp, 2005.
Hannerz, Ulf. *Foreign News: Exploring the World of Foreign Correspondents*. Chicago: University of Chicago Press, 2004.
Harvey, David. *Paris, Capital of Modernity*. New York: Routledge, 2003.
Herf, Jeffrey. *Divided Memory: The Nazi Past in the Two Germanys*. Cambridge, MA: Harvard University Press, 1997.
Heyland, Carl. *Das Widerstandsrecht des Volkes*. Tübingen: J. C. B. Mohr, 1949.
Hobbes, Thomas. *Leviathan*. Cambridge: Cambridge University Press, 1998.
Höfling, Wolfram, editor. *Transplantationsgesetz: Kommentar*. Berlin: Schmidt, 2003.
Holman, Edwin J. "The Time Lag between Medicine and Law." 9 *Lex et Scientia* 102 (1972), reprinted in W. Curran and E. Shapiro, *Law, Medicine, and Forensic Science*. Boston: Little, Brown, 1982.
Jasanoff, Sheila. *Designs on Nature*. Princeton: Princeton University Press, 2005.
———, editor. *States of Knowledge: The Coproduction of Science and Social Order*. London: Routledge, 2004.
Kant, Immanuel. *Metaphysik der Sitten*. In *Werke*, vols. 5 and 6. Berlin: Akademie Ausgabe, 1969 (1797).
———. "Was ist Aufklärung?" In Michel Foucault, *The Politics of Truth*, 7–20. New York: Semiotext(e), 1997.
Kass, Leon. "The Wisdom of Repugnance." *The New Republic*, 216, no. 22 (1997): 17–26.
Kipphoff, Petra. "Verweile Nicht, du bist so schön!" *Die Zeit*, June 30, 1995.
Knabe, Hubertus. *Die Unterwanderte Republik: Stasi im Westen*. Berlin: Propyläen, 1999.
Konrad, Monica. "From Secrets of Life to the Life of Secrets: Tracing Genetic Knowledge as Genealogical Ethics in Biomedical Britain." *Journal of the Royal Anthropological Institute* 9 (2003): 339–58.
Lachmund, Jens. "Knowing the Urban Wasteland: Ecological Expertise as Local Process." In *Earthly Politics: Local and Global in Environmental Governance*, edited by Sheila Jasanoff and Marybeth Long Martello. Cambridge, MA: MIT Press, 2004.
Latour, Bruno. *Science in Action: How to Follow Scientists and Engineers Through Society*. Cambridge, MA: Harvard University Press, 1987.
Latour, Bruno, and Steve Woolgar. *Laboratory Life: The Construction of Scientific Facts*. Princeton: Princeton University Press, 1986.
Leicht, Robert. "Auch der Bürger Kohl hat Rechte; Ein Kanzler als Opfer: Warum

die Stasi-Abhörprotokolle unter Verschluss bleiben müssen; Eine Antwort an Marianne Birthler," *Die Zeit*, February 2, 2001.

———. "Vom Reichstag fallen die Hüllen; Aufbruch zur Neuen Republik." *Die Zeit*, July 7, 1995.

Leinemann, Jürgen. "Das Neue ist die Größe," *Der Spiegel* 36 (1999): 46.

Macrakis, Kristie. *Seduced by Secrets: Inside the Stasi's Spy-Tech World*. Cambridge: Cambridge University Press, 2008.

Maier, Charles. *The Unmasterable Past: History, Holocaust, and German National Identity*. Cambridge, MA: Harvard University Press, 1988.

Merton, Robert K. "The Normative Structure of Science." In *The Sociology of Science: Theoretical and Empirical Investigations*, edited by Robert K. Merton. Chicago: University of Chicago Press, 1973.

Morlok, Martin. "Glaubens-, Gewissens- und Bekenntnisfreiheit, Kriegsdienstverweigerung." In *Grundgesetz, Kommentar*, edited by Horst Dreier, vol. 1, 294–343. Tübingen: Mohr Siebeck, 1996.

Nader, Laura. "Up the Anthropologist—Perspectives Gained From Studying Up." In *Reinventing Anthropology*, edited by Dell Hymes, 284–311. New York: Vintage Books, 1974.

Niederberghaus, Thomas. "Abbau West: Die Ostsee ist das beliebteste Kurzreiseziel der Westdeutschen." *Die Zeit*, July 25, 2002.

Nikolow, Sibylle. "Der statistische Blick auf Krankheit und Gesundheit: 'Kurvenlandschaften' in Gesundheitsausstellungen am Beginn des 20. Jahrhunderts in Deutschland." In *Infografiken, Medien, Normalisierung: Zur Kartografie politisch-sozialer Landschaften*, by Ute Gerhard, Jürgen Link, and Ernst Schulte-Holtey. Heidelberg: Synchron, 2001.

Novas, Carlos, and Nikolas Rose. "Genetic Risk and the Birth of the Somatic Individual." *Economy & Society* 29, no. 4 (2000): 485–513.

Paul, Diane B. "Eugenic Anxieties, Social Realities, and Political Choices." In *Are Genes Us? The Social Consequences of the New Genetics*, edited by Carl F. Cranor, 142–54. New Brunswick, NJ: Rutgers University Press, 1994.

Peine, Franz-Josef, and Dieter Heinlein. *Einführung in das Recht des Berufsbeamtentums*. Heidelberg: C. F. Müller, 1999.

Petersen, Alan. "The Best Experts: The Narratives of Those Who Have a Genetic Condition." *Science & Social Medicine*. 63 (2006): 32–42.

Polanyi, Michael. *Personal Knowledge: Towards a Post-Critical Philosophy*. Chicago: University of Chicago Press, 1974 (1958).

Rabinow, Paul. *French Modern: Norms and Forms of the Social Environment*. Chicago: University of Chicago Press, 1995.

Rheinberger, Hans-Jörg. *Toward a History of Epistemic Things: Synthesizing Proteins in the Test Tube*. Stanford: Stanford University Press, 1997.

Rogers, Annie. *A Shining Affliction: A Story of Harm and Healing in Psychotherapy*. New York: Viking Penguin, 1992.

Röper, Erich. "Allzeit verschiedene Deutsche." *Kritische Vierteljahreszeitschrift für Gesetzgebung und Rechtswissenschaft* 84 (2001): 5–23.

Roth, Martin. "Menschenökonomie, oder Der Mensch als technisches und künstlerisches Meisterwerk." In *Der Gläserne Mensch: Eine Sensation*, edited by Rosmarie Beier and Martin Roth. Stuttgart: Verlag Gerd Hatje, 1990.

Sahlins, Marshall. *Islands of History*. Chicago: University of Chicago Press, 1997
———. "The Sadness of Sweetness: The Native Anthropology of Western Cosmology." *Current Anthropology* 37 (1996): 395–415.
Samerski, Silja. *Die verrechnete Hoffnung*. Münster: Westfälisches Dampfboot, 2002.
Sandel, Michael. "The Case against Perfection." *Atlantic Monthly*, April 2004, 51–62.
Schicktanz, Silke, and Jörg Naumann, editors. *Bürgerkonferenz: Streitfall Gendiagnostik*. Wiesbaden: VS-Verlag für Sozialwissenschaften, 2003.
Schiller, Friedrich. "Brief an Friedrich Christian von Augustenburg, 13. Juli 1793." In *Schillers Werke*, vol. 6, *Nationalausgabe*, edited by Edith Nahler and Horst Nahler, page 26, Z.8–14. Weimar, 1992.
Schlink, Bernhard. "An der Grenze des Rechts." *Der Spiegel*, January 17, 2005.
Schröder, Gerhard. "Der Neue Mensch—Beitrag zur Gentechnik." *Die Woche*, December 20, 2000.
Schulzki-Haddouti, Christina. *Im Netz der Inneren Sicherheit: Die Neuen Methoden der Überwachung*. Hamburg: Europäische Verlagsanstalt, 2004.
Shapin, Steven, and Simon Schaffer. *Leviathan and the Airpump*. Princeton: Princeton University Press, 1985.
Silver, Lee. *Remaking Eden*. New York: Harper Perennial, 1998.
Sperling, Stefan. "Converting Ethics into Reason: German Stem Cell Policy between Science and the Law." *Science as Culture* 17, no. 4 (2008): 363–75.
———. "Managing Potential Selves: Stem Cells, Immigrants, and German Identity." *Science and Public Policy* 31, no. 2 (2004): 139–49.
Steffani, Winfried. *Parlamentarismus ohne Transparenz*. Opladen: Westdeutscher Verlag, 1971.
Stödter, Rolf. *Deutschlands Rechtslage*. Hamburg: Rechts- und Staatswissenschaftlicher Verlag, 1948.
Stölb, Marcus. "Undurchsichtige Tricks mit der Gläsernen Kuh." *Spiegel Online*, September 17, 2003. http://www.spiegel.de/panorama/0,1518,265493,00.html (accessed April 25, 2012).
Strathern, Marilyn. "The Tyranny of Transparency." *British Educational Research Journal* 26, no. 3 (2000): 309–21.
Wagner, Matthias. *Das Stasi-Syndrom: Über den Umgang mit den Akten des MfS in den 90er Jahren*. Berlin: Edition Ost, 2001.
Weber, Max. "Politik als Beruf." In *From Max Weber: Essays in Sociology*, edited by H. H. Gerth and C. Wright Mills, 77–128. London: Routledge 1991.
"Wir können mit Grenzen umgehen." *Die Welt*, May 30, 2001.
Wolff, Gerhard. Eugenik und genetische Beratung—Ethische Probleme humangenetischer Diagnostik. *Medizinische Genetik* no. 2 (1990): 14–20.
Zimmer, René. *Begleitende Evaluation der Bürgerkonferenz "Streitfall Gendiagnostik."* Karlsruhe: Fraunhofer Institut für Systemtechnik und Innovationsforschung, 2002.

Index

abortion law in Germany: compromise law in 1992, 231; content of the applicable Penal Code, 178–79; counseling agencies' role in abortions, 178, 179, 306n42; East and West German laws addressing abortions, 230–31, 309nn42–44; logic of the law, 179–80; relationship between the state and the woman regarding, 178–79, 180–81; unintended consequences of the law, 232
Ach, Johann, 49, 50, 65
Adenauer, Konrad, 196
Advanced Cell Technology (ACT), 23, 42, 266
Afghanistan war debate, 181–83, 306n43
Akademie der Wissenschaften der DDR, 31
Aktion Mensch, 297n3
Allgemeine Rundfunkanstalten Deutschland (ARD), 8–9, 97

antidiscrimination law (*Antidiskriminierungsgesetz*), 72
architecture: approach to governance expressed in, 7–10, 295–96nn6–10; juxtaposition of new and old buildings in Berlin, 10; Paul-Löbe-Haus, 1–2, 4, 295n1; symbolism of the EK building, 66; symbolism of the executive and parliamentary buildings, 2, 4, 11, 25
Archive of the German Representatives (*Archiv der Deutschen Abgeordneten*) (artwork), 4
ARD (Allgemeine Rundfunkanstalten Deutschland), 8–9, 97
Aufarbeitung, 201
Auf dem Weg zur liberalen Eugenik (Habermas), 102
Aufklärung (education), 135, 138, 140
Ausschuss für Bildung und Forschung (Committee for Education and Research), 33

awareness of what is lawful (*Rechtsbewusstsein*), 224

Barnstone, Deborah Ascher, 295n6
Basic Law: constitutional right to refuse military service, 173–74; disabled rights in Article 3, 72; freedom of conscience and, 146, 147–48, 151, 304n1; individual protections afforded by, 151–52, 283; prohibition of a cost-benefit calculation involving lives, 169; public reminders of, 6, 11, 21; reinterpreted after reunification, 200; state's role in setting ethical criteria and, 273–74
Baumgarten, Klaus-Dieter, 309n37
Bayertz, Kurt, 79
BBAW (Berlin-Brandenburgische Akademie der Wissenschaften), 30
Beamte: conflict between conscience and duty, 171; employment conditions for, 162–63; history up to WWII, 158–59; post-WWII prosecution by the Allies, 159–60; presumption of acting on "best knowledge and conscience," 163–64
Becker, Wolfgang, 243
Beck-Gernsheim, Elisabeth, 39
Beek, Cornelia, 35, 65, 75, 95–96, 107–8, 298n24
Benda, Ernst, 297n17
Benda Commission, 37, 297n17
Bentham, Jeremy, 115, 116
Berlin-Brandenburgische Akademie der Wissenschaften (BBAW), 30
Berliner Rede, 41, 282, 298n22
Berlin Museum for Communication, 212, 213
Berth, Hendrik, 202
Bildung and education, 119–20
Bildungsbürgertum (educated elite), 120
bioethics: citizen conference on (*see* citizen conference on genetic testing); commissions establishment to facilitate debate (*see* ethics commissions); Council on Bioethics, U.S., 48–49, 91, 190; idea that bioethics is a value-free *Saalordner*, 52; invoking of Kantian ethics during debates on, 156–57; nature of in Germany, 14–15, 296n14; perceived problem of false labeling as a science, 51–52; prejudice against East German bioethics opinions, 233; public bioethics discussions in East Germany, 238–39; sense of rightness that pervades the German discourse on bioethics, 61–62; social engineering basis of bioethics in East Germany, 235–36; start of in West Germany, 234; stem cell controversy (*see* stem cell debate); thematic treatment of, 15–16; transparency of for the public (*see* Bundestag debate over stem cell importation; citizen conference on genetic testing)
Bioethics Convention (Convention on Human Rights and Biomedicine), 27, 52
Birnbacher, Dieter, 79
Birthler, Marianne, 217–19, 220, 221
Bismarck, Otto von, 151
BMBF (Ministry of Education and Research), 251–53
Boas, Franz, 112
Böckenförde, Ernst-Wolfgang, 298n20
Boltanski, Christian, 4–5
Bolz, Norbert, 286
border guards (*Mauerschützen*), 223, 224
Borges, Jorge Luis, 275
Borneman, John, 197
Bourdieu, Pierre, 300n20
Boyer, Dominic, 120
Brandenburg Gate, 193–94, 195
Brandt, Willi, 221
break in the dam (*Dammbruch*), 168, 284–87

Brezan, Jurij, 236
Brüstle, Oliver, 41, 53, 63, 247, 268, 269, 280, 287, 310n8
Bundesamt für die Anerkennung ausländischer Flüchtlinge (Federal Authority for the Recognition of Foreign Refugees), 209
Bundesgerichtshof (Federal Court of Justice), 161
Bundesgerichtshof in Zivilsachen (Federal Court of Civil Affairs), 161
Bundesgleichstellungsgesetz, 72
Bundesrat, 5
BundesStasiUnterlagenBehörde (BStU). *See* Federal Stasi Files Authority
Bundestag debate over stem cell importation: assumption of a connection between wealth and ethics, 56; belief that the ethical order is tied to social order, 60; codification of (*see* Stem Cell Law); EK argument for selectively allowing research, 54–55; ethics linked to national sovereignty, 56; FDP argument for allowing all research, 55; Green Party argument for prohibiting research, 55–56; illustration of the German ideals of rational debate, 58; implication of this being a vote for the future of humankind, 57–58; issues being debated, 54; NER's large public profile, 299n26; proposals for an importation law, 53; role of the EK and NER in the debate, 58, 59; visitor rules, 53–54; voting process, 57
Bundestag ticker service, 77–78
Bundesverfassungsgericht (Federal Constitutional Court), 101, 161, 170
Bundeszentrale für gesundheitliche Aufklärung (Central Federal Office for Health Education), 230

Bürgerämter (citizen services offices), 127
Bürgerkonferenz. *See* citizen conference on genetic testing
Buruma, Ian, 167
Bush, George W., 42, 91, 190, 254
Busquin, Philippe, 257

categorical imperative by Kant, 154, 155, 156, 187
Catenhusen, Wolf-Michael, 33
CDU (Christian Democratic Union), 42, 216
Center for Art and Media Technology (ZKM), 101
Central Ethics Commission for Stem Cell Research. *See* ZES
Central Federal Office for Health Education (Bundeszentrale für gesundheitliche Aufklärung), 230
Chancen und Risiken der Gentechnologie (Enquete Kommission on Opportunities and Risks of Gene Technology), 31–33, 234
Checkpoint Charlie Museum, 214
Christian Democratic Union (CDU), 42, 216
Christo, 23
citizen conference on genetic testing (*Bürgerkonferenz*): aim of the conference, 121, 122; conference impact on the participants, 140; details of time spent on specific tasks, 134–35; experts' reactions to the citizens' opinion, 139–40; final opinion on genetic research and testing, 137–39, 303nn57–59; formal evaluation of, 133, 137, 140, 303n55; location (*see* German Hygiene Museum); participants (*see* citizen participants at the citizen conference); process of educating citizens (*see* citizen education at the citizen conference); questions about how the lay public would be engaged, 121–22; structure of, 121

citizen education at the citizen conference: audience members' interest in the process and not the content, 133; details of time spent on specific tasks, 134–35; disabled persons presentations' impact, 136; experts' focus on science and ethics facts, 135; participants' comments on adequacy of their preparation, 133–34; participants' comments on how their opinions were shaped, 134; participants' initial confusion about their role, 131–32; questionable success in generating dialogue between citizens and experts, 132, 302–3nn43–44

citizen participants at the citizen conference: chosen topics, 130; composition of the panel, 128–29; conference impact on the participants, 140; critique of the moderator, 130; impartiality requirement, 129–30; organizers' control of the citizen-expert dialogue, 130; perceived success at representing all citizens, 129; process of finding participants, 127–28, 302n35; trust placed in the composition of a deliberative body, 129; vote against PGD, 128

citizen services offices (*Bürgerämter*), 127

Civilian Service Law of 1983, 174–75

clarity of law (*Rechtsklarheit*), 265

Clement, Wolfgang, 41

Cohen, Randy, 299n26

Comaroff, Jean and John, 115

Committee for Education and Research (Ausschuss für Bildung und Forschung), 33

competent (*mündig*), 302n42

conscience (*Gewissen*), 145, 148, 150, 154. *See also* decisions of conscience; political debates and conscience

"conscience of science" (*wissenschaftliches Gewissen*), 45

conscientious (*gewissenhaft*), 178

conscientious objectors: argument that individual conscience is more important than the needs of the state, 175; certification process in mid-1980s, 172; Civilian Service Law of 1983 and, 174–75; constitutional right to refuse military service, 173–74; form letters detailing how to justify a pacifist position, 175–77; growth in civilian service numbers before suspension of mandatory service, 172–73; institutionalization of creating a reasoned position, 175–78; reasons people had for avoiding military service, 172; suspension of mandatory military service in 2011, 173

Convention on Human Rights and Biomedicine (Bioethics Convention), 27, 52

Council on Bioethics, U.S., 48–49, 91, 190

Dammbruch (break in the dam), 168, 284–87

Daschner, Wolfgang, 165, 166, 170–71, 283

Day of the German Protestant Church, 41

decisions of conscience: abortion law and (*see* abortion law in Germany); Afghanistan war support debate, 181–83, 306n43; argument that citizens' do not have a right or the ability to resist the state, 160–62; Basic Law and freedom of conscience, 146, 147–48, 151, 304n1; case of alleged torture as an example of public versus private reason (*see* torture case); changes in the meaning of conscience, 149–50, 154; claims by East Germans that they had been justly loyal to the state, 228–29; conflict between law and morality, 170; conscientious objectors (*see* conscientious objectors); court cases dealing with individual accountability for official actions, 223; culturally inculcated shared moral standard of conscience in Germany, 154–55; history of German constitutions, 150–51; individual conscience versus legal authority in the context of the greater good, 169–71, 306nn37–38; individual protections afforded by the Basic Law, 151–52; Kantian ethics' enduring importance in Germany, 155–57; Kant's distinction between public and private reason, 157–58; Kant's inner moral law of conscience, 152–54, 170; legal category of "crimes against humanity," 159–60; political debates and (*see* political debates and conscience); private and public aspects of conscience, 145–47; question of how the state helps develop a citizen's conscience, 224–26; question of individual justification for violating collective norms, 282–83; requirement for reasoned exercise of conscience, 190–91; requirement that conscience be transparent and based on universally shared action, 154; restrictions on religious minorities in past German law, 148–49; state workers and (see *Beamte*)

Defense Council, East German (Nationaler Verteidigungsrat), 223

Degussa, 205, 287

Deleuze, Gilles, 154

democracy and transparency: connection between, 112, 113, 117, 141, 279; example of democracy made transparent (*see* citizen conference on genetic testing [*Bürgerkonferenz*])

Der (im)perfekte Mensch (*The [Im]perfect Human*), 127

Der Mensch als Industriepalast (The Human as a Palace of Industry), 124

"Der Neue Mensch" ("New Human") (Schröder), 28

Der Vorleser (*The Reader*) (Schlink), 11–12

Deutscher Ethikrat (German Ethics Council), 296n15

Deutsch Forschungsgemeinschaft (DFG), 257

DFG (German Research Society), 40–41

Dienstweg (official channels), 73–74

Die Tageszeitung (newspaper), 39

Die Zeit (newspaper), 217, 299n26

disabled persons: inclusion in the ethics discussion, 38, 71–72; ineffectiveness of disabled persons' lobbying efforts, 72–73; legitimacy of a commission member based on his disability, 71; presentations' impact at the citizens' conference, 136; reality of poor handicapped accessibility in the parliament building, 72; rights in the Basic Law, 72

Dreier, Horst, 281

Dumont, Louis, 120

East and West German relations: abortion policy differences (*see* abortion law in Germany); architecture and, 8–10; argument that retributive justice is necessary in a democracy, 198; asymmetry of the power of the West over the East, 227; authors' and artists' unofficial function of creating a vision for an East German society, 236–37; Basic Law after reunification, 200; Brandenburg Gate history, 193–94,

East and West German relations (*cont.*) 195; claims by East Germans that they had been justly loyal to the state, 228–29; concerns over birthrate decline, 229; Constitutional Court ruling that past East German law violated human rights, 223–24; contradictions in the retelling of history, 194–95; court cases dealing with individual accountability for official actions, 223; difficulty in reconciling the new Germany with old actions, 228–29; discrimination of the East by the West, 199–201; East and West's attempts to define Germany, 197; East German people's revolution, 201; East German secret police (*see* Stasi); East Germany anger over the marginalization of the uprising, 202; East Germany framed as an unlawful state (see *Unrechtsstaat*); East Germany's antifascist ideology in practice, 197; economic decline in the East after the Wall fell, 203; forming of commissions to study East Germany, 196; *Gaterslebener Begegnungen*, 237–39, 241–42, 309n55; legal proceedings against military leaders, 221–22, 226–27, 309n37; legal strategy in *Mauerschützen* trials questioned, 227; marginalization of East German scientists during reunification, 240–41; memorial for the murdered Jews, 204–6; number of attempted escapes from East Germany, 214, 222–23, 308n24, 308n33; official East German reaction to public bioethics discussions, 238–39; post-reunification changes experienced by East Germans, 203–4, 307nn10–11; prejudice against East German bioethics opinions, 233; process of reunification and, 198–99; public sphere of bioethics in East Germany, 239; question of how the state helps develop a citizen's conscience, 224–26; reconciling with the Stasi after reunification (*see* Federal Stasi Files Authority); reunified Germany's commitment to privacy in communications, 212–13; scientific losses in transition to a Western Research model, 241; signing of the Unification Treaty, 201–2; social engineering basis of bioethics in East Germany, 235–36; societal influences on percentages of childless women, 229–30; start of bioethics in West Germany, 234; transparency between science and the public in the East, 235–36; treatment of East German judges, 200, 226; treatment of some former East German professionals, 232–34; West German courts' treatment of East German state defendants, 222; West Germans' visits to East Germany, 202–3, 204; West Germany's approach to denazification, 196–97; West Germany's dominance after reunification, 287–89

East German parliament (Volkskammer), 207

East German Socialist Unity Party (Sozialistische Einheitspartei Deutschlands) (SED), 27

educated elite (*Bildungsbürgertum*), 120

education (*Aufklärung*): Hygiene Museum mission of educating citizens, 124, 125–26; value placed on literacy, 11–13. *See also* citizen education at the citizen conference

Eisenman, Peter, 204

EK. *See* Enquete Kommission on Law and Ethics in Modern Medicine

Embryo Protection Law (1991), 27, 41
Enola Gay (airplane), 306n38
Enquete Kommission on Law and Ethics in Modern Medicine (EK), 9, 14; concerns of disability rights groups over the Bioethics Convention, 27; distress over a breach of confidentiality, 93–94; events at Karlsruhe, 104; guest speaker choice debate for report presentation, 102–3; impetus for the formation of, 16, 26–27; information outreach to other countries, 298n24; intern experience (*see* intern experience at the EK); legacy of the early gene technology commission, 31–33; mandate for, 27–28; media treatment of (*see* media and the EK); panel membership debate for report presentation, 103–4; paperwork associated with (*see* writing bioethics); public relations issues, 35–36; published report on stem cell research, 43, 44; purpose of, 22; questions about its "democratic legitimation," 36–37; rivalry with the NER, 33–34; role of the EK and NER in the debate, 58, 59; scheduled duration, 28; source of its authority, 21; text production (*see* text production in the EK); transparency and (*see* transparency of the commissions)
Enquete Kommission on Opportunities and Risks of Gene Technology (Chancen und Risiken der Gentechnologie), 31–33, 234
Eppelmann, Rainer, 196
equal positioning law (*Gleichstellungsgesetz*), 72
Ersatz service, 173
escape helpers (*Fluchthelfer*), 214
ethical sand dune (*ethische Wanderdüne*), 256

"Ethicist, The" (news column), 299n26
ethics commissions: churches' protests against embryo research, 41–42; claim to function as *Saalordner*, 44–46; contention over "democratically legitimated" phrase, 36–37; cutoff date considerations, 34; disabled persons representation on each, 38; dislike of the term "bioethics" in Germany, 50, 52; each commission's belief that it brings order to the debate, 47–48; EK (*see* Enquete Kommission on Law and Ethics in Modern Medicine); financing of each, 38; metaphor of the Rubicon, 42; NER (*see* Nationaler Ethikrat); parliamentary debate on proposals (*see* Bundestag debate over stem cell importation); sense that the work of ethics lags research, 43; transparency and (*see* transparency of the commissions); unauthorized negotiation to import Israeli stem cell lines, 41
ethische Wanderdüne (ethical sand dune), 256
eugenics, 125, 297n2
European Convention on Human Rights, 223, 224
European Court of Human Rights, 228
Ezrahi, Yaron, 113–14

FAZ (*Frankfurter Allgemeine Zeitung*), 98
FDJ (Freie Deutsche Jugend), 173
FDP (Free Democratic Party), 42, 55
Federal Authority for the Recognition of Foreign Refugees (Bundesamt für die Anerkennung ausländischer Flüchtlinge), 209
Federal Constitutional Court (Bundesverfassungsgericht), 101, 161, 170

Federal Court of Civil Affairs (Bundesgerichtshof in Zivilsachen), 161
Federal Court of Justice (Bundesgerichtshof), 161
Federal Stasi Files Authority (BStU): Birthler's advocacy of a widespread investigation, 221; Birthler's promotion of expanding use of the files to allow accountability, 217–18; central aim of the Stasi Files Law, 208, 307–8n17; difficulty in reconciling notions of privacy and data protection with file transparency, 218; difficulty in separating the victims from the perpetrators, 216; discovery of informal collaborators in the files, 209; estimate of time needed for reconstruction, 210–11; implications the volume of files had for East German society, 215–16; Leicht's advocacy of a targeted response, 220, 221; Leicht's interpretation of transparency, 219–20; political scandal that challenged file transparency, 216–17; procedure for reconstructing torn documents, 209–10; profiling used by West Germany to find guilty individuals, 213–14; proposal for computer-based reconstruction, 210; scope of work involved in checking the files, 208–9; the Stasi and (see Stasi); symbolism of the different understandings of the law's role in reunified Germany, 220–21; use of the terms victim and perpetrator, 214–15
Fischer, Andrea, 29, 255, 281, 298n19
Flach, Ulrike, 55
Flierl, Thomas, 240
Fluchthelfer (escape helpers), 214
Forschungsausschuss, 255
Fortpflanzungsmedizingesetz (Law on Reproductive Medicine), 29

Foster, Norman, 7–8, 116, 295n7
Foucault, Michel, 113, 116, 150, 157
Foundations of a Metaphysics of Morals (Kant), 152
Fragen an die deutsche Geschichte (*Questions to German History*), 24, 72
Fraktionszwang (party pressure), 182, 186, 187
Frankfurter Allgemeine Zeitung (FAZ), 98
Franz, Wolfgang: application process, 248, 272–73; research background, 271–72; restrictions placed on conducting his research, 272; view of why stem cell research in Germany is limited, 273
Free Democratic Party (FDP), 42
freedom of conscience and the Basic Law, 146, 147–48, 151, 304n1
Freie Deutsche Jugend (FDJ), 173
Fristenlösung, 230–31
Funder, Anna, 210
"Funes, The Memorious" (Borges), 275–76

Gäfgen, Magnus, 165
Ganten, Detlev, 40
Gaterslebener Begegnungen (Gatersleben Encounters), 237–39, 241–42, 309n55
Gauck, Joachim, 208, 214–15, 216, 307n16
Gauck Authority, 208
Geissler, Erhard, 236, 238
Gene Technology Law (*Gentechnikgesetz*), 32
genetic testing: citizens' conference on (see citizen conference on genetic testing); inclusion of disabled people in the ethics discussion, 38, 71–72; pre-implantation genetic diagnosis, 29, 46, 91–92, 128; selection prohibition, 250, 310n2; Stem Cell Law and (see Stem Cell Law)

Gene Worlds (*Genwelten*), 126
Gentechnikgesetz (Gene Technology Law), 32
German Embryo Protection Law, 42
German Ethics Council (Deutscher Ethikrat), 296n15
German Hygiene Museum: careful management of the notion of "citizen," 127; guiding purpose of, 123, 302n28; human body presented as biological and social, 124–25; location, 120–21, 122, 124; mission of educating citizens, 124; role as a center for health education, 125–26; Transparent Human centerpiece, 126; used by the National Socialists to espouse eugenics, 125
German Ideology (Dumont), 120
German Research Society (DFG), 40–41
Germany: act of setting limits seen as an ethical act, 280–82; Afghanistan war debate, 181–83, 306n43; approach to governance expressed in architecture, 7–10, 295–96nn6–10; East-West relations (*see* East and West German relations); EK hearing meant to position Germany in the world of bioethics, 62–63; executive ethics commission (*see* Nationaler Ethikrat); framings of the *Dammbruch* argument, 284–87; juxtaposition of new and old buildings in Berlin, 10; new Parliamentoffice building (*see* Paul-Löbe-Haus); parliamentary ethics commission (*see* Enquete Kommission on Law and Ethics in Modern Medicine); potentiality consideration in the stem cell debates, 283–84; question of individual justification for violating collective norms, 282–83; reminders of Berlin's ruination in WWII, 9–10; reputation as a place of order, 17–18; role of the future in the stem cell debate, 284; Schröder's goal of a new direction for the country, 101; sense of rightness that pervades the German discourse on bioethics, 61–62; significance of November 9, 276–77; symbolism of the executive and parliamentary buildings, 25; transparency in (*see* transparency); value placed on literacy, 11–13
Geron, 266
Gesetzgebung (legislation), 85
Gesundheitsausschuss, 255
Gewissen. *See* conscience
gewissenhaft (conscientious), 178. *See also* conscientious objectors
Gläserne Fabrik (transparent factory), 114–15, 124
Gläserne Kuh (transparent cow), 114
Gläserner Bürger (transparent citizen), 118
Gläserner Mensch (Transparent Human), 124, 125, 126, 143
Gläserner Motor (glass motor), 124
Gläserner Patient (transparent patient), 118
Gläsernes Labor (transparent laboratory), 115
Gleichstellungsgesetz (equal positioning law), 72
Golin, Simon, 34–35, 297n16
Goodbye Lenin! (film), 243–45
Gorbachev, Mikhail, 227
Göring, Hermann, 5
Göring, Peter, 228
Green Party, 55–56, 216
Grüber, Katrin, 183–84

Habermas, Jürgen, 102, 156
Härtel, Ingo, 62, 63, 65, 72, 97
Harvey, David, 7
Haussmann, Georges, 7
Helmholtz Institute, 8–9
Herf, Jeffrey, 196, 198, 201, 287
Herzog, Roman, 298n22

Hescheler, Jürgen: investigation of rhythm of heart cells, 267–68; national research regulation's inhibiting of his work, 268–69; research application process, 248, 269–70; research background, 259, 266–67; start of interest in cultural differences in ethics, 266; stem cell research with Wobus, 267; view of the Stem Cell Law, 270

Heyland, Carl, 160–61

Hildebrandt, Alexandra, 223

Hobbes, Thomas, 149

Höfling, Wolfram, 300n12

Hohlfeld, Rainer, 236, 241

Holocaust, 15, 147, 159, 160, 277, 287, 289

Honecker, Erich, 226

Honnefelder, Ludger, 23, 88, 102

Huch, Ricarda, 11

"Human as a Palace of Industry, The" (*Der Mensch als Industriepalast*), 124

human embryonic stem cells. *See* citizen conference on genetic testing; stem cell debate; Stem Cell Law

Human Genome Project, U.S. National Institute of Health, 239

Humboldt University, 9

Iconoclash (art exhibit), 105

(Im)perfect Human, The (*Der [im]perfekte Mensch*), 127

Indikationsregelung, 231

Institute for Neuropathology, 247

Institute for Systems Technology and Innovation Research, 133, 137

Institut für Pflanzengenetik und Kulturpflanzenforschung (IPK), 237

Institut Mensch, Ethik, Wissenschaft (IMEW), 183

International Hygiene Exhibition (1911), 123

intern experience at the EK: beginning of author's association with the EK, 62–63; building housing the EK, 64, 66 (*see also* Paul-Löbe-Haus); bureaucratic complexity of the internship reauthorization procedure, 73–74; contrast between insiders' and outsiders' access to the building, 66–67; disabled rights in the Basic Law, 72; importance of using proper administrative channels, 73–74; inclusion of disabled people in the ethics discussion, 71–72; ineffectiveness of disabled persons lobbying efforts, 72–73; internship wrap-up, 108–9; ironies in the security process, 65; legitimacy of a member based on his disability, 71; missing transparency about non-public matters, 69; observations of a myopic focus on the wording of the text, 69; observations of procedural allowances, 70, 299n3; observations of the amount of paperwork (*see* writing bioethics); observations that the concept of transparency did not apply to behind-the-scenes discussions, 69; parliamentarians' posturing for the media during commission meetings, 68; practical application of "transparency," 65–66; problem of combining informality with hierarchy, 74–75, 299n5; reminder of the role of rules in a bureaucracy, 107–8, 300n20; social rituals surrounding meals, 74, 299n4; sociological goal of the internship, 67–68; staff members' resistance to the process being made public, 92–94; start of internship, 64–65; strategy meeting attendance, 69; symbolism of the transparency of the building, 66

IPK (Institut für Pflanzengenetik und Kulturpflanzenforschung), 237

Itskovitz-Eldor, Joseph, 247, 268

INDEX

Jähn, Sigmund, 243
Jakob-Kaiser-Haus, 5–6
Jeanne-Claude, 23
Jewish bioethics, 266

Kahn, Fritz, 124
Kaiser Wilhelm Memorial Church, 296n10
Kaminer, Ariel, 299n26
Kant, Immanuel: enduring importance of his ethics in Germany, 155–57; on Enlightenment and reason, 157–58; inner moral law described by, 152–54, 170, 187–88
Kanzleramt, 8, 10, 24–25
Kass, Leon, 190
Kentenich, Heribert, 281
Knabe, Hubertus, 221
Knoche, Monika, 55, 101, 104
Kohl, Helmut, 24, 202, 208, 217–18
Köhler, Horst, 297n13
Kollek, Regine, 33, 40, 298n21
Krabat oder die Verwandlung der Welt (*Krabat, or the Transformation of the World*) (Brezan), 236
Krenz, Egon, 227, 228, 309n37
Kriegsdienstverweigerer aus Gewissensgründen. See conscientious objectors
Kühlungsborn Kolloquia, 236

Lachmund, Jens, 241
Latour, Bruno, 95, 105
Law for Aerial Security (*Luftsicherheitsgesetz*), 169, 170, 282
Law for the Reinstitution of the *Beamten* Status, 159
lawful state. See *Rechtsstaat*
Law on Reproductive Medicine (*Fortpflanzungsmedizingesetz*), 29
legislation (*Gesetzgebung*), 85
Leibniz, Gottfried Wilhelm von, 31
Leibniz Academy, 31
Leicht, Robert, 217, 218–20, 220, 221

Letters on Aesthetic Education (Schiller), 119
Leviathan (Hobbes), 149
Leviathan and the Airpump (Schaffer and Shapin), 301n7
Limbach, Jutta, 102
Lingner, Karl August, 123
Löbe, Paul, 295n1
Lohkamp, Christiane, 38
Luftsicherheitsgesetz (Law for Aerial Security), 169, 170, 282
Luther, Ernst, 86, 232–33, 234
Luther, Martin, 148

Madness and Civilization (Foucault), 150
Mann, Thomas, 3
manslaughter (*Totschlag*), 227
Marg, Volkwin, 287
Markl, Hubert, 41
Marquardt, Odo, 103
Marx, Karl, 67
Mauermuseum (Wall Museum), 214
Mauerschützen (border guards), 223, 224, 227
meaning of life exhibits, 2–3
media and the EK: ARD reporters explanation for their level of coverage, 97–98; coverage of NER's first open meeting, 99–100; goal of autonomy of ethical judgments, 97; media treatment of the commission's reports, 94–95; NER's courting of the media, 98–99; perception of prioritizing the NER over the EK, 96–97
Memorial to the Murdered Jews of Europe, 204–6
Merkel, Angela, 56, 57, 187
Mertonian ethos, 241
Metzler, Jakob, 165
Mielke, Erich, 227
Ministry of Education and Research (BMBF), 251–53
Momper, Walter, 101
Motor der Republik (Motor of the Republic), 4

"Motor of the Republic," 68
mündig (competent), 302n42
mündige Bürger (responsible citizens), 131, 302n42

Nachvollziehbarkeit. See transparency
Nader, Laura, 296n13
National Bioethics Advisory Council, U.S. (NBAC), 27
Nationaler Ethikrat (NER), 16; claim of ethical sovereignty, 42; courting of the media, 98–99; criticism of a lack of balance on the commission, 39–40; cutoff date suggestion for stem cell imports, 34; director's presentation to the U.S. Council on Bioethics, 48–49; first public meeting on its framing of the ethics questions, 46–47, 99–100; function perceived as handing down ethics to the people, 38–39; large public profile, 59, 299n26; membership and focus, 30; nominating process and nominees, 39–40; organizational hierarchy emphasis, 75, 299n6; proposal for a national ethics council, 29–30; published report on stem cell research, 43–44; purpose of, 22, 38–39; questions about the council's independence, 37–38; questions about the first director's salary and authority, 34–35, 297n16; rivalry with the EK, 33–34; role of the EK and NER in the debate, 58, 59; Schröder's argument against research regulation, 28–29; selective admiration for the UK legal system, 100–101; Simitis's presentation to the U.S. Council on Bioethics, 48–49; source of its authority, 21; symbolism of location at the BBAW, 30–31; transparency and (*see* transparency of the commissions)

"Nationaler Ethikrat" (news column), 299n26
Nationaler Verteidigungsrat (Defense Council, East German), 223
National Socialists, 5, 7, 11, 125, 161
Nazi Germany, 5–6, 8, 159, 209, 222
NBAC (National Bioethics Advisory Council, U.S.), 27
NER. *See* Nationaler Ethikrat
Neues Forum, 201
"New Human" ("Der Neue Mensch") (Schröder), 28
no punishment without law principle (*Nulla poena sine lege*), 223
Nuremberg trials, 160
Nüsslein-Volhard, Christiane, 40

official channels (*Dienstweg*), 73
180 Grad Berlin, 193–94
Opfer (victims), 215
order within the state (*Ordnung im Staate*), 161
organ transplantation and ethics discussion, 85–87, 300n12
Outline of a Theory of Practice (Bourdieu), 300n20

parliamentary debate. *See* Bundestag debate over stem cell importation
parliament building. *See* Paul-Löbe-Haus; Reichstag
parliament's starry hour (*Sternstunde des Parlaments*), 57–58, 249
Party of German Socialism (Partei des Deutschen Sozialismus) (PDS), 27
party pressure (*Fraktionszwang*), 182, 186, 187
Paul-Löbe-Haus: architecture, 1–2, 4, 295n1; democratic philosophy expressed by, 2, 4, 8, 11; new home of the EK, 68; reasoning behind the design choices, 6–7; walkway letters theme, 2–3, 295n3
Paulskirchenverfassung, 150

PDS (Party of German Socialism [Partei des Deutschen Sozialismus]), 27, 71
Peace of Westphalia, 148–49, 150
Penal Code, 178–79
Perruche case, 69, 299n2
Personal Knowledge (Polanyi), 300n13
PGD. *See* pre-implantation genetic diagnosis
Pietrass, Richard, 237
Planck, Max, 9
Polanyi, Michael, 300n13
political debates and conscience: Afghanistan decision and individual conscience, 182–83, 306n43; commissions' deciding of questions of conscience while connecting public and private reasoning, 188–90; conflicts between individual and sovereign conscience in parliament, 184–85; debate over Germany's participation in Afghanistan, 181–82; election process in Germany, 306–7n44; ethics commissions' view of their role, 45; *Fraktionszwang* in parliament, 186, 187; importance given to transparency in (*see* Bundestag debate over stem cell importation); Kant's inner moral law and, 187–88; parallels between personal autonomy and political sovereignty, 183; question of conscience in the political sphere, 183–84; requirement to be reasonable, 186–87
PPL Therapeutics, 79
pre-implantation genetic diagnosis (PGD): seeking a visual definition of the "beginning of life," 91–92; suspending of a restrictive law, 29; vote against at citizens' panel, 46, 128. *See also* bioethics
Preuß, Hugo, 150–51
Preussisches Landrecht (1794), 150
Primor, Avi, 205

Pro Familia, 179, 306n41
profiling (*Rasterfahndung*), 213–14
Propping, Peter, 40

Questions to German History (Fragen an die deutsche Geschichte), 24

Radbruch, Gustav, 223
Radbruch formula, 223, 224
Radtke, Peter, 38
Rasterfahndung (profiling), 213–14
Rau, Johannes, 41, 282, 298n22
Reader, The (Der Vorleser) (Schlink), 11–12
Reason (*Vernunft*), 125
Rechtsbewusstsein (awareness of what is lawful), 224
Rechtsklarheit (clarity of law), 265
Rechtsstaat (lawful state): citizens' rights and, 152; concept of, 6, 15, 198, 223, 226; inviolability of West German laws, 200, 287; Stasi Files Law and, 214, 219, 220, 221; West versus East power asymmetry, 227
Reformation, 148
Regeln für den Menschenpark (Sloterdijk), 102
Reich, Jens, 253–54, 282
Reichstag: Basic Law glass plates, 6; boxes art installation, 4–5; glass dome, 7–8, 116, 117, 295n7; Jakob-Kaiser-Haus and tunnels, 5–6; plenary hall, 5; symbolism of, 25; veiling and unveiling of the building, 23–24; view from the observation deck, 25–26; Weimar Republic representatives' memorial, 11, 296n11
Remaking Eden (Silver), 309n47
Renesse, Margot von: appointment to head the EK, 28; on conscience and elected representatives, 184–88; focus on the process, 61–62; importation compromise offered, 54–55; leadership in

Renesse, Margot von (cont.)
meetings, 81–82; on the legitimacy of NER, 37; on the logic of the abortion law, 179–80; on the taint attached to "bioethics" term, 52; writing of the Stem Cell Law, 249, 252
responsible citizens (*mündige Bürger*), 131, 302n42
Rheinberger, Hans-Jörg, 310n6
Ricken, Friedo, 156
Riedel, Ulrike, 76, 254, 282
Robert Koch Institute (RKI), 248, 250, 258
Röper, Erich, 199

Saalordner (neutral provider of order), 44–46, 54
Sahlins, Marshall, 112
Schaffer, Simon, 114, 301n7
Schiller, Friedrich, 119, 141
Schily, Otto, 213–14
Schirrmacher, Frank, 298n20
Schlink, Bernhard, 11, 169, 170, 283
Schmidt, Ulla, 29, 298n19
Schmölling, Klaus, 31–33
Schreiber, Jürgen, 166
Schröder, Gerhard, 22, 28–29, 56, 181–82, 297n13, 306n43
Schröder, Richard, 189, 285
Second International Hygiene Exhibition (1930), 123
SED (East German Socialist Unity Party [Sozialistische Einheitspartei Deutschlands]), 27
Seifert, Ilja, 38, 70–71
Settling Accounts (Borneman), 197
Shapin, Steven, 114, 301n7
Siep, Ludwig: acknowledgment that ethics was made subservient to the law by RKI, 262; appointment to the ZES, 261–62; exercise of public versus private reason, 265; exercise of public versus private uses of reason, 188–89
Silver, Lee, 236, 309n47
Simitis, Spiros, 46–47

Sinn und Form, 236, 237
Sloterdijk, Peter, 102, 104
Socialist Unity Party (SED), 201, 309n37
Sonnenallee (film), 242–43
Spirit and System (Boyer), 120
Spurzheim, Johann Kaspar, 150
Staatsratsgebäude, 25
Stammzelldebatte. See stem cell debate
Stammzellgesetz. See Stem Cell Law
Stasi: basis of its role in Germany, 206; destruction of files before the Wall fell, 207; law dealing with the files (*see* Federal Stasi Files Authority); museums about, 211–14; provisions for preserving and investigating the saved files, 207–8
Stasi Files Law, 208, 307–8n17
Stasiland (Funder), 210
Stasi Museum, 211–12
stem cell debate (*Stammzelldebatte*): about, 14, 19–20; ethics commissions established for (*see* Enquete Kommission on Law and Ethics in Modern Medicine; Nationaler Ethikrat); goal of initiating dialogue before topics become controversies, 23; in parliament (*see* Bundestag debate over stem cell importation); perception of a "law lag" between science and laws, 21–22, 297n5; popular media's portrayal of the issues, 21–22, 297n3; potentiality consideration in the stem cell debates, 283–84; purpose of a second ethics commission establishment, 20–21; role of the future in the stem cell debate, 284; "stem cells" term use, 296n16
Stem Cell Law (*Stammzellgesetz*): allowances for transparency, 252, 265; application approval process, 250–51, 310n6; central contradiction, 250, 252, 255–56;

criteria for ethically acceptable determination, 251; criteria for implementation of the law's imperatives, 249; cutoff date considerations, 34, 254–55, 256; dearth of applications for importation, 310n5; decision to allow the importation of stem cells, 42–43, 247; defining ethical German research, 256–57; impact on researchers (*see* Franz, Wolfgang; Hescheler, Jürgen); implication that ethical acceptability is equated with scientific merit, 259; importance of debate to transparency of laws, 260–61; lawyers' role as the agents of collective conscience, 259; paradoxes and ambiguities contained in, 253–54; parliamentary debate (*see* Bundestag debate over stem cell importation); potentiality consideration in the stem cell debates, 283–84; purpose of the law, 249; question of its constitutionality, 260; rationale for the law, 251–53; role of the future in the stem cell debate, 284; scientists' view of the law's requirements, 258; selection prohibition, 250, 310n2; separation of duties between the ZES and the RKI, 250, 253, 258; shifting meanings of "high priority" and "absence of alternatives," 259–60; small number of applications received, 248; the state's role in setting ethical criteria, 273–74; ZES and application approval (*see* ZES)
Stödter, Rolf, 160
Strathern, Marilyn, 112
supernumerary embryos (*überzählige Embryonen*), 88–89

Taupitz, Jochen, 260
text production in the EK: analogy between translating language and fitting ethics into an idiom, 87–88; delicate nature of "official translations," 89; difficulty in crafting language that would be valid for emerging technologies, 83–84; emergent order of the proposed texts, 84–85; *erzeugt* versus *gezeugt* example, 82–83; juridical language viewed as clear and precise, 90–91; process used to decide grammar and word choices, 82–83; rationale for removing legal and ethical terminology from the glossary, 89–91; seeking a visual definition of the "beginning of life," 91–92; struggle with the concept of supernumerary embryos, 88–89
Thierse, Wolfgang, 205
Thomson, James, 267
Tibbets, Paul, 306n38
torture case: allegations of torture by the suspect, 165, 168; allowed treatment of suspects, 165–66; decision to threaten the suspect, 166, 170–71; Germany's view of torture, 167–68; outcome of the case, 168; publication of the case, 166–67; publics' reaction to the story, 167, 305n33
Totschlag (manslaughter), 227
transparency (*Nachvollziehbarkeit*): allowances for in the Stem Cell Law, 252, 265; citizen agency and the state's performance of transparency, 280; conscience and, 154; difficulty in reconciling notions of privacy and data protection with file transparency, 218; ethics commissions and (*see* transparency of the commissions); functioning of in Germany, 277–78; Hygiene Museum mission and (*see* German Hygiene Museum); importance of debate to transparency of laws, 260–61; museums used to support the

transparency (*cont.*)
reunified state's transparency through pedagogy, 211–14; requirement that conscience be transparent and based on universally shared action, 154; Stasi files and (*see* Federal Stasi Files Authority [BStU]); theme in architecture of German politics, 7–8, 295n6; transparency and democracy linked through culture, 279; work involved in creating transparency, 278

transparency of the commissions: concerns over transparency of genetic information, 118–19, 301nn18–20; connection with democracy, 112, 113, 117; contingencies of transparency of the state, 117–18; example of democracy made transparent (*see* citizen conference on genetic testing); examples of the State's commitment to transparency, 114–15, 301n9; extremes of theories of political visibility, 113–14; horizontal solidarity preserved through vertical control, 118; importance of debate to transparency of laws, 260–61; legitimating role of transparency, 111, 278–79; official rhetoric about, 112, 300n1; practical application of "transparency," 65–66, 115–16; preoccupation with educating citizens and the concept of *Bildung*, 119–20; role in democracy, 141

transparent citizen (*Gläserner Bürger*), 118
transparent cow (*Gläserne Kuh*), 114
transparent factory (*Gläserne Fabrik*), 114–15, 124
Transparent Human (*Gläserner Mensch*), 124, 125, 126, 143
transparent laboratory (*Gläsernes Labor*), 115
transparent patient (*Gläserner Patient*), 118
turn (*Wende*), 195, 196

überzählige Embryonen (supernumerary embryos), 88–89
unconditional solidarity (*uneingeschränkte Solidarität*), 181
UN Convention against Torture, 167, 168
Understanding (*Verstand*), 125
Unification Treaty, 200, 201–2, 207, 227, 240
Unrechtsstaat (unlawful state): concept of, 15, 198, 288; disposability of East German laws, 200, 287; Germany defined by the Allies as, 160; proving that East Germany had been an *Unrechtsstaat* (*see* Stasi); question of where conscience comes from in, 225; Stasi Files Law and, 220; treatment of East German judges, 226; treatment of East German state defendants, 222; West versus East power asymmetry, 227
"Up the Anthropologist—Perspectives Gained from Studying Up" (Nader), 296n13
Urania, 45

Van den Daele, Wolfgang, 33, 46, 298n21
Vernunft (Reason), 125
Verstand (Understanding), 125
victims (*Opfer*), 215
Volkskammer (East German parliament), 207
Voltra, Manfred, 237
Von Augustenburg, Christian Friedrich, 119
Von Weizsäcker, Richard, 72, 208

Wages of Guilt (Buruma), 167
Wagner, Matthias, 215, 308n27
Waigel, Theo, 224

Wall, the, 23–24, 126, 172, 194, 199, 201, 203–4, 214, 222–23, 225, 227, 244
Wall Museum (Mauermuseum), 214
Watson, James, 238–39
Weber, Max, 150–51, 158
Weimar constitution, 151
Wende (turn), 195, 196
West Germany. *See* East and West German relations
WiCell, 272
Winnacker, Ernst-Ludwig, 33, 40, 282
"Wisdom of Repugnance, The" (Kass), 190
wissenschaftliches Gewissen ("conscience of science"), 45
Wobus, Anna, 237, 238, 241, 255, 261, 267
Wodarg, Wolfgang, 28, 88
Wolf, Christa, 237
Wolf, Eckhard, 23
Wolfschmidt, Matthias, 49–50, 65
writing bioethics: archive contents, 77; focus on maintaining the authoritative nature of the texts, 80–81; quantities of paper used, 75–76; reality of ethical arguments attaining a routine status, 79–80; sources of information, 77–78; tendency to view any new information as relevant and vital, 78–79, 300n8; text production (*see* text production in the EK)

ZES (Zentrale Ethikkommission für Stammzellforschung) (Central Ethics Commission for Stem Cell Research): application requirements, 250–51, 264, 310n6; appointment of Siep, 261–62; composition of the commission, 262–63; cutoff date considerations, 34, 255; decision-making process, 188–90; ethics made subservient to the law by RKI, 262, 263; exercise of public versus private reason, 188–89; importation approval, 247, 248; paucity of applications, 263–65; sense of transparency mandated by the law, 265; separation of duties between the ZES and the RKI, 250, 258, 263
ZKM (Center for Art and Media Technology), 101
Zyklon B, 205